High Temperature/High Performance Composites

MATERIALS RESEARCH SOCIETY SYMPOSIUM PROCEEDINGS   VOLUME 120

High Temperature/ High Performance Composites

Symposium held April 5-7, 1988, Reno, Nevada, U.S.A.

EDITORS:

F.D. Lemkey
United Technologies Research Center, East Hartford, Connecticut, U.S.A.

S.G Fishman
Office of Naval Research, Arlington, Virginia, U.S.A.

A.G. Evans
University of California, Santa Barbara, California, U.S.A.

J.R. Strife
United Technologies Research Center, East Hartford, Connecticut, U.S.A.

MRS   MATERIALS RESEARCH SOCIETY
Pittsburgh, Pennsylvania

CODEN: MRSPDH

Copyright 1988 by Materials Research Society.
All rights reserved.

This book has been registered with Copyright Clearance Center, Inc. For further information, please contact the Copyright Clearance Center, Salem, Massachusetts.

Published by:

Materials Research Society
9800 McKnight Road, Suite 327
Pittsburgh, Pennsylvania 15237
Telephone (412) 367-3003

Library of Congress Cataloging in Publication Data

High temperature/high performance composites : symposium held April 5-7, 1988, Reno, Nevada, U.S.A. / editors, F.D. Lemkey, S.G. Fishman, A.G. Evans, J.R. Strife.

    p.   cm. — (Materials Research Society symposium proceedings, ISSN 0272-9172 ; v. 120)
   Bibliography: p.
   ISBN 0-931837-90-1
   1. Composite materials—Congresses.   2. Heat resistant materials—Congresses.
   I. Lemkey, F.D.   II. Materials Research Society   III. Series: Materials Research Society symposium proceedings ; v. 120.

TA418.9.C6H545	1988	88-30704
620.1'1817—dc19		CIP

Manufactured in the United States of America

Manufactured by Publishers Choice Book Mfg. Co.
Mars, Pennsylvania 16046

These proceedings are dedicated to

Tony Kelly FEng, FRS

scientist, administrator, scholar
and pioneer of composite materials

Contents

PREFACE   xi

MATERIALS RESEARCH SOCIETY SYMPOSIUM PROCEEDINGS   xiii

PART I: NOVEL PROCESSING METHODS FOR METAL BASED COMPOSITES

*NEW PATHWAYS TO PROCESSING COMPOSITES   3
Robert Mehrabian

HIGH-ENERGY HIGH-RATE PROCESSING OF HIGH-TEMPERATURE
METAL-MATRIX COMPOSITES   23
C. Persad, S. Raghunathan, B.-H. Lee, D.L. Bourell, Z. Eliezer, and H.L. Marcus

XD^{TM} TITANIUM ALUMINIDE COMPOSITES   29
L. Christodoulou, P.A. Parrish, and C.R. Crowe

MICROMECHANICAL STUDIES OF MODEL MATRIX COMPOSITES   35
H.N.G. Wadley, F.S. Biancaniello, and R.B. Clough

EFFECT OF TEMPERATURE ON THE MECHANICAL PROPERTIES AND
MICROSTRUCTURES OF IN SITU FORMED Cu-Nb AND Cu-Ta
COMPOSITES   45
W.A. Spitzig, P.D. Krotz, L.S. Chumbley, H.L. Downing, and J.D. Verhoeven

PRELIMINARY INVESTIGATIONS OF ALUMINA-FIBER REINFORCED
Ni_3Al MATRIX COMPOSITES   51
B. Moore, A. Bose, R.M. German, and N.S. Stoloff

HIGH TEMPERATURE INTERMETALLIC COMPOSITES   57
D.L. Anton

PART II: DEFORMATION MECHANISMS IN METAL MATRIX COMPOSITES

*MECHANISMS AND MODELS OF HIGH TEMPERATURE DEFORMATION OF
COMPOSITES   67
Malcolm McLean

CRACK-TIP SHIELDING IN METAL-MATRIX COMPOSITES:
MODELLING OF CRACK BRIDGING BY UNCRACKED LIGAMENTS   81
Jian Ku Shang and R.O. Ritchie

ELEVATED TEMPERATURE SLOW PLASTIC DEFORMATION OF
$NiAl/TiB_2$ PARTICULATE COMPOSITES   89
R.K. Viswanadham, J. Daniel Whittenberger, S.K. Mannan, and B. Sprissler

*Invited Paper

TOUGHENING MECHANISMS IN INTERMETALLIC γ-TiAl ALLOYS
CONTAINING DUCTILE PHASES   95
   C.K. Elliott, G.R. Odette, G.E. Lucas, and
   J.W. Sheckherd

MICROSTRUCTURAL EFFECTS ON DUCTILE PHASE TOUGHENING
OF Nb-Nb SILICIDE COMPOSITES   103
   J.J. Lewandowski, D. Dimiduk, W. Kerr, and
   M.G. Mendiratta

INTERNAL FRICTION OF CAST GRAPHITE-MAGNESIUM COMPOSITES   111
   J.H. Armstrong, S.P. Rawal, and M.S. Misra

THE ELEVATED TEMPERATURE RESPONSE OF SILICON CARBIDE
AND BORON REINFORCED ALUMINUM AND TITANIUM METAL
MATRIX COMPOSITES   121
   M.S. Madhukar, A. Fareed, J. Awerbuch, and
   M.J. Koczak

THERMAL, VISCOPLASTIC ANALYSIS OF COMPOSITE LAMINATES   129
   E. Krempl and K.D. Lee

ENHANCED PLASTICITY OF MECHANICALLY ALLOYED ALUMINUM
IN90211   137
   T.R. Bieler, T.G. Nieh, J. Wadsworth, and
   A.K. Mukherjee

PART III: CERAMIC COMPOSITE MICROSTRUCTURAL DEVELOPMENT

*GLASS AND CERAMIC MATRIX COMPOSITES PRESENT AND
FUTURE   145
   Karl M. Prewo

PROCESSING OF POLYMERIC PRECURSOR, CERAMIC MATRIX
COMPOSITES   157
   R.J. Diefendorf and R.P. Boisvert

FRACTURE BEHAVIOR OF 3-D BRAIDED NICALON/SILICON
CARBIDE COMPOSITE   163
   J.-M. Yang, J.-C. Chou, and C.V. Burkland

MICROSTRUCTURAL CHARACTERIZATION OF A SIC WHISKER-
REINFORCED HIPped REACTION-BONDED SI_3N_4 169
   S.C. Farmer, P. Pirouz, and A.H. Heuer

SUSPENSION PROCESSING OF SiC WHISKER-REINFORCED
COMPOSITES   175
   Michael D. Sacks, Hae-Weon Lee, and Oswaldo E. Rojas

THEORETICAL ANALYSIS OF CHEMICAL VAPOR INFILTRATION
IN CERAMIC/CERAMIC COMPOSITES   185
   Nyan-Hwa Tai and Tsu-Wei Chou

FIBERS AND GRIDS BY INTEGRATED CIRCUIT TECHNOLOGY   193
   James E. Steinwall and H.H. Johnson

*Invited Paper

SYNTHESIS OF TITANIUM AND BORON CONTAINING POLYMERS: POTENTIAL PRECURSORS FOR ADVANCED CERAMICS   199
     Kenneth E. Gonsalves and K.T. Kembaiyan

PYROELECTRIC AND DIELECTRIC PROPERTIES OF POLYMER-CERAMIC COMPOSITES   205
     D.K. Das-Gupta and M.J. Abdullah

PART IV: CERAMIC COMPOSITE MECHANICAL PERFORMANCE

*THE MECHANICAL PERFORMANCE OF FIBER REINFORCED CERAMIC MATRIX COMPOSITES   213
     A.G. Evans and D.B. Marshall

THE FRICTIONAL RESISTANCE TO SLIDING OF A SiC FIBER IN A BRITTLE MATRIX   247
     T.P. Weihs, C.M. Dick, and W.D. Nix

EFFECT OF THERMAL EXPANSION MISMATCH ON FIBER PULL-OUT IN GLASS MATRIX COMPOSITES   253
     U.V. Deshmukh, A. Kanei, S.W. Freiman, and D.C. Cranmer

ROLE OF FIBER-MATRIX INTERFACIAL SHEAR STRESS ON THE TOUGHNESS OF REINFORCED OXIDE MATRIX COMPOSITES   259
     Raj N. Singh

THE MECHANICAL PROPERTIES AT HIGH TEMPERATURES OF SiC WHISKER-REINFORCED ALUMINA   265
     Kenong Xia and Terence G. Langdon

CREEP BEHAVIOR OF AN Al_2O_3-SIC COMPOSITE   271
     P. Lipetzky, S.R. Nutt, and P.F. Becher

FRACTURE MECHANISMS IN SiC-WHISKER REINFORCED ALUMINA   279
     Christophe H. Boulanger, Yih-Cherng Chiang, Azar P. Majidi, and Tsu-Wei Chou

IMPACT BEHAVIOR OF FIBER REINFORCED GLASS MATRIX COMPOSITES   285
     D.F. Hasson and S.G. Fishman

PART V: COMPOSITE INTERFACIAL EFFECTS

*STRUCTURE AND CHEMISTRY OF METAL/CERAMIC INTERFACES   293
     M. Rühle and A.G. Evans

INTERFACIAL CHEMISTRY-STRUCTURE AND FRACTURE OF CERAMIC COMPOSITES   313
     L.H. Schoenlein, R.H. Jones, C.H. Henager, C.H. Schilling, and F. Gac

THERMAL OXIDATION OF Al_2O_3-SiC WHISKER COMPOSITES: MECHANISMS AND KINETICS   323
     F. Lin, T. Marieb, A. Morrone, and S. Nutt

*Invited Paper

THE INFLUENCE OF HEAT TREATMENT UPON FIBER PULL-OUT IN A CERAMIC COMPOSITE M.D. Thouless, O. Sbaizero, E. Bischoff, and E.Y. Luh	333
ULTRASONIC PROPAGATION AT CYLINDRICAL METAL-CERAMIC INTERFACES IN COMPOSITES H.N.G. Wadley, J.A. Simmons, and E. Drescher-Krasicka	341
HREM CHARACTERIZATION OF THE INTERFACE IN A SiC FIBERS/Ti MATRIX COMPOSITE M. Lancin, J.S. Bour, J. Thibault-Desseaux	351
PHASE STABILITY AND INTERFACE REACTIONS IN THE Al-SiC SYSTEM Doh-Jae Lee, Mark D. Vaudin, Carol A. Handwerker, and Ursula R. Kattner	357

PART VI: COMPOSITE STRUCTURES

NOVEL METHOD FOR CONSTRUCTING TETRAHEDRAL FRAMES John J. Gilman	369
AUTHOR INDEX	379
SUBJECT INDEX	381

Preface

Composites--both high temperature and high performance--continue to be at the forefront of materials research and development, fundamentally as well as with a view toward future generations of aerospace structures, propulsion devices, and energy conversion systems. The aim of this three-day symposium was to bring together researchers from diverse disciplines to compare their latest results concerning the synthesis, structure-property relationships, and mechanics of metal, intermetallic, glass, and ceramic matrix composites.

This volume includes papers on high temperature/high performance composites invited or submitted for oral presentation at the Materials Research Society Spring Meeting held in Reno, Nevada April 5-8, 1988. MRS has contributed to this subject for a number of years by presenting symposia on in situ composites (see Volume 12 in the MRS Symposium Proceedings Series) and related topics (see Volumes 37, 39, 56, 64, 73, 78, 81, and 114).

The papers are arranged in six sections: (1) novel processing methods for metal-based composites; (2) deformation mechanisms in metal matrix composites; (3) ceramic composite microstructural development; (4) ceramic composite mechanical performance; (5) composite interfacial effects; and (6) novel composite structures. The authors are listed alphabetically at the back of the volume for convenient reference.

The editors and participants are indebted to the Defense Advanced Research Projects Agency for financial support of the symposium; however, no official endorsement is implied by this source of support. A large measure of credit for the success of the symposium must go to the invited speakers: R. Mehrabian, M. McLean, K. Prewo, A. Evans, and J. Gilman (presented at Symposium X) for their delivery of provocative and illuminating papers.

Special thanks for typing and secretarial services are due to Ms. Joyce Hurlburt. Her assistance, provided cheerfully and efficiently under trying circumstances, has assured uniformity of format to these proceedings.

September, 1988

Frank Lemkey
Tony Evans
Steve Fishman
Jim Strife

MATERIALS RESEARCH SOCIETY SYMPOSIUM PROCEEDINGS

ISSN 0272 - 9172

Volume 1—Laser and Electron-Beam Solid Interactions and Materials Processing, J. F. Gibbons, L. D. Hess, T. W. Sigmon, 1981, ISBN 0-444-00595-1

Volume 2—Defects in Semiconductors, J. Narayan, T. Y. Tan, 1981, ISBN 0-444-00596-X

Volume 3—Nuclear and Electron Resonance Spectroscopies Applied to Materials Science, E. N. Kaufmann, G. K. Shenoy, 1981, ISBN 0-444-00597-8

Volume 4—Laser and Electron-Beam Interactions with Solids, B. R. Appleton, G. K. Celler, 1982, ISBN 0-444-00693-1

Volume 5—Grain Boundaries in Semiconductors, H. J. Leamy, G. E. Pike, C. H. Seager, 1982, ISBN 0-444-00697-4

Volume 6—Scientific Basis for Nuclear Waste Management IV, S. V. Topp, 1982, ISBN 0-444-00699-0

Volume 7—Metastable Materials Formation by Ion Implantation, S. T. Picraux, W. J. Choyke, 1982, ISBN 0-444-00692-3

Volume 8—Rapidly Solidified Amorphous and Crystalline Alloys, B. H. Kear, B. C. Giessen, M. Cohen, 1982, ISBN 0-444-00698-2

Volume 9—Materials Processing in the Reduced Gravity Environment of Space, G. E. Rindone, 1982, ISBN 0-444-00691-5

Volume 10—Thin Films and Interfaces, P. S. Ho, K.-N. Tu, 1982, ISBN 0-444-00774-1

Volume 11—Scientific Basis for Nuclear Waste Management V, W. Lutze, 1982, ISBN 0-444-00725-3

Volume 12—In Situ Composites IV, F. D. Lemkey, H. E. Cline, M. McLean, 1982, ISBN 0-444-00726-1

Volume 13—Laser-Solid Interactions and Transient Thermal Processing of Materials, J. Narayan, W. L. Brown, R. A. Lemons, 1983, ISBN 0-444-00788-1

Volume 14—Defects in Semiconductors II, S. Mahajan, J. W. Corbett, 1983, ISBN 0-444-00812-8

Volume 15—Scientific Basis for Nuclear Waste Management VI, D. G. Brookins, 1983, ISBN 0-444-00780-6

Volume 16—Nuclear Radiation Detector Materials, E. E. Haller, H. W. Kraner, W. A. Higinbotham, 1983, ISBN 0-444-00787-3

Volume 17—Laser Diagnostics and Photochemical Processing for Semiconductor Devices, R. M. Osgood, S. R. J. Brueck, H. R. Schlossberg, 1983, ISBN 0-444-00782-2

Volume 18—Interfaces and Contacts, R. Ludeke, K. Rose, 1983, ISBN 0-444-00820-9

Volume 19—Alloy Phase Diagrams, L. H. Bennett, T. B. Massalski, B. C. Giessen, 1983, ISBN 0-444-00809-8

Volume 20—Intercalated Graphite, M. S. Dresselhaus, G. Dresselhaus, J. E. Fischer, M. J. Moran, 1983, ISBN 0-444-00781-4

Volume 21—Phase Transformations in Solids, T. Tsakalakos, 1984, ISBN 0-444-00901-9

Volume 22—High Pressure in Science and Technology, C. Homan, R. K. MacCrone, E. Whalley, 1984, ISBN 0-444-00932-9 (3 part set)

Volume 23—Energy Beam-Solid Interactions and Transient Thermal Processing, J. C. C. Fan, N. M. Johnson, 1984, ISBN 0-444-00903-5

Volume 24—Defect Properties and Processing of High-Technology Nonmetallic Materials, J. H. Crawford, Jr., Y. Chen, W. A. Sibley, 1984, ISBN 0-444-00904-3

MATERIALS RESEARCH SOCIETY SYMPOSIUM PROCEEDINGS

Volume 25—Thin Films and Interfaces II, J. E. E. Baglin, D. R. Campbell, W. K. Chu, 1984, ISBN 0-444-00905-1

Volume 26—Scientific Basis for Nuclear Waste Management VII, G. L. McVay, 1984, ISBN 0-444-00906-X

Volume 27—Ion Implantation and Ion Beam Processing of Materials, G. K. Hubler, O. W. Holland, C. R. Clayton, C. W. White, 1984, ISBN 0-444-00869-1

Volume 28—Rapidly Solidified Metastable Materials, B. H. Kear, B. C. Giessen, 1984, ISBN 0-444-00935-3

Volume 29—Laser-Controlled Chemical Processing of Surfaces, A. W. Johnson, D. J. Ehrlich, H. R. Schlossberg, 1984, ISBN 0-444-00894-2

Volume 30—Plasma Processing and Synthesis of Materials, J. Szekely, D. Apelian, 1984, ISBN 0-444-00895-0

Volume 31—Electron Microscopy of Materials, W. Krakow, D. A. Smith, L. W. Hobbs, 1984, ISBN 0-444-00898-7

Volume 32—Better Ceramics Through Chemistry, C. J. Brinker, D. E. Clark, D. R. Ulrich, 1984, ISBN 0-444-00898-5

Volume 33—Comparison of Thin Film Transistor and SOI Technologies, H. W. Lam, M. J. Thompson, 1984, ISBN 0-444-00899-3

Volume 34—Physical Metallurgy of Cast Iron, H. Fredriksson, M. Hillerts, 1985, ISBN 0-444-00938-8

Volume 35—Energy Beam-Solid Interactions and Transient Thermal Processing/1984, D. K. Biegelsen, G. A. Rozgonyi, C. V. Shank, 1985, ISBN 0-931837-00-6

Volume 36—Impurity Diffusion and Gettering in Silicon, R. B. Fair, C. W. Pearce, J. Washburn, 1985, ISBN 0-931837-01-4

Volume 37—Layered Structures, Epitaxy, and Interfaces, J. M. Gibson, L. R. Dawson, 1985, ISBN 0-931837-02-2

Volume 38—Plasma Synthesis and Etching of Electronic Materials, R. P. H. Chang, B. Abeles, 1985, ISBN 0-931837-03-0

Volume 39—High-Temperature Ordered Intermetallic Alloys, C. C. Koch, C. T. Liu, N. S. Stoloff, 1985, ISBN 0-931837-04-9

Volume 40—Electronic Packaging Materials Science, E. A. Giess, K.-N. Tu, D. R. Uhlmann, 1985, ISBN 0-931837-05-7

Volume 41—Advanced Photon and Particle Techniques for the Characterization of Defects in Solids, J. B. Roberto, R. W. Carpenter, M. C. Wittels, 1985, ISBN 0-931837-06-5

Volume 42—Very High Strength Cement-Based Materials, J. F. Young, 1985, ISBN 0-931837-07-3

Volume 43—Fly Ash and Coal Conversion By-Products: Characterization, Utilization, and Disposal I, G. J. McCarthy, R. J. Lauf, 1985, ISBN 0-931837-08-1

Volume 44—Scientific Basis for Nuclear Waste Management VIII, C. M. Jantzen, J. A. Stone, R. C. Ewing, 1985, ISBN 0-931837-09-X

Volume 45—Ion Beam Processes in Advanced Electronic Materials and Device Technology, B. R. Appleton, F. H. Eisen, T. W. Sigmon, 1985, ISBN 0-931837-10-3

Volume 46—Microscopic Identification of Electronic Defects in Semiconductors, N. M. Johnson, S. G. Bishop, G. D. Watkins, 1985, ISBN 0-931837-11-1

MATERIALS RESEARCH SOCIETY SYMPOSIUM PROCEEDINGS

Volume 47—Thin Films: The Relationship of Structure to Properties, C. R. Aita, K. S. SreeHarsha, 1985, ISBN 0-931837-12-X

Volume 48—Applied Materials Characterization, W. Katz, P. Williams, 1985, ISBN 0-931837-13-8

Volume 49—Materials Issues in Applications of Amorphous Silicon Technology, D. Adler, A. Madan, M. J. Thompson, 1985, ISBN 0-931837-14-6

Volume 50—Scientific Basis for Nuclear Waste Management IX, L. O. Werme, 1986, ISBN 0-931837-15-4

Volume 51—Beam-Solid Interactions and Phase Transformations, H. Kurz, G. L. Olson, J. M. Poate, 1986, ISBN 0-931837-16-2

Volume 52—Rapid Thermal Processing, T. O. Sedgwick, T. E. Seidel, B.-Y. Tsaur, 1986, ISBN 0-931837-17-0

Volume 53—Semiconductor-on-Insulator and Thin Film Transistor Technology, A. Chiang. M. W. Geis, L. Pfeiffer, 1986, ISBN 0-931837-18-9

Volume 54—Thin Films—Interfaces and Phenomena, R. J. Nemanich, P. S. Ho, S. S. Lau, 1986, ISBN 0-931837-19-7

Volume 55—Biomedical Materials, J. M. Williams, M. F. Nichols, W. Zingg, 1986, ISBN 0-931837-20-0

Volume 56—Layered Structures and Epitaxy, J. M. Gibson, G. C. Osbourn, R. M. Tromp, 1986, ISBN 0-931837-21-9

Volume 57—Phase Transitions in Condensed Systems—Experiments and Theory, G. S. Cargill III, F. Spaepen, K.-N. Tu, 1987, ISBN 0-931837-22-7

Volume 58—Rapidly Solidified Alloys and Their Mechanical and Magnetic Properties, B. C. Giessen, D. E. Polk, A. I. Taub, 1986, ISBN 0-931837-23-5

Volume 59—Oxygen, Carbon, Hydrogen, and Nitrogen in Crystalline Silicon, J. C. Mikkelsen, Jr., S. J. Pearton, J. W. Corbett, S. J. Pennycook, 1986, ISBN 0-931837-24-3

Volume 60—Defect Properties and Processing of High-Technology Nonmetallic Materials, Y. Chen, W. D. Kingery, R. J. Stokes, 1986, ISBN 0-931837-25-1

Volume 61—Defects in Glasses, F. L. Galeener, D. L. Griscom, M. J. Weber, 1986, ISBN 0-931837-26-X

Volume 62—Materials Problem Solving with the Transmission Electron Microscope, L. W. Hobbs, K. H. Westmacott, D. B. Williams, 1986, ISBN 0-931837-27-8

Volume 63—Computer-Based Microscopic Description of the Structure and Properties of Materials, J. Broughton, W. Krakow, S. T. Pantelides, 1986, ISBN 0-931837-28-6

Volume 64—Cement-Based Composites: Strain Rate Effects on Fracture, S. Mindess, S. P. Shah, 1986, ISBN 0-931837-29-4

Volume 65—Fly Ash and Coal Conversion By-Products: Characterization, Utilization and Disposal II, G. J. McCarthy, F. P. Glasser, D. M. Roy, 1986, ISBN 0-931837-30-8

Volume 66—Frontiers in Materials Education, L. W. Hobbs, G. L. Liedl, 1986, ISBN 0-931837-31-6

Volume 67—Heteroepitaxy on Silicon, J. C. C. Fan, J. M. Poate, 1986, ISBN 0-931837-33-2

Volume 68—Plasma Processing, J. W. Coburn, R. A. Gottscho, D. W. Hess, 1986, ISBN 0-931837-34-0

Volume 69—Materials Characterization, N. W. Cheung, M.-A. Nicolet, 1986, ISBN 0-931837-35-9

Volume 70—Materials Issues in Amorphous-Semiconductor Technology, D. Adler, Y. Hamakawa, A. Madan, 1986, ISBN 0-931837-36-7

MATERIALS RESEARCH SOCIETY SYMPOSIUM PROCEEDINGS

Volume 71—Materials Issues in Silicon Integrated Circuit Processing, M. Wittmer, J. Stimmell, M. Strathman, 1986, ISBN 0-931837-37-5

Volume 72—Electronic Packaging Materials Science II, K. A. Jackson, R. C. Pohanka, D. R. Uhlmann, D. R. Ulrich, 1986, ISBN 0-931837-38-3

Volume 73—Better Ceramics Through Chemistry II, C. J. Brinker, D. E. Clark, D. R. Ulrich, 1986, ISBN 0-931837-39-1

Volume 74—Beam-Solid Interactions and Transient Processes, M. O. Thompson, S. T. Picraux, J. S. Williams, 1987, ISBN 0-931837-40-5

Volume 75—Photon, Beam and Plasma Stimulated Chemical Processes at Surfaces, V. M. Donnelly, I. P. Herman, M. Hirose, 1987, ISBN 0-931837-41-3

Volume 76—Science and Technology of Microfabrication, R. E. Howard, E. L. Hu, S. Namba, S. Pang, 1987, ISBN 0-931837-42-1

Volume 77—Interfaces, Superlattices, and Thin Films, J. D. Dow, I. K. Schuller, 1987, ISBN 0-931837-56-1

Volume 78—Advances in Structural Ceramics, P. F. Becher, M. V. Swain, S. Sōmiya, 1987, ISBN 0-931837-43-X

Volume 79—Scattering, Deformation and Fracture in Polymers, G. D. Wignall, B. Crist, T. P. Russell, E. L. Thomas, 1987, ISBN 0-931837-44-8

Volume 80—Science and Technology of Rapidly Quenched Alloys, M. Tenhover, W. L. Johnson, L. E. Tanner, 1987, ISBN 0-931837-45-6

Volume 81—High-Temperature Ordered Intermetallic Alloys, II, N. S. Stoloff, C. C. Koch, C. T. Liu, O. Izumi, 1987, ISBN 0-931837-46-4

Volume 82—Characterization of Defects in Materials, R. W. Siegel, J. R. Weertman, R. Sinclair, 1987, ISBN 0-931837-47-2

Volume 83—Physical and Chemical Properties of Thin Metal Overlayers and Alloy Surfaces, D. M. Zehner, D. W. Goodman, 1987, ISBN 0-931837-48-0

Volume 84—Scientific Basis for Nuclear Waste Management X, J. K. Bates, W. B. Seefeldt, 1987, ISBN 0-931837-49-9

Volume 85—Microstructural Development During the Hydration of Cement, L. Struble, P. Brown, 1987, ISBN 0-931837-50-2

Volume 86—Fly Ash and Coal Conversion By-Products Characterization, Utilization and Disposal III, G. J. McCarthy, F. P. Glasser, D. M. Roy, S. Diamond, 1987, ISBN 0-931837-51-0

Volume 87—Materials Processing in the Reduced Gravity Environment of Space, R. H. Doremus, P. C. Nordine, 1987, ISBN 0-931837-52-9

Volume 88—Optical Fiber Materials and Properties, S. R. Nagel, J. W. Fleming, G. Sigel, D. A. Thompson, 1987, ISBN 0-931837-53-7

Volume 89—Diluted Magnetic (Semimagnetic) Semiconductors, R. L. Aggarwal, J. K. Furdyna, S. von Molnar, 1987, ISBN 0-931837-54-5

Volume 90—Materials for Infrared Detectors and Sources, R. F. C. Farrow, J. F. Schetzina, J. T. Cheung, 1987, ISBN 0-931837-55-3

Volume 91—Heteroepitaxy on Silicon II, J. C. C. Fan, J. M. Phillips, B.-Y. Tsaur, 1987, ISBN 0-931837-58-8

Volume 92—Rapid Thermal Processing of Electronic Materials, S. R. Wilson, R. A. Powell, D. E. Davies, 1987, ISBN 0-931837-59-6

MATERIALS RESEARCH SOCIETY SYMPOSIUM PROCEEDINGS

Volume 93—Materials Modification and Growth Using Ion Beams, U. Gibson, A. E. White, P. P. Pronko, 1987, ISBN 0-931837-60-X

Volume 94—Initial Stages of Epitaxial Growth, R. Hull, J. M. Gibson, David A. Smith, 1987, ISBN 0-931837-61-8

Volume 95—Amorphous Silicon Semiconductors—Pure and Hydrogenated, A. Madan, M. Thompson, D. Adler, Y. Hamakawa, 1987, ISBN 0-931837-62-6

Volume 96—Permanent Magnet Materials, S. G. Sankar, J. F. Herbst, N. C. Koon, 1987, ISBN 0-931837-63-4

Volume 97—Novel Refractory Semiconductors, D. Emin, T. Aselage, C. Wood, 1987, ISBN 0-931837-64-2

Volume 98—Plasma Processing and Synthesis of Materials, D. Apelian, J. Szekely, 1987, ISBN 0-931837-65-0

Volume 99—High-Temperature Superconductors, M. B. Brodsky, R. C. Dynes, K. Kitazawa, H. L. Tuller, 1988, ISBN 0-931837-67-7

Volume 100—Fundamentals of Beam-Solid Interactions and Transient Thermal Processing, M. J. Aziz, L. E. Rehn, B. Stritzker, 1988, ISBN 0-931837-68-5

Volume 101—Laser and Particle-Beam Chemical Processing for Microelectronics, D.J. Ehrlich, G.S. Higashi, M.M. Oprysko, 1988, ISBN 0-931837-69-3

Volume 102—Epitaxy of Semiconductor Layered Structures, R. T. Tung, L. R. Dawson, R. L. Gunshor, 1988, ISBN 0-931837-70-7

Volume 103—Multilayers: Synthesis, Properties, and Nonelectronic Applications, T. W. Barbee Jr., F. Spaepen, L. Greer, 1988, ISBN 0-931837-71-5

Volume 104—Defects in Electronic Materials, M. Stavola, S. J. Pearton, G. Davies, 1988, ISBN 0-931837-72-3

Volume 105—SiO_2 and Its Interfaces, G. Lucovsky, S. T. Pantelides, 1988, ISBN 0-931837-73-1

Volume 106—Polysilicon Films and Interfaces, C.Y. Wong, C.V. Thompson, K-N. Tu, 1988, ISBN 0-931837-74-X

Volume 107—Silicon-on-Insulator and Buried Metals in Semiconductors, J. C. Sturm, C. K. Chen, L. Pfeiffer, P. L. F. Hemment, 1988, ISBN 0-931837-75-8

Volume 108—Electronic Packaging Materials Science II, R. C. Sundahl, R. Jaccodine, K. A. Jackson, 1988, ISBN 0-931837-76-6

Volume 109—Nonlinear Optical Properties of Polymers, A. J. Heeger, J. Orenstein, D. R. Ulrich, 1988, ISBN 0-931837-77-4

Volume 110—Biomedical Materials and Devices, J. S. Hanker, B. L. Giammara, 1988, ISBN 0-931837-78-2

Volume 111—Microstructure and Properties of Catalysts, M. M. J. Treacy, J. M. Thomas, J. M. White, 1988, ISBN 0-931837-79-0

Volume 112—Scientific Basis for Nuclear Waste Management XI, M. J. Apted, R. E. Westerman, 1988, ISBN 0-931837-80-4

Volume 113—Fly Ash and Coal Conversion By-Products: Characterization, Utilization, and Disposal IV, G. J. McCarthy, D. M. Roy, F. P. Glasser, R. T. Hemmings, 1988, ISBN 0-931837-81-2

Volume 114—Bonding in Cementitious Composites, S. Mindess, S. P. Shah, 1988, ISBN 0-931837-82-0

Volume 115—Specimen Preparation for Transmission Electron Microscopy of Materials, J. C. Bravman, R. Anderson, M. L. McDonald, 1988, ISBN 0-931837-83-9

MATERIALS RESEARCH SOCIETY SYMPOSIUM PROCEEDINGS

Volume 116—Heteroepitaxy on Silicon: Fundamentals, Structures, and Devices, H.K. Choi, H. Ishiwara, R. Hull, R.J. Nemanich, 1988, ISBN: 0-931837-86-3

Volume 117—Process Diagnostics: Materials, Combustion, Fusion, A. K. Hays, A.C. Eckbreth, G.A. Campbell, 1988, ISBN: 0-931837-87-1

Volume 118—Amorphous Silicon Technology, A. Madan, M.J. Thompson, P.C. Taylor, P.G. LeComber, Y. Hamakawa, 1988, ISBN: 0-931837-88-X

Volume 119—Adhesion in Solids, D.M. Mattox, C. Batich, J.E.E. Baglin, R.J. Gottschall, 1988, ISBN: 0-931837-89-8

Volume 120—High-Temperature/High-Performance Composites, F.D. Lemkey, A.G. Evans, S.G. Fishman, J.R. Strife, 1988, ISBN: 0-931837-90-1

Volume 121—Better Ceramics Through Chemistry III, C.J. Brinker, D.E. Clark, D.R. Ulrich, 1988, ISBN: 0-931837-91-X

Volume 122—Interfacial Structure, Properties, and Design, M.H. Yoo, W.A.T. Clark, C.L. Briant, 1988, ISBN: 0-931837-92-8

Volume 123—Materials Issues in Art and Archaeology, E.V. Sayre, P. Vandiver, J. Druzik, C. Stevenson, 1988, ISBN: 0-931837-93-6

Volume 124—Microwave-Processing of Materials, M.H. Brooks, I.J. Chabinsky, W.H. Sutton, 1988, ISBN: 0-931837-94-4

Volume 125—Materials Stability and Environmental Degradation, A. Barkatt, L.R. Smith, E. Verink, Jr., 1988, ISBN: 0-931837-95-2

Volume 126—Advanced Surface Processes for Optoelectronics, S. Bernasek, T. Venkatesan, H. Temkin, 1988, ISBN: 0-931837-96-0

MATERIALS RESEARCH SOCIETY CONFERENCE PROCEEDINGS

Tungsten and Other Refractory Metals for VLSI Applications, R. S. Blewer, 1986; ISSN 0886-7860; ISBN 0-931837-32-4

Tungsten and Other Refractory Metals for VLSI Applications II, E.K. Broadbent, 1987; ISSN 0886-7860; ISBN 0-931837-66-9

Ternary and Multinary Compounds, S. Deb, A. Zunger, 1987; ISBN 0-931837-57-x

Tungsten and Other Refractory Metals for VLSI Applications III, Victor A. Wells, 1988; ISSN 0886-7860; ISBN 0-931837-84-7

Atomic and Molecular Processing of Electronic and Ceramic Materials: Preparation, Characterization and Properties, Ilhan A. Aksay, Gary L. McVay, Thomas G. Stoebe, 1988; ISBN 0-931837-85-5

Materials Futures: Strategies and Opportunities, R. Byron Pipes, U.S. Organizing Committee, Rune Lagneborg, Swedish Organizing Committee, 1988; ISBN 0-55899-000-3

PART I

Novel Processing Methods for Metal Based Composites

NEW PATHWAYS TO PROCESSING COMPOSITES

by

Robert Mehrabian
Materials Department
College of Engineering
University of California, Santa Barbara
Santa Barbara, CA 93106

ABSTRACT

Compositing routes are reviewed for the fabricaiton of metal and metal-ceramic matrices combined with ceramic reinforcements and/or ductilizing phases. The important role of micromechanics in elucidating microstructural design principles to guide the processing "pathways" is emphasized. Specific processing techniques are described including incorporation of particles and fibers into melts, melt infiltration into preforms, powder metallurgy and melt oxidation.

I. INTRODUCTION

Structural metal and metal-ceramic matrix composites exhibit unique set of microstructures/properties not available in either monolitic ceramic or metallic materials. Potential new applications range from reciprocal and turbine engines to aerospace and space structures. Resurgence of interest in this field has evolved from two significant recent developments. First, a new approach for the design of composites based on micromechanics has emerged, and second, a variety of new processing techniques have been developed that permit production of tailored microstructures based on these design principles.

The central thrust of the microstructural design effort is to exploit micromechanics models to guide the materials processing "pathways" for the production of composites. This approach has emerged primarily through the DARPA University Research Initiative (1). Figure 1 shows the relationship between processing, characterization, mechanical measurements/observations, and micromechanics models and microstructural design. Obviously this approach to material processing is an iterative one where each step of the cycle is continuously revised and fine tuned based on knowledge-base generated in the previous steps. Nevertheless, the microstructural design principles provide the primary guidance for the processing of useful components and products embodying the use of novel microstructures. For example, research concerning the toughening of brittle (ceramic) matrices by the incorporation of ductile phases and brittle fibers and whiskers has provided important perspectives on property optimization in high temperature intermetallic systems.

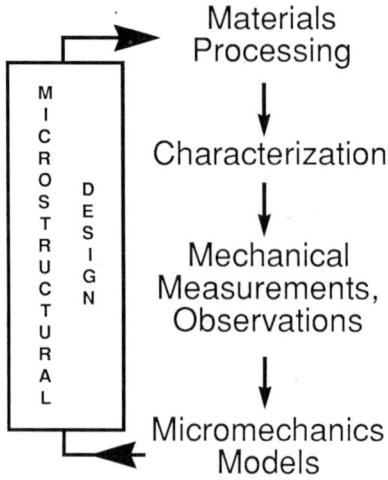

Figure 1. Current approach to materials processing based on microstructural design principle.

In particular, it is now accepted that brittle matrix composites having high toughness require strongly bonded interfaces for ductile reinforcements and weakly bonded interfaces for brittle (e.g. fiber) reinforcements. Ductile phases can thus contribute most effectively to toughening by bridging contributions from the deforming particles. Examples of improved toughness in intermetallic γ-TiAl alloy by incorporation of ductile Nb particles are given by C.K. Elliott et al in this volume (2). On the other hand, toughening by brittle fibers and whiskers requires debonding along the fiber/matrix interface at the crack front, which is governed by the inherent bond strength of the interface. Interface debonding thus allows the fibers to remain intact in the immediate crack wake, and subsequent fiber pull-out contributes to the toughening. An example of the latter, an Al/Al_2O_3 matrix composite containing coated Nicalon™ fibers produced by the Lanxide™ process, is shown in Figure 2.

Measured room temperature toughness of this composite is 29 MPa-m$^{1/2}$. These examples not only provide evidence of success for the new microstructural design approach, but also show the critical role of interfacial phenomena in multi-material composite systems.

The combination of advances in micromechanics and the availability of broader processing "pathways", due to development of new techniques, has provided a unique opportunity in this field. In general, each compositing technique is confined to a thermodynamic/kinetic space within which specific "pathways" can be mapped-out, through control of composition and process variables, to achieve the desired composite microstructures. Obviously, this field is too broad for coverage in a single review article. In this paper emphasis is placed on compositing processes which yield either a metal or a metal-ceramic matrix incorporating particle, whisker, and fiber reinforcements. Examples are drawn from three generic processing approaches which include:

Liquid Metal Processing,
Powder Processing,
and Melt Oxidation.

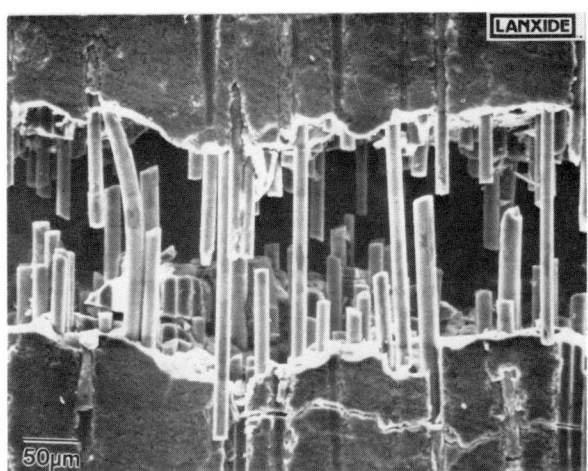

Figure 2.   Aluminum alloy/Al_2O_3 matrix composite reinforced with Nicalon™ fibers produced by the Lanxide™ process. Note extensive fiber pull-out that contributes to toughening. From A.W. Urquhart, Lanxide Corporation.

II. LIQUID METAL PROCESSING

Liquid metal processing techniques can be divided into two general categories; those in which particles, whiskers and fibers are introduced into a melt, and processes where a liquid metal infiltrates fiber bundles and/or sintered preforms. There are a number of variations in the latter category vis-a-vis the use and magnitude of pressurization.

Incorporation of Particles/Fibers into Melts

A significant body of work done in the metal-matrix composites field deals with introduction of ceramic particles and discontinuous fibers both to agitated partially solid slurries and to completely liquid alloys. Most of this work has utilized aluminum alloy matrices into which SiC and Al_2O_3 particulates (3 to 150 µm in size) and Al_2O_3 fibers (3 to 6mm long) have been introduced (3-6).*

---

*Martin Marietta's proprietary XD™ process is not included in this review. In this process small particles (<1 µm) of hard phases such as TiB_2 and TiN are formed in a liquid matrix.

A requirement for ductile matrix reinforced composites is a strong bond to the fibers to permit load transfer from matrix to fiber. In the Al/Al$_2$O$_3$ system, simple exposure of the fiber to liquid unalloyed aluminum does not yield "wetting" unless T ≥ 1173K. One approach is to alloy the aluminum with an element which can interact chemically with the fiber to produce a new phase at the interface which is readily "wet". In two studies (4,5), discontinuous Al$_2$O$_3$ fibers were introduced into partially solid slurries and completely liquid alloys. In both cases it was found that "wetting" and bonding of the fibers could be achieved by these techniques and that agitation was essential. Induced convection of the melt permits disruption of contamination films or absorbed layers and produces intimate contact between the fiber and the melt, so that interface interaction is facilitated. Microscopic examination of the composite interfaces revealed the existance of an altered microstructure around the Al$_2$O$_3$ fibers which consisted of a fine multi-phase material. Features common to all the structures were the existence of an intimate bond, the absence of voids at the fiber boundary and the presence of fine polycrystalline α-Al$_2$O$_3$ in the interaction zone. In the case of Al-Mg alloys, the interaction zone also contained MgAl$_2$O$_4$ (4,7). It was suggested that the MgAl$_2$O$_4$ spinel, Figure 3, formed by reaction between magnesium, which was in solution, and both the Al$_2$O$_3$ fiber and fine α-Al$_2$O$_3$, which resulted from oxidation of the melt. Reactions [1] and [2] are equivalent.

$$Mg + 2Al + 2O_2 = MgAl_2O_4 \qquad [1]$$

$$Mg + 4/3 Al_2O_3 = MgAl_2O_4 + 2/3\, Al \qquad [2]$$

The fibers are thought to provide suitable substrate for the growth of the spinal. In the Al-Cu-Mg alloy both Mg and Cu enrichments were detected around the fibers. The interaction zone most likely consisted of MgAl$_2$O$_4$, α-Al$_2$O$_3$ and CuAl$_2$O$_4$ (4).

Although the Al$_2$O$_3$ fibers were successfully incorporated into both the partially solid and completely liquid aluminum alloys, some problems still existed in the fabrication processes. First, fiber damage was observed, especially when partially solid slurries were used. Second, and more importantly, interface interactions are a function of time and temperature of fabrication and the control of an adequate matrix/fiber interfacial bond, without extensive interphase formation is difficult. This problem will of course be exacerbated when high temperature alloy matricies and/or more reactive fibers (e.g. SiC) are used.

An important finding of one of these studies (5) was that planar random orientation and increased volume fraction of discontinuous fibers in the composite could be achieved by squeeze casting the original composites in a die containing a porous ceramic filter. Figure 4 shows a schematic illustration of the forging dies and the ceramic filter used. Pressurization of the composite in this arrangement results in infiltration of some of the liquid metal into the porous structure of the filter with concurrent alignment of the fibers in a random two-dimensional mat.

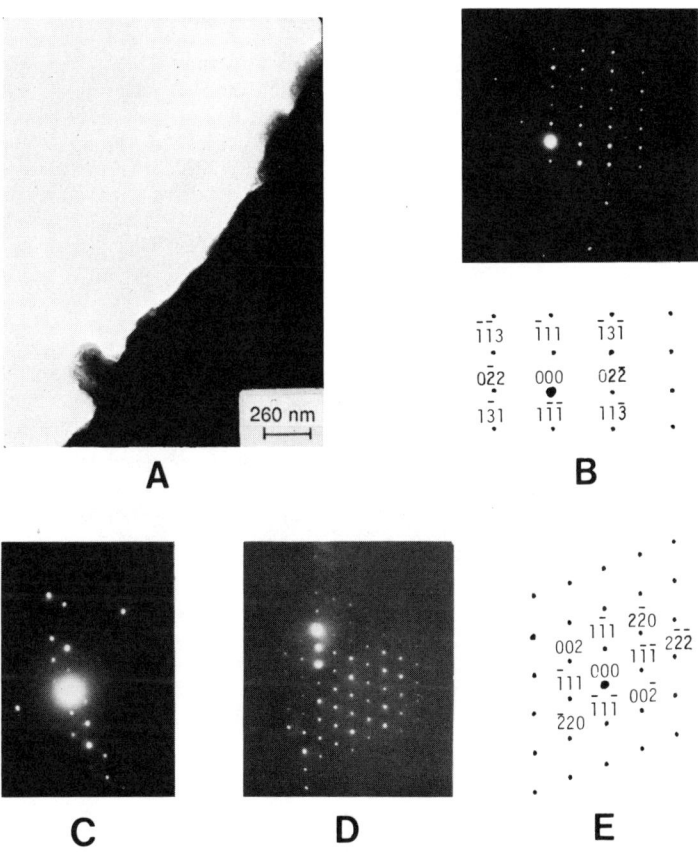

Figure 3. Observations relating to spinel diffraction patterns in Al-4wt%Mg alloy/Dupont FP I α-Al$_2$O$_3$.
- (a) Profile of fiber isolated from the matrix.
- (b) Diffraction pattern from one point on fiber edge with indexing as spinel, Z.A. [211].
- (c) Diffraction pattern from another point on fiber edge.
- (d) Diffraction pattern from MgAl$_2$O$_4$ powder.
- (e) Indexing of D and C as spinel, Z.A. [110]. 002 reflections are the result of double diffraction.

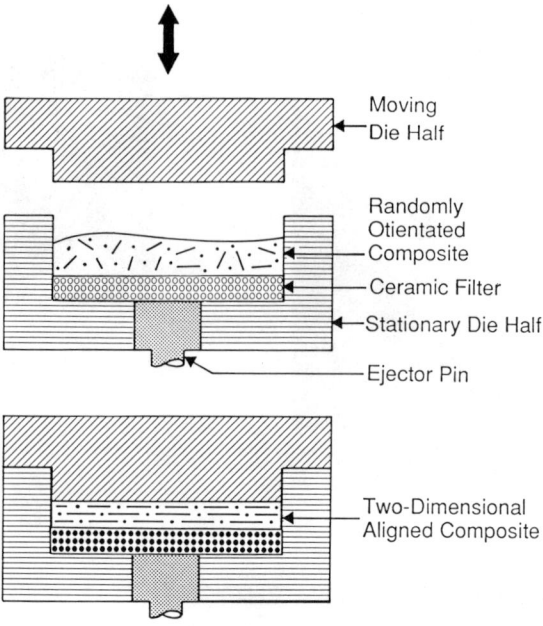

Figure 4. Schematic of the forging dies and ceramic filter used to two-dimensionally (planar-random) orient the fibers in Al/Al$_2$O$_3$ composites.

Figure 5 shows an SEM micrograph of the vertical section of the composite which was electroetched to dissolve away approximately 50 μm of the matrix.

Melt Infiltration into Preforms

Infiltration processes are carried out with and without application of external pressure. The "wetting" characteristic of the ceramic preform materials by the liquid alloy is an important parameter since it affects infiltration by capillary action and/or externally applied pressure. The variables influencing "wettability" include surface and interfacial energies which are in turn influenced by alloy composition, ceramic preform material, surface treatments, surface geometry, interfacial interactions, atmosphere, temperature and time. The "wettability" is typically measured by the equilibrium contact angle of the liquid alloy on the solid ceramic and may be described by the Dupre equation:

$$\cos \theta = (\gamma_{SV} - \gamma_{LS}) / \gamma_{LV} \qquad [3]$$

where θ is the contact angle, γ_{LV} is the liquid-alloy/gas-phase surface energy, γ_{SV} is the ceramic-preform/gas-phase surface energy and γ_{LS} is the liquid-alloy/ceramic-preform interfacial energy.

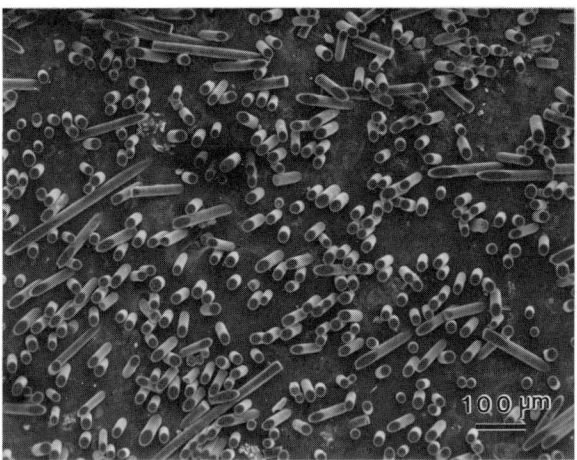

Figure 5. Electroetched vertical section of Al/ α-Al$_2$O$_3$ planar-random oriented fiber composite. ~50 μm of the matrix has been dissolved away.

Figure 6 shows schematic presentations of the surface energies involved and the contact angle for "wetting" and "non-wetting" conditions in a ceramic-preform capillary. When $\gamma_{SV} < \gamma_{LS}$, θ > 90° the system is characterized as "non-wetting". The surface energy of the ceramic-preform/gas-phase is less than the liquid/preform interfacial energy. Therefore, by depression of the liquid in the capillary below the melt top, the surface area of the ceramic exposed to the gas phase is increased at the expense of the liquid/preform area. Similarly, when $\gamma_{SV} > \gamma_{LS}$, θ < 90° the system is characterized as "wetting" and the liquid rises in the capillary to form a liquid-solid interface at the expense of the solid surface area. The pressure differential, ΔP in Figure 6, can be related to hydrostatic pressure in the capillary by:

$$\Delta P = \rho g H = 2\gamma_{LV} \cos \theta / r \qquad [4]$$

where H is height of the liquid column, g is acceleration due to gravity, r is radius of the capillary, and ρ is the density of the liquid.

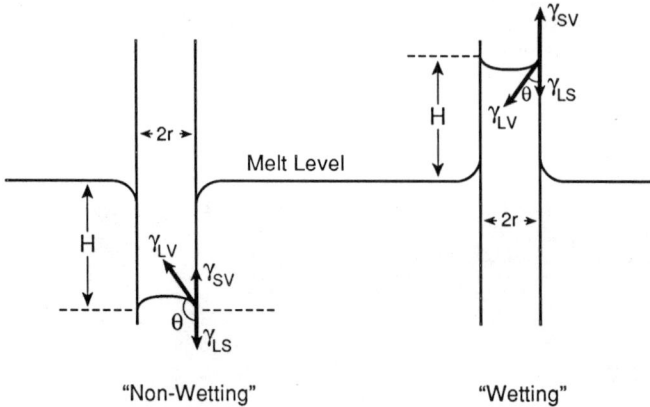

Figure 6. Schematic illustration of capillary rise and depression in the interstice of a ceramic preform in contact with the liquid alloy.

Equation [4] also describes the mininum pressure for melt entry, initiation of infiltration, into the capillaries of a porous fiber/particulate preform. An additional pressure gradient in the melt is required for the infiltration process to compensate for the resistance to flow caused by viscous forces in the fine interstices of the porous medium. Estimates and measurements of these pressures have been made for specific fiber/particulate geometrics, volume fractions ceramic and alloy/preform compositions. [8-11].

As examples, consider infiltration of liquid aluminum into a two-dimensional square fiber array, simulating planar-random orientation of FP $\alpha - Al_2O_3$ fibers and inorganic binder (SiO_2) cemented Saffil™ product. The former, ~20 μm in diameter is primarily pure $\alpha - Al_2O_3$, while the latter, ~3 μm in diameter, is $\delta - Al_2O_3$ + 3 to 4wt% SiO_2 cemented together with an additional ~5wt% SiO_2 binder. The surface energy of liquid aluminum as a function of temperature is given by the following relationship [12]:

$$\gamma_{LV} = 914 - 0.35 \, (T - 933) \qquad [5]$$

For a square fiber array equation [4] becomes:

$$\Delta P = 2\gamma_{LV} \cos\theta \, / \, \left\{ R\left[(\pi / 2V_f)^{1/2} - 1\right] \right\} \qquad [6]$$

when R is fiber radius and V_f is volume fraction of fibers.

From equation 5, the surface energy of pure liquid aluminum, at 1100K is ~855 mJ/m². Sessile drop experiments with aluminum on Al_2O_3 indicate reported contact angles of up to 140° at this temperature [4]. The calculated pressures for generating the meniscus curvatures at the initiation of infiltration thus become ~0.1MPa for 10 μm radius FP Al$_2$O$_3$ and 0.68 MPa for 1.5 μm radius Saffil™ when $V_f = 0.3$.

Similar calculations can be readily carried out for the initiation of infiltration in ceramic preforms consisting of sintered particulates. In this case the capillary radius in equation [4] is replaced with the hydraulic radius [13]. For spherical particles of radius R it is:

$$r_h = \frac{R(1 - V_f)}{3V_f} \quad [7]$$

The pressure drop necessary to overcome the viscous flow resistance in the preform can be similarly calculated using the Blake-Kozeny equation [13]. Calculations [8] and measurements [11] for laboratory sized specimens show these pressures to be of the same order of magnitude as those necessary for infiltration initiation. Infiltration times are proportional to the applied external pressure while the pressure itself is inversely proporational to the square of the dimensions of the interstices in the preform. A further complication arises when the preform is below the melting temperature of the alloy and concurrent solidification in these interstices has to be taken into account.

Shortcomings of pressure infiltration processes include fiber damage, preform compression, microstructural nonuniformity and fiber to fiber contact. For example, Mortenson et al [14] studied the solidification characteristics of Al-4.5wt%Cu matrix alloy pressure infiltrated (1000psi) into bundles of 140 μm SiC fibers. Reduced microsegragation was found in the matrix, within the fiber interstices, for long solidification times when compared to conventional solidification. The technique used, however, produced inhomogeneous microstructures with extensive fiber to fiber contact in the center of the specimens while pheripheral regions were free of fibers. One proposed method to address some of these issues is to use inorganic binders to lend rigidity to the preform [8,9]. It is expected that these composites would have low toughness. In another approach, fine SiC whiskers are distributed among coarser carbon fibers to keep them separated during liquid infiltration and control microstructural uniformity [15]. Finally, a new process developed by the author permits uniform infiltration of two-dimensionally random-oriented mats of discontinuous fibers, Figure 7.

Pressure infiltration is not necessary if melt/preform compositions and temperatures are controlled such that "wetting" conditions are achieved and the melt permeates, "wicks", into the preform. Aksay and his co-workers have studied the fabrication of microdesigned metal-ceramic composites using capillarity and reaction thermodynamics to advantage [16].

Figure 8 shows measured contact angle of Al on B$_4$C. "Wetting" is readily achieved at high temperatures through the formation of new phases as a function of time (16). High and low volume fraction B$_4$C powder compacts were prepared by dispersed and flocculated aqueous suspensions, respectively. Subsequent sintering in the 2073 to 2473K temperature range resulted in B4C skeletons ready for infiltration. These heat treatments not only control the preform and composite microstructures, Figure 9, but also influence reaction rates and products during infiltration (16). Infiltration is achieved by the capillary suction of the liquid metal into the interstices of the preform. Variation of time and temperature of infiltration can result in clean interfaces, such as that

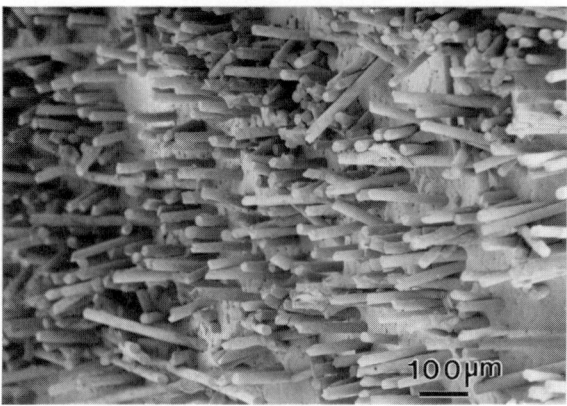

Figure 7. Microstructure of Al-Mg/FP Al2O3 planar-random composite produced by a new process. The matrix has been electro-etched to reveal fiber distribution.

shown in Figure 10, or the formation of compounds. The latter can also be formed by subsequent heat treatment of the composite. Properties can thus be tailored by both the volume fraction and distribution of the aluminum and the B4C, and the extent and composition of the chemical reaction products in the microstructure (16).

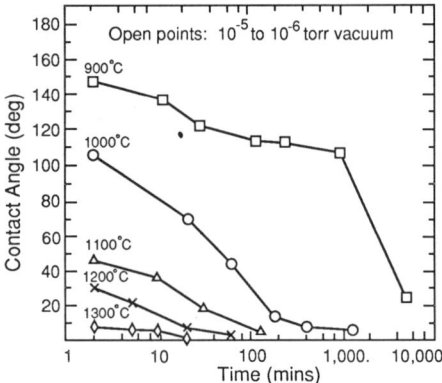

Figure 8 Contact angle of molten aluminum on B4C as a function of temperature and time. From reference [16].

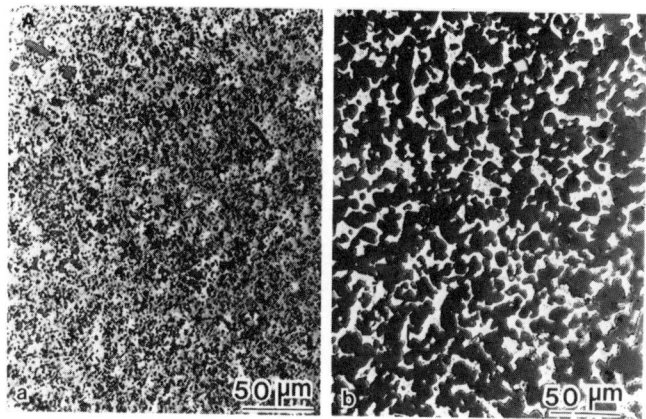

Figure 9.   Al/B4C composites prepared by the capillary infiltration approach. [a]. Without B4C pre-heat treatment, [b] with B4C pre-heat treatment. From reference [16].

III. POWDER PROCESSING

Atomization (rapid solidification) has provided "pathways" to new microstructures for metals and intermetallics not available in conventional ingot metallurgy (17,18). These microstructures are presently being exploited as composite matrices into which strengthening and/or ductilizing materials are incorporated.

In one process pioneered by General Electric, rapid solidification by plasma deposition (RSPD) is used for direct production of composites. Powders of the matrix alloy are fed into an RF plasma gun and deposited onto a rotating mandrel in vacuum. Filaments of the desired fiber are wound on the mandrel prior to deposition to produce monotapes. The advantage of this process is that it combines rapid solidification of molten droplets (on the mandrel substrate) with the incorporation of continuous fibers. Subsequent closure of porosity in the composite may be facilitated by hot isostatic pressing (HIP).

In more conventional processes, particulates and whiskers of the reinforcement are blended with rapidly solidified matrix alloy powders and consolidated. One such approach, primarily used for aluminum alloys, is schematically presented in Figure 11. The blended composite is cold compacted, degassed, hot pressed and extruded into final shape. Figure 12 shows representative microstructures of aluminum matrix composites containing 0.15 to 0.2 volume fraction SiC particles produced by this technique. Mechanical properties of an extruded composite containing $V_f = 0.15$ SiC particles are given in Table I. Note the very respectable fracture toughness of this composite. Fatigue properties of the same composite at room temperature and 450K are superior to the unreinforced matrix alloy, Figure 13.

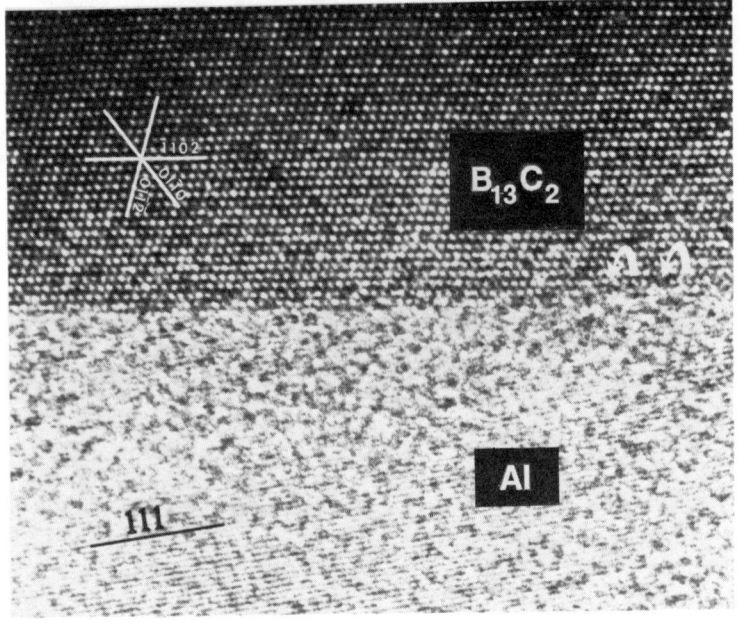

Figure 10. High resolution transmission electron micrograph revealing no interphase formation in Al/B4C composite. From reference [16].

TABLE I
Properties for 2000 series P/M aluminum alloy matrix/15% SiC composite.
From T. B. Gurganus, ALCOA.

	Orientation	
	L, L-T	T, T-L
Ultimate tensile strength, MPa	524	482
Tensile yield strength, MPa	372	358
Elastic modulus GPa	103	-
Elongation, %	7.5	4
Reduction of area, %	9.4	5.2
Fracture toughness, K_{IC}, MPa.m$^{1/2}$	36	27.5
Fatigue limit, MPa ($N=10^7$, $R=-1$)	241	

(all properties taken in longitudinal direction from 25mm X 89mm extruded bar stock)

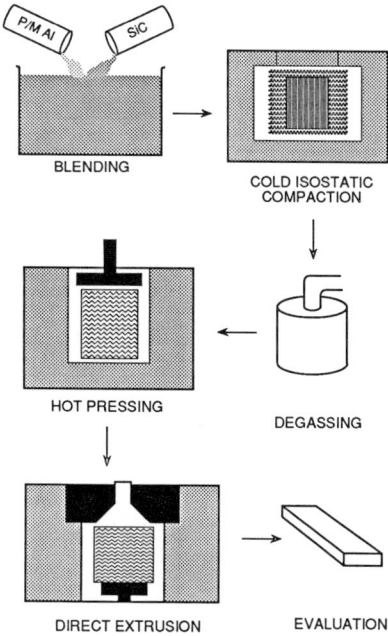

Figure 11. Schematic description of processing route for P/M Al/SiC particulate composites. From T.B. Gurganus, ALCOA.

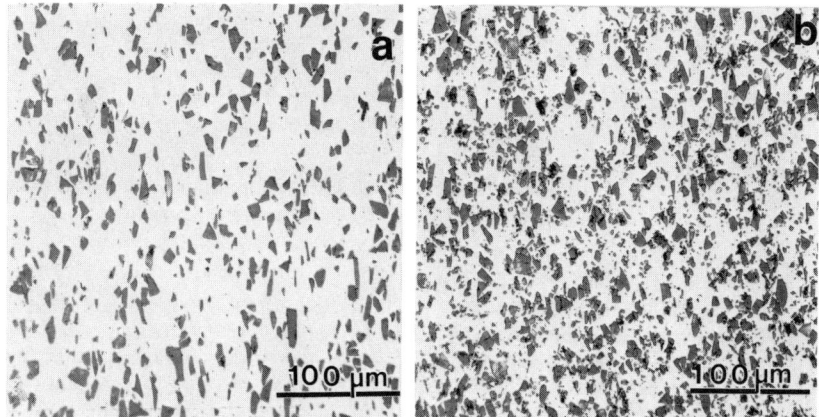

Figure 12. Microstructures of 2000 series P/M aluminum alloy matrix/ SiC particulate composites. [a]. Volume fraction of SiC $V_f = 0.15$, [b] $V_f = 0.2$. From T.B. Gurganus. ALCOA.

Reported mechanical properties of SiC whisker reinforced PM aluminum matrix composites are generally higher than the particulate reinforced ones listed in Table I (19).

Recent interest in high temperature intermetallics, e.g. titanium aluminides, has focused efforts in both the development of designed microstructures and the processing of same. Fiber reinforcements are considered for high temperature creep improvement. On the other hand, improvements in room temperature toughness has been demonstrated through ductile phase compositing. For example, γ-TiAl powders produced by rapid solidification were mixed with 0.2 volume fraction of Nb powders and hot isostaticaly pressed to full consolidation to improve toughness (2). One of the difficulties involved in processing from powders concerns interface interactions during densification. Figure 14 shows the evolution of an interphase zone between Nb and γ-TiAl. Since service temperatures are generally lower than the consolidation temperatures, it is important to establish conditions of pressure, temperature and time that permit full densification without appreciatable reaction.

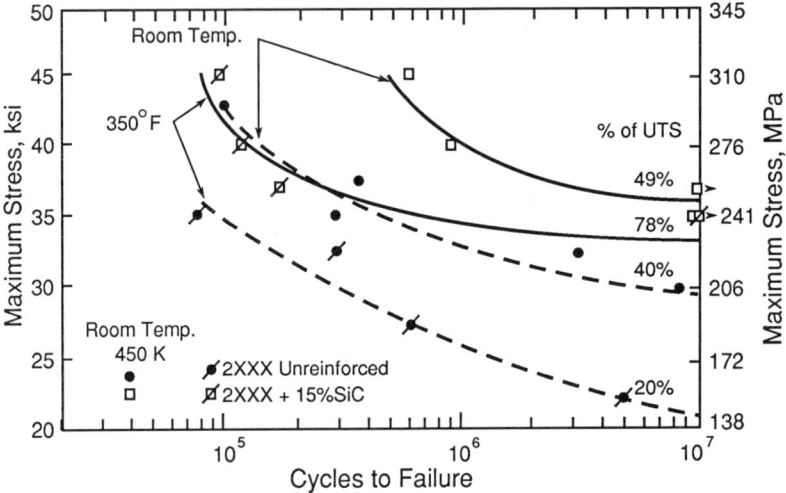

Figure 13. S-N fatigure behavior of 2000 series aluminum alloy matrix/ 15% SiC composite (R=-1, 25 Hz). From T.B. Gurganus, ALCOA.

The consolidation of powders into dense bodies by HIPing can be expressed by means of densification rate maps developed by Ashby [20,21]. Densification rate is predicted as a function of material properties and process parameters and depicted in maps, where density is plotted as a function of pressure for constant temperature, or vice versa. Figure 15 shows a preliminary map for densification of γ-TiAl calculated by Ashby, assuming an initial packing density of 62% and a temperature of 1480K [22]. The map is divided into regions representing the range of dominance of the different densification mechanisms; Nabaro-Herring creep at low pressures, power law creep at intermediate pressures and plastic yielding at high pressures. The isochronal curves in Figure 15 represent the density attainable at a given pressure in the time indicated. The critical role of increased pressure in reducing time at temperature to reach equivalent densification is clearly evident thus lending further credence to the need for rapid high pressure consolidation processes such as shock compaction and the Ceracon™ process [23].

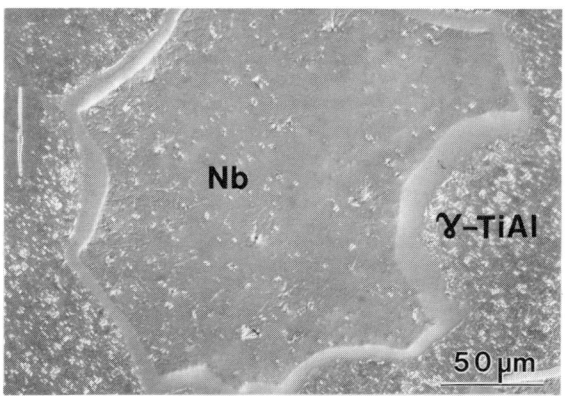

Figure 14. Development of interaction zone between ductile niobium particle and γ-TiAl matrix during HIP consolidation.

A promising approach yet to be investigated is to combine thermodynamic and kinetic information on interface interactions, between matrix and reinforcing phases, with densificiation maps to generate optimum "pathways" to full consolidation and controlled interface interactions. A first step in this process is to develop densification maps for composite systems where the role of reinforcements, especially particles/fibers stronger than the matrix, in the densification rates are first predicted.

III MELT OXIDATION

This is a new and novel technique for the fabrication of ceramic-metal matrix composites which utilizes a ceramic (particulate,fiber,whisker) preform that is continuously infiltrated by a molten alloy as it undergoes oxidation reaction with a gas phase. The high temperature oxidation of the molten alloy in the interstices of the ceramic preform produces a matrix material composed of a mixture of the oxidation reaction product and unreacted metal alloy. The growth of the oxide occurs continuously away from the initial molten-alloy/ceramic preform interface while fresh alloy is supplied to the gas/metal reaction interface by the fine microscopic channels between the grains of the oxide product [23]. The final composite consists of the ceramic preform in an interconnected matrix (ceramic reaction product and residual metal alloy). The primary advantages of the process include its ability to form relatively complex, fully dense composite shapes [24] with the requisite microstructures designed by emerging micromechanics models. As example, the unreacted metal alloy serves as the ductile phase while duplex interfaces containing a weak interfacial bond between the reaction product and the ceramic preform may provide the additional toughening by interfacial sliding and fiber pull-out mechanisms, Figure 2. Exceptional mechanical properties have been reported for these composites, Table II.

ROOM TEMPERATURE PROPERTIES
OF SELECTED LANXIDE™ COMPOSITES
From A.W. Urquhart, Lanxide Corporation

MATRIX/REINFORCEMENT	4-pt MOR	TOUGHNESS
aluminum-alumina/alumina particles	500MPa	9 MPa-m$^{1/2}$
aluminum-alumina/silicon carbide particles	523	8
aluminum-alumina/Nicalon™ fibers	997	29

The compositing process [23] permits fabrication of uniform microstructures through continuous "wicking" of the molten alloy, by capillary suction; to the oxidation reaction interface and continuous availability of the gas phase necessary for the reaction. Specific dopants are added to the molten alloy or externally introduced to the alloy/preform interface to affect interfacial energetics and the oxidation process. For aluminum alloys the process is carried out at relatively high temperatures, ~1000K to ~1700K. As example, typical composite fabrication temperatures for aluminum alloy -Al_2O_3 matrix composites with particulate Al_2O_3 preforms are ~1400K to ~1650K with dopant additions of Mg-Si, Mg-Ge, Mg-Sn and Mg-Pb [23]. Typical oxidation reaction rates for Al_2O_3 formation are 2-3 cm in 24 hours using air as the oxidant.

Figure 16 shows microstructures of aluminum alloy -Al_2O_3 matrix composites containing particulate and fiber SiC. The light areas denote unoxidized metal through which molten alloy is continuously transported to the oxidation front.

CONCLUDING REMARKS

The historical approach to materials processing through the exploitation of processing/microstructure/property-performance relationships must be suplemented with microstructural design principles if we are to continue making progress in the commercialization of complex multimaterial high performance composites. The micromechanics models have progressed to a point where an iterative approach to process and micromechanics modelling/experiments is most valuable. Further, cross-cultivation of ideas in the design and processing of interfaces in ceramic and metallic matrix composites is also needed.

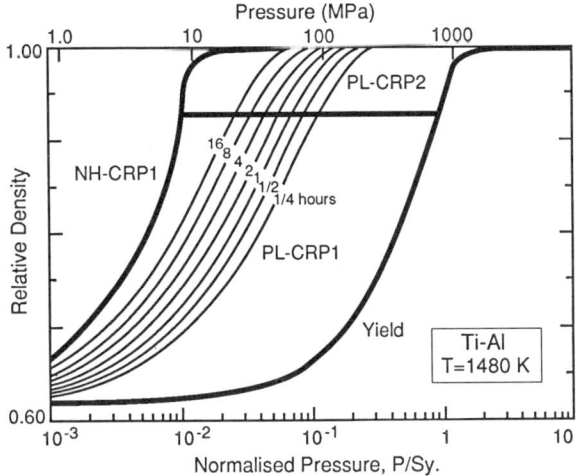

Figure 15. HIP diagram for TiAl powders with an initial packing density of 0.62 at a constant temperature of 1480K. Regions are labeled after the dominating densification mechanism (PL-CRP for power-law creep and NH-CRP for Nabarro-Herring creep). Numbers after mechanism refer to stage 1 and 2 of the densification process.

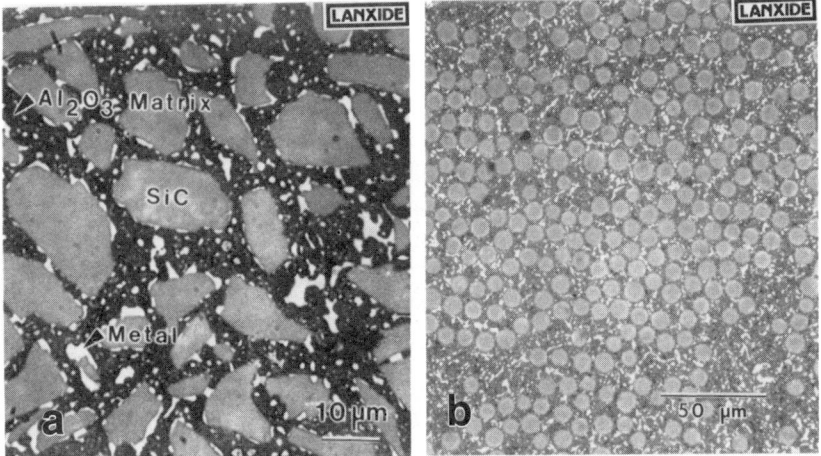

Figure 16. Aluminum alloy Al$_2$O$_3$ matrix/SiC composites produced by melt oxidation. [a] Sic is in particulate form, [b] SiC is Nicalon™ fibers. From A.Q. Urquhart, Lanxide Corporation

ACKNOWLEDGEMENTS

The author is grateful to Drs. I.A. Aksay of Washington University, T.B. Gurganus of ALCOA and A.W. Urquhart of Lanxide Corporation for enlightening discussions, background materials and a large fraction of the figures and data presented in this review. The support of the Defense Advanced Research Projects Agency (DARPA) through a University Research Initiative Grant N00014-86-K-0753, supervised by Dr. P.A. Parrish and monitored by Dr. S.G. Fishman of the Office of Naval Research, is also gratefully acknowledged.

REFERENCES

1. A.G. Evans and R. Mehrabian: URI Annual Reports: Contract N00014-86-K-0753

2. C.K. Elliott, G.R. Odette, G.E. Lucas and J.W. Sheckherd, in this proceeding.

3. A. Sato and R. Mehrabian, Met. Trans. 5, 1974, p. 1899

4. C.G. Levi, G.J. Abbaschian and R. Mehrabian, Met Trans. 9A, 1978, p. 697

5. B.F. Quigley, G.J. Abbaschian, R. Wunderlin and R. Mehrabian, Met Trans. 13A, 1982, p. 93

6. F.M. Hoskings, F. Folgar Portillo, R. Wunderlin and R. Mehrabian, Journal of Materials Science 17, 1982, p. 477

7. A. Munitz, M. Metzger and R. Mehrabian, Met. Trans. 13A, 1982, p. 93

8. T.W. Clyne and J.F. Mason, Met. Trans. 18A, 1987, p. 1519

9. T.W. Clyne, M.G. Bader, G.R. Cappleman, P.A. Hubert, Journal of Materials Science, 20, 1985, p.85

10. S.Y. Oh, J.A. Cornie and K.C. Russell, Ceramic Engineering and Science Proceedings of the 11th Conference on Composites, Adv. Ceram. Materials, 1987, p. 912

11. J.A. Cornie, A. Mortensen and M.C. Flemings, Proceedings of the Sixth International Conference on Composite Materials, ICCM and ECCM, Editors, F.L. Mathews, N.C.R. Buskell, J.M. Hodgkinson and J. Morton, 1987, p. 2.297

12. Smithers Metals Reference Book, 6th Edition, Edited by Eric A. Brandes, Butterworths, London, 1983

13. D.H. Geiger and D.R. Poirier, Transport Phenomena in Metallurgy, Addison Wesley, Reading, Mass., 1973, p. 93

14. A. Mortensen, J.A. Cornie and M.C. Flemings, Met. Trans. 19A, 1988, p. 709

15. Sin-ichi Towata, Hajime Ikuno and Sen-ichi Yamada, Proceedings Sixth International Conference on Composite Materials, ICCM and ECCM, Editors F.L. Mathews, W.C.R. Buskell, J.M. Hodgkinson and J. Morton, 1987, p. 2.412

16. A.J. Pyzik, I.A. Aksay and M. Sarikaya, in Ceramic Microstructures '86, Role of Interfaces, Editors J.A. Pask and A.G. Evans, Plenum Press, N.Y. 1987, p. 45

17. Rapid Solidification Processing: Principle and Technologies IV, Editors R. Mehrabian and P.A. Parrish, Claitor's Publishing Div., Baton Rouge, Louisiana, 1988

18. Processing of Structural Metals by Rapid Solidification, Editors, F.H. Froes and S.J. Savage, ASM International, 1987

19. P. Niskanen and W.R. Mohn, Advanced Mats. and Processes, Metals Progress 3, 1988, p. 39

20. A.S. Helle, K.E. Easterling and M.F. Ashby, Acta Metall. 33, 1985, p. 2163

21. M.F. Ashby, submitted for publication

22. M.F. Ashby, private communication

23. M.S. Newkirk, A.W. Urquhart, H.R. Zwicker, and E. Breval: Journal of Materials Research 1 (1), 1986, p. 81

24. A.W. Urquhart, private communication

HIGH-ENERGY HIGH-RATE PROCESSING OF HIGH-TEMPERATURE METAL-MATRIX COMPOSITES

C. Persad, S. Raghunathan, B.-H. Lee, D.L. Bourell, Z. Eliezer and H.L. Marcus

Center for Materials Science and Engineering
The University of Texas at Austin,
Austin, Texas 78712.

ABSTRACT

Advances in kinetic energy storage devices have opened up a new approach to powder processing of High Temperature Composites. The processing consists of internal heating of a customized powder blend by a fast electrical discharge of a homopolar generator. The high-energy high-rate "1MJ in 1s" pulse permits rapid heating of a conducting powder in a cold wall die. This short time at temperature approach offers the opportunity to control phase transformations and the degree of microstructural coarsening not readily possible using standard powder processing approaches. This paper will describe the consolidation results of two high temperature composite materials, (W-Ni-Fe)/B_4C and (Ti_3Al+Nb)/SiC. The focus of this study was the identification of the reaction products formed at the matrix/reinforcement interface as a function of input energy and applied stress. Input energies beyond a threshold value for each system were required to produce detectable reaction products. In the (W-Ni-Fe)/B_4C system, the reaction products formed at 4000 kJ/kg input energy under 420 MPa applied stress were a series of complex carbides and borides including W_2C, FeWB, Fe_3C, Fe_6W_6C and Ni_4B_3. The intermetallic Fe_7W_6 was also observed. In the (Ti_3Al+Nb)/SiC system, the reaction products observed at 3400 kJ/kg and 210 MPa were TiC and $TiSi_2$.

INTRODUCTION

The new (Ti_3Al + Nb) - based matrices reinforced with silicon carbide represent a class of high temperature metal matrix composites (MMC) with attractive specific modulus and potential for high temperature service. It has been shown that the (Ti_3Al + Nb) /SiC composites have significantly higher strength/density values over the wrought superalloys in the 500K to 1400K temperature range [1]. These materials are expected to find application in the next generation of high performance aircraft and turbines. The Ti-6Al-4V/SiC MMC shows reduced tensile strength after high temperature exposure. This reduction in strength has been correlated with the formation and growth of deleterious reaction products at the matrix-reinforcement interface [2]. It has been suggested that lowering of the consolidation temperature will result in a reduced amount of reaction product [3]. Similar reaction products may be responsible for the 3 μm thick reacted layer observed in the hot-pressed (Ti_3Al + Nb)/ SiC composites observed by Brindley [1].

A completely different group of high temperature applications utilizing high hot hardness are anticipated for the cobalt-free substitutes for WC + Co composites. The B_4C -reinforced W-Ni-Fe composites are a prototypical system [4,5]. The hot hardness of the 2.5 B_4C - 97.5 W-Ni-Fe [95.0 W - 3.5 Ni - 1.5 Fe] is 10 to 20% higher than the premium grade of WC + Co at 1073K. When produced by hot pressing at 1733K, the B_4C is fully reacted and phases with a variety of chemistries are produced. These include graphite, tungsten carbide, and

Fe-Ni and W-B-C phases. Further analysis revealed the onset of the series of reactions to be at 1173K [4].

For both these systems an alternative approach to control of the reaction zone is by limiting the duration of the high temperature exposure during processing. This is the thrust of the high-energy high-rate consolidation processing approach applied to these materials systems and described in this paper.

EXPERIMENTAL PROCEDURE

Materials

(Ti_3Al + Nb)/SiC

The "Alpha 2" titanium aluminide intermetallic was supplied in powder form by United Technologies Pratt and Whitney Division, FL. These Nb-stabilized compositions were RS processed and were classified to - 80 mesh (< 177 µm). The as-received phase structure of the Alpha 2 composition was determined by XRD and consisted of the Nb-stabilized $\beta + \beta_2$ (bcc structure with B2 ordering).

The SiC particulate was supplied by the Superior Graphite Corp, Grade HSC-ROF95. These +325 mesh (> 44 µm) ß-SiC particulates were deliberately supplied with a graphite-enriched surface layer to enhance local electrical conductivity. This layer extends 1-10 µm below the surface, depending upon particle size (6). The mass of SiC was 20 percent of the total composite mass, corresponding to a volume fraction of 30 %.

W-Ni-Fe/B_4C

The tungsten-based composite was supplied as a premixed powder blend by Los Alamos National Laboratories, NM. The metallic matrix elemental powders W:Ni:Fe were mixed in a mass ratio of 93:5:2. The mass of B_4C was 1.3 or 1.65 percent of the total composite mass, corresponding to a volume fraction of 8% or 10%. Detailed characteristics of such powder blends have been reported previously by Sheinberg [5].

Processing

Advances in kinetic energy storage devices have opened up a new approach to powder processing of High Temperature Composites. The processing consists of internal heating of a customized powder blend by a fast electrical discharge of a homopolar generator. The high-energy high-rate "1MJ in 1s" pulse permits rapid heating of a conducting powder in a cold wall die. This short time at temperature approach offers the opportunity to control phase transformations and the degree of microstructural coarsening not readily possible using standard powder processing approaches.

The underlying fundamental approaches to high-energy high-rate processing have been described elsewhere [7]. The general details of the experimental apparatus and the pulse characteristics have also been reported[8], and the technique has been employed in the processing of Al-SiC composites [6,9]. In this study, 50g quantities of each of the composite powder blends were loaded into insulated die cavities between copper electrodes. Pressures between 210 and 420 MPa were applied, and the compact was rapidly heated by a homopolar generator pulse discharge. Disks with diameters of 30 mm or 50 mm were produced. Specific energy inputs ranged from 2000 kJ/kg to 7000 kJ/kg.

Microstructure Evaluation

Standard metallographic procedures were employed in the preparation of radial cross sections of each of the consolidated composite materials. The Ti_3Al/SiC required the use of water-jet machining. Vickers microhardness measurements were performed at room temperature to determine the degree of structural homogeneity and to aid in the detection of the reaction zones.

For the $(Ti_3Al + Nb)/SiC$ system, heat treatments of 24h to 96h at 1473K were used to follow the growth of the reaction zone. Unetched and etched cross-sections were examined in an optical microscope fitted with a Nomarski interference contrast system. This permits the polishing relief developed due to phases of different hardnesses to become readily apparent in color micrographs[10]. From such micrographs, average zone width measurements corresponding to each heat treatment were made. These measurements are not absolute values since they are uncorrected for magnification due to random angle sectioning.

X-ray diffraction analyses were performed on a Phillips diffractometer fitted with a Cu tube and a graphite monochromator. The powder samples were held onto a glass slide with double-sided adhesive tape. Sufficient powder was used so that no signal was detected from the glass slide. A JEOL 35M SEM with EDS and WDS capability was employed to verify the chemical discontinuities associated with the formation of reaction products at the matrix-reinforcement interface.

RESULTS AND DISCUSSION

Consolidation Parameters

The processing parameters used for these initial consolidation experiments on each of these composite systems were derived from parameter sets previously developed for the unreinforced matrix materials. The inherent assumption is that beyond a set low electrical conductivity threshold, the pulse resistive heating under pressure would provide rapid densification. The most identifiable solid state densification mechanism is powder forging. Densities greater than 95% of the calculated theoretical values were obtained. The parameters for the $(Ti_3Al + Nb)/SiC$ system were: Applied Pressure = 30 ksi (210 MPa), Specific energy input: 2000 to 4000 kJ/kg. For the $W-Ni-Fe/B_4C$ system they were: Applied Pressure = 45 - 60 ksi (315 - 420 MPa), Specific energy input: 3000 to 7000 kJ/kg.

Microstructure and Chemistry

Figures 1 and 2 provide typical overviews of the rapidly densified microstructures developed in the $W-Ni-Fe/B_4C$ system and the $(Ti_3Al + Nb)/SiC$ system respectively. The metal matrices are well consolidated, and excellent geometric conformability to the darker reinforcing phases is evident in both systems. The reinforcement appears well dispersed.

$(Ti_3Al + Nb)/SiC$

In Figure 2 from the $(Ti_3Al + Nb)/SiC$, which was heavily etched in HCl, preferential chemical attack occured at the SiC/matrix interface where a reaction zone was observed. Consolidation appears to have occured in the solid state with the energy input of 3200 kJ/kg under 210 MPa applied pressure. XRD analyses of the phases present in the processed composite reveal the likely reaction products to be TiC and $TiSi_2$. The $TiSi_2$ phase appears to be associated with the regions where localized matrix melting at higher specific energy inputs has occured. The $TiSi_2$ was not observed in the lower energy consolidations.

Phases with corresponding chemistries could be identified in the vicinity of interfaces by EDS in the SEM as shown in Figure 3. The diagonal interface separates a SiC particle (lower half of photo) from the matrix. The light-colored islands on the SiC particle surface are rich in Ti

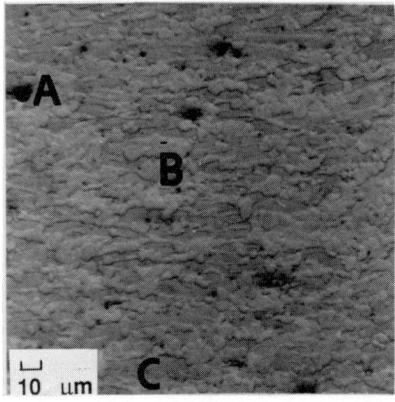

Fig.1. Optical photomicrograph of an etched radial cross-section of a (W-Ni-Fe)/B_4C composite specimen produced with an energy input of 4000 kJ/kg under 420 MPa applied pressure. The macroscopic features are A-B_4C, B-W, and C-W-Ni-Fe eutectic.

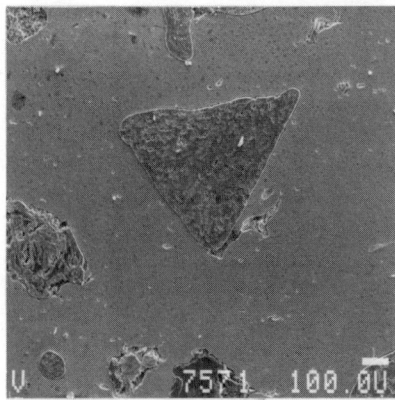

Fig.2. SEM photomicrograph of a heavily etched radial cross-section of a (Ti_3Al+Nb)/SiC composite specimen produced with an energy input of 3200 kJ/kg under 210 MPa applied pressure. Tight microencapsulation of the SiC particles is evident.

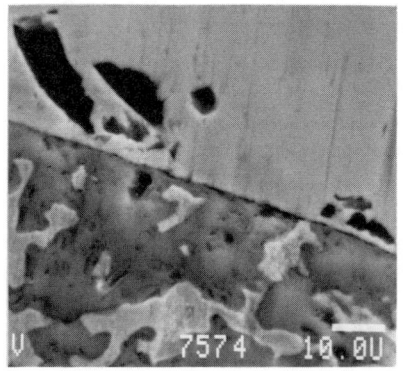

Fig.3. SEM photomicrograph of an unetched radial cross-section of a (Ti_3Al+Nb)/SiC composite specimen in which localized melting has occured. The region above the diagonal interface is the matrix with dark TiC reaction products. The lower section is a SiC particle on which islands rich in Ti and Si appear as light-colored areas.

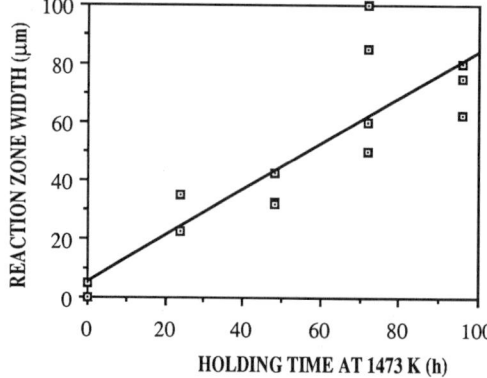

Fig. 4. Average apparent interface reaction zone width measurements corresponding to heat treatments of a (Ti_3Al + Nb)/SiC composite for 24h to 96h at 1473K. These measurements are not absolute values. They are uncorrected for magnification due to random angle sectioning.

and Si and have very little carbon, consistent with the XRD observation of $TiSi_2$. EDS analysis of the dark phases visible above the diagonal interface show these to be TiC.

The multiphase reaction zones were grown by heat treatments at 1473K for 24 to 96h. The apparent width of the reacted layer measured from a series of etched cross-sections is plotted in Figure 4. Additional unidentified phases were produced during these heat treatments. TEM investigations of the phases present are currently underway.

(W-Ni-Fe)/B_4C

In Figure 1, the apparent orientation of the W-Ni-Fe matrix is attributed to flow of the material under pressure during the transient liquid-phase-assisted consolidation. The three microscopic features observed are the B_4C, W and the W-Ni-Fe eutectic.

XRD analyses of the phases present in the high-density processed composite reveal the reaction products to be a series of complex carbides and borides, when sufficient energy is delivered to induce liquid-phase assisted consolidation. The lines indexable in a 10 to 80 degree two-theta scan for the W-Ni-Fe/B_4C system after liquid-phase assisted consolidation includes W (110),(200),(211), W_2C (021),(002),(121),(102),(321),(302), FeWB (001),(112), W_2B (110),(002),(200),(211), Fe_3C (101), Fe_6W_6C (660),(822) and Ni_4B_3 (211),(410),(403),(013). The intermetallic Fe_7W_6 (119) was also observed. The elemental Fe and Ni were fully reacted. These phases were not observed in the material consolidated completely in the solid state, where the maximum applied pressure of 420 MPa was insufficient to produce full densification.

Several of the carbide forming reactions are highly exothermic, and once triggered, these reactions occurring at multiple and distributed interface nodes can be expected to maintain the high temperature developed in the early stages of the processing. Indeed this processing cycle then takes on characteristics similar to those observed in Self-Propagating High-Temperature Synthesis of ceramic phases [11]. The reaction products so derived effectively transform the nature of the composite material introducing the attributes of the in-situ composites, in which the reinforcing phases are produced during processing.

As a clearer understanding develops of the complex physical metallurgy of the high-temperature matrix materials, in particular the new RSP aluminides, alternative approaches to the selection of reinforcement phases become crucial. Some guidelines for such approaches have been discussed by Fine et al [12], and are being applied to our continued effort in high-temperature metal matrix composites.

ACKNOWLEDGMENT

We thank Ralph Anderson and Sandy Shuleshko of Pratt and Whitney, Read Stewart of Superior Graphite, and Haskell Sheinberg of Los Alamos National Laboratories who provided the powders. Assistance with processing was provided by Ted Aanstoos and Jim Allen at CEM-UT. Mike Schmerling, H. Tello, H.-G. Chun, C.J. Lund, and Ming-Jy Wang assisted in the structure evaluation. This research was supported by DARPA/ARO Contract DAAL03-87-K-0073.

REFERENCES

1. P. K. Brindley in High-Temperature Ordered Intermetallic Alloys II, edited by N. S. Stoloff, C. C. Koch, C. T. Liu, and O. Izumi (Mater. Res. Soc. Proc. 81,Pittsburgh, PA 1987) pp. 419 - 424.

2. A. G. Metcalfe in Interfaces in Metal Matrix Composites, Volume 1, A. G. Metcalf (ed.),(Academic Press, New York, NY, 1974) pp. 67-123.

3. C. G. Rhodes, A. K. Ghosh, and R. A. Spurling, Metallurgical Transactions A, Volume 18A, Dec. 1987, pp. 2151-2126.

4. H. Sheinberg, Int. J. of Refractory and Hard Metals, March 1983, pp.17 - 26.

5. H. Sheinberg, Int. J. of Refractory and Hard Metals, December 1986, pp. 230 - 237.

6. G. Elkabir, PhD Dissertation, The University of Texas at Austin, August 1987.

7. H. L. Marcus, D. L. Bourell, Z. Eliezer , C. Persad and W. F. Weldon, Journal of Metals, December 1987, pp. 6-10.

8. G. Elkabir , L. K. Rabenberg, C. Persad and H. L. Marcus, Scripta Metallurgica, 20 (10),1986 pp. 1411-1416.

9. H. L. Marcus, L. K. Rabenberg, L. D. Brown, G. Elkabir, and Y. M. Cheong, in ICCM VI/ECCM 2-Composite Materials, eds. F. L. Matthews, N. C. R. Buskell, J. M. Hodgkinson and J. Morton (Elsevier Applied Science, London, 1987) p. 2.459.

10. C.E. Price in Metals Handbook, 9th. edition, V. 9, Metallography and Microstructures (ASM, Metals Park, OH, 1985) p. 151.

11. J. B. Holt and D. D. Kingman in Emergent Process Methods for High-Technology Ceramics, eds.R. F.Davis, H. Palmour III, and R. L. Porter (Plenum Press, New York, NY, 1984) p. 167.

12. M. E. Fine, D. L. Bourell, Z. Eliezer , C. Persad, and H. L. Marcus , submitted to Scripta Met., March 1988.

XD™ TITANIUM ALUMINIDE COMPOSITES

L. CHRISTODOULOU*, P.A. PARRISH**, and C.R. CROWE***
* Martin Marietta Laboratories, 1450 South Rolling Road, Baltimore, MD 21227
** Defense Advanced Research Agency, 1400 Wilson Blvd., Rosslyn, VA 22209
*** Naval Research Laboratory, Washington, DC 20375

ABSTRACT
The advantages of reinforcing metals with ceramic particles to produce metal matrix composites are well known. The behavior of discontinuously reinforced intermetallic compounds, however, has not been extensively studied. Martin Marietta Laboratories has produced a new generation of discontinuously reinforced titanium aluminide composites using a proprietary casting process known as XD™ technology. These new materials possess enhanced properties at room and elevated temperatures and may be cast, extruded, or forged. The effects of matrix composition, reinforcing phase, and thermal mechanical processing on properties have been studied using optical and various electron microscopy and mechanical and physical property measurement techniques to characterize the alloys. To date, most work has been done on a two-phased lamellar Ti-45 a/o Al alloy reinforced with TiB_2 ceramic having an equiaxed morphology. Data on temperature dependence of the dynamic Young's modulus, coefficient of thermal expansion, deformation and fracture behavior, and microstructure are presented.

INTRODUCTION
Intermetallic compounds, especially various aluminides, are known for their excellent high-temperature properties which, in combination with their low density, rigidity, and resistance to oxidation, make them and their composites suitable candidates for components in gas turbine engines and structures for hypersonic vehicles. To date, their use has been limited because of their poor ductility and fracture toughness at room temperature.

Following earlier studies [1,2] these materials are only coming to maturity now via renewed efforts to develop both ingot and powder metallurgy techniques. The recently developed Martin Marietta XD™ technology is one example of a way to develop new discontinuously reinforced titanium aluminide composite materials for hot structure aerospace applications.

The XD™ approach has concentrated on the attainment of complex "designed" or "engineered" microstructures that incorporate a range of reinforcements to achieve the combinations of properties required in structural materials (Fig. 1). This is achieved by forming reinforcements in-situ during the base alloy preparation. For example, hard phases can yield improved compressive properties depending on their size, morphology, and volume fraction. Fine particles can enhance tensile strength, whereas larger asymmetric shapes can substantially improve creep properties. Soft phase additions, on the other hand, can improve fracture toughness and damage tolerance in brittle matrix materials. When the XD™ technology is used, many different ceramic reinforcements, with morphologies ranging from whiskers to platelets, can be introduced to achieve the predicted benefits.

The XD™ technology has the further advantage that it is compatible with casting and fabrication techniques, and wrought products such as extrusions, forgings, and rolled sheet have been obtained as well as investment castings. Some of the findings of the XD™ development program for the TiB_2-γ Ti aluminide system are summarized in this communication.

ALLOY THEORY

From a fundamental physics viewpoint, very little is known about the factors that control strength, ductility, and fracture of ordered intermetallics. Recent advances in solid state theory have provided some tools to calculate stable crystal structures, the important bonding features associated with deformation behavior, and the effects of third element alloying additions on the crystal structure. Figure 2 shows total energy vs c/a ratio for the $L1_0$ TiAl crystal structure plotted by Klein, Papaconstantopoulou, and Chub at NRL. These results predict both the c/a ratio and the value of the lattice spacing, a, to within 1% of the experimental value. A similar analysis for VAl in the same crystal structure was also made. Comparing the equilibrium configuration with the density of states reveals a correlation between the minumum total energy and a low density of states at the Fermi level which implies that the addition of a small amount of transition element heavier than Ti shifts the crystal structure to a more stable state.

Mark Eberhart at Los Alamos National Labs (LANL) and Dimitri Vvedensky from Imperial College, London, have been using molecular orbital calculations to model the TiAl/TiB$_2$ and TiAl/Ti$_3$Al interfaces. According to their work, the bonding between TiAl and Ti$_3$Al is very strong between the closely packed planes of Ti$_3$Al and the basal plane of TiB$_2$; that between other crystal orientations is far less favorable. Furthermore, bond strength decreases with increasing Al content. Addition of Nb introduces lattice strain into the metal layer, weakening the interfacial bonding.

MICROSTRUCTURE

The TiB$_2$/TiAl cast ingot consists of equiaxed colonies of Ti$_3$Al and TiAl distributed in a lamellar morphology (Fig. 3). Orientation relationships between the phases in the microstructure show that the close packed planes of the α_2 and γ are parallel according to $\{111\}_\gamma$ $\{0001\}_{\alpha_2}$ and $<110>_\gamma$ $<1120>_{\alpha_2}$; the results agree with Blackburn [3]. The TiB$_2$ particle size distribution is centered around one micron and both intergranular and intragranular particles are observed (Fig. 3). Transmission electron microscopy (TEM) evaluation reveals that although the reinforcements are occasionally faulted, they are single TiB$_2$ crystals and exhibit clean particle/matrix interfaces (Fig. 4). No reaction zones between the matrix and particles have been observed. Heat treatment of the as-cast alloy does not markedly alter the microstructure, whereas thermomechanical processing induces the matrix microstructure to be transformed from a lamellar to an equiaxed morphology. The transformation occurs with no TiB$_2$ break-up (Fig. 5). This microstructure appears more stable under creep conditions than the lamellar structure and also exhibits increased room-temperature elongation, Table I.

PHYSICAL PROPERTIES

Density, measured using a displacement method, is 3.99 g/cm^3. Specific modulus as a function of temperature has also been measured from room temperature to 500°C using the PUCOT techniques previously used by Harmouche and Wolfenden to study B2 transition metal aluminides [4]. E/ρ fits the relation: $E/\rho = [43.305 - 0.01015 \times T] \times 10^{-6}$m, where T is temperature in °C.

The coefficient of thermal expansion (CTE) as a function of temperature has also been measured from room temperature to 800°C. The CTE at 20°C equals 1.0×10^{-5}/°C and increases to 1.7×10^{-5}/°C at 800°C. The $\Delta L/L$ curves show a slight hysteresis during heating and cooling and fall midway between the curves for titanium and TiAl.

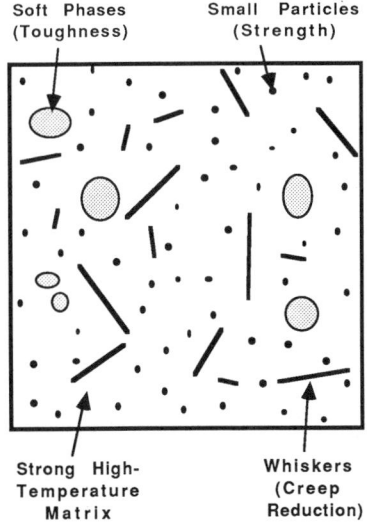

Figure 1. Schematic of a "Designed" microstructure exhibiting several features considered important in the development of intermetallic alloys (Adapted from DARPA URI Workshop, La Jolla California 1986).

Figure 2. Total energy calculations as a function of c/a ratio for the $L1_0$ TiAl structure

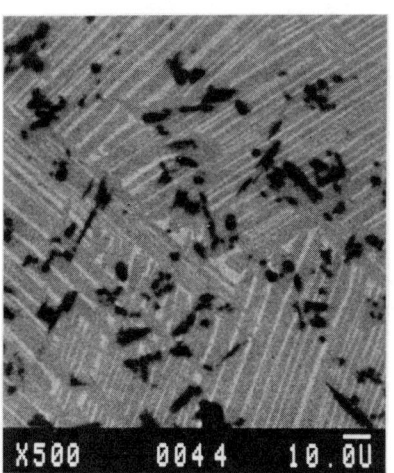

Figure 3. Backscattered electron image of an TiB_2-reinforced XD™ Ti-45 at% Al alloys. TiB_2 particulates (black) are uniformly dispersed in a α_2 (bright) / γ (grey) lamellar matrix.

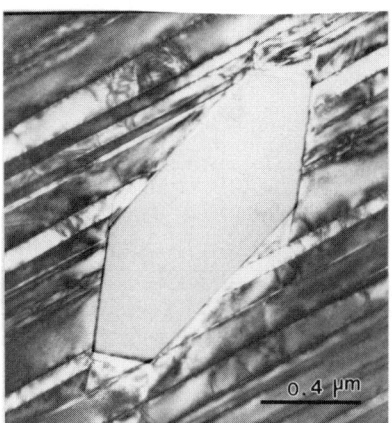

Figure 4. Bright field transmission electron image of a single crystal TiB_2 reinforcement in XD™ Ti-45 at% Al.

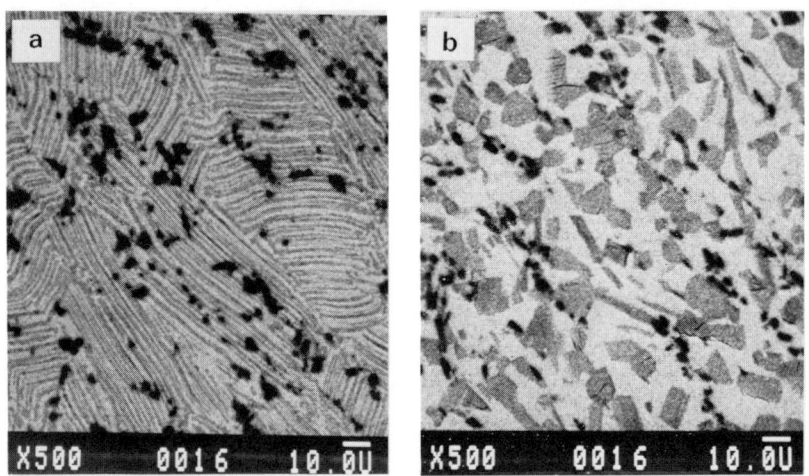

Figure 5.(a,b) As-extruded (a) and extruded and heat treated (b) microstructures of XD™ Ti-45 at% Al alloy. The transformed "equiaxed" structure exhibits increased room temperature elongation and improved stability under creep.

Table I

Properties of Ti-45Al-75% TiB$_2$

	20°C			800°C		
	Yield Strength (MPa)	Ultimate Strength (MPa)	% Elong. (Plastic)	Yield Strength (MPa)	Ultimate Strength (MPa)	% Elong. (Plastic)
As Extruded	-	793	0	448	710	11
As Extruded and Heat Treated	793	862	0.5	427	600	20

MECHANICAL PROPERTIES

Some of the mechanical properties of the XD™ material in various states are listed in Table I. The deformation behavior of the composite alloys has been studied at ambient and elevated temperatures. Fracture surfaces of the specimens tested have shown a change in fracture mode with increasing temperature. The microstructural features identified on the fracture surface by stereo microscopy have been correlated with those seen through metallography.

The fracture behavior of the alloys appears as cleavage at room temperature and gradually changes to microvoid coalescence at elevated temperatures (Fig. 6). Examination of the material below the fracture surface by a technique that preserves fine detail reveals that fracture proceeds both among and across lamellar boundaries (Fig. 7). Fracture of TiB_2 particles was never observed. Although the fracture initiation sites have not yet been identifed, the reinforcement appears to be effective at deflecting the advancing crack. Fracture toughness measurements on the lamellar structure yield values ranging from 11-13 ksi in.

Creep rates at 760°C are being measured by P.L. Martin at LANL and by V. Sikka at ORNL. Initial results at a 69-MPa load show steady-state creep rates of $2 \times 10^{-7} s^{-1}$. This compares to $9 \times 10^{-7} s^{-1}$ in non-reinforced alloys at the same load and temperature.

SUMMARY

Titanium aluminide composites based on a two-phase microstructure of Ti_3Al and TiAl reinforced with a ceramic dispersoid are being developed for high-temperature applications. TiB_2-particulate reinforced titanium-45 a/o aluminum base alloy has been made recently by Martin Marietta Laboratories. The microstructure and phases present in this alloy have been analyzed using scanning electron microscopy (SEM), TEM, and other methods. The TiB_2/TiAl cast ingot consists of equiaxed grains of TiAl with lamellae of Ti_3Al and TiAl. The TiB_2 particles are observed not only inside the grains but also at the grain boundaries. Mechanical property measurements, metallography, and analytical microscopy studies were conducted to determine the sensitivity of these microstructures to thermal and mechanical processing and alloying additions. Results indicate that strength and creep resistance in the composites are substantially improved without degradation of other properties.

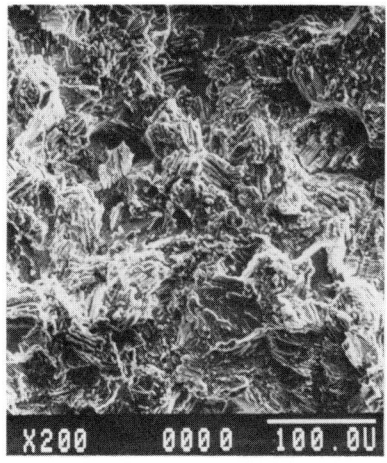

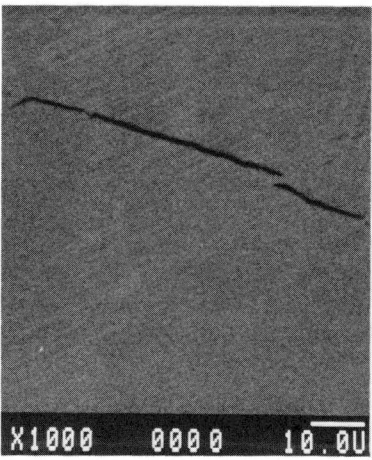

Figure 6. Room temperature fracture surface of a lamellar XD™ Ti-45Al alloy showing cleavage-type fracture.

Figure 7. Subcritical crack in base unreinforced alloy Ti-45Al showing fracture across and along α_2 / γ lamellae.

ACKNOWLEDGEMENTS

The work reported in this communication is a small fraction of a large effort currently underway in the Composites Group at Martin Marietta Laboratories and the Material Science and Technology Division at NRL, and at LANL and ORNL. The authors are grateful for the dedicated efforts of all contributors - too many to mention by name. Funding by the Defense Advanced Projects Agency and USAF under contract N00014-86-C-2277 and the Martin Marietta Corporation Emerging Technology program is gratefully acknowledged.

REFERENCES

1. H.A. Lipsitt, D. Shechtman, and R.E. Schafrik, Met. Trans. 6A, 1991 (1975).

2. M.J. Blackburn and M.P. Smith, United Technologies Corporation, Pratt and Whitney Group, AFML-TR79-4056 (May 1979).

3. M.J. Blackburn in The Science, Technology and Applications of Titanium (Pergamon Press, London 1970) p. 633.

4. M.R. Harmouche and A. Wolfenden, in High Temperature Ordered Intermetallic Alloys, (Mater. Res. Soc. Proc. 39 Pittsburgh, PA 1985) pp. 343-348.

MICROMECHANICAL STUDIES OF MODEL METAL MATRIX COMPOSITES

H. N. G. WADLEY, F. S. BIANCANIELLO, and R. B. CLOUGH
National Bureau of Standards, Gaithersburg, MD 20899

INTRODUCTION

The fiber-matrix interface is believed to play a central role in the load-supporting and crack-growth resisting capabilities of fiber-reinforced high temperature composites. In metal matrix composites, this interface can be very complex: new phases may be present due to reactions between the matrix and reinforcement, compositional gradients due to interdiffusion occur, and the outer fiber surface is usually coated to avoid handling damage, improve fiber wetability, and/or control matrix reactivity. Attempts at measurements of fiber and interface strength from studies of actual composite samples are thwarted in part by the difficulty of determining the local distribution of stress and strain in these highly inhomogeneous materials. Furthermore, the inability to determine the point, during loading, where failure occurs in the interfacial zone also complicates the problem in these opaque materials. It is not surprising, therefore, that the fundamental relationships between the structure and properties of interfaces, on the one hand, and the load-supporting/crack-resisting properties of bulk metal matrix composities, on the other, have not been determined to date.

We are exploring the use of a more direct approach to study the influence of the fiber-matrix interface upon the micromechanics of composite failure, by making strength measurments of individual failure events. In order to determine the local stress/strain distributions, we use a simple cylindrical geometry tensile sample with a single fiber centrally aligned along the sample axis. Such samples can be conveniently prepared using a single crystal growth technique, and by varying fiber coating, melt temperature, and liquid metal contact time, it is possible to systematically vary the interface microstructure.

To determine the location and load at which failure occurs, we are investigating the use of acoustic emission measurements during these mechanical tests. Acoustic emission at this point serves to give an acoustic indication of failure since we cannot "see" into these materials as we can e.g. in polymer matrix systems. It also shows promise for locating the fracture site to within 30 μm, and eventually providing fundamental insight into the micromechanics of failure.

EXPERIMENTAL

Specimen Preparation [1,2]. Single crystal/single fiber tensile samples having a dumbell geometry (with a reduced section 57 mm in length and 4 mm in diameter), were prepared from 99.99% Al and 140 μm diameter AVCO[1] SiC fibers with both unmodified and C-rich (SCS-2) surfaces. A Bridgman technique with a melt temperature of 900°C was used for sample growth. Samples with shallow (~1 μm) reaction zones of Al_4C_3 were grown with 140 s solidification time, while samples with extensive (~10 μm) reaction zones of Al_4C_3 were grown with 6840 s (1.9 h) solidification time. Samples with no fibers were also grown so that the effect of fiber reinforcement could be independently observed and separated from matrix behavior.

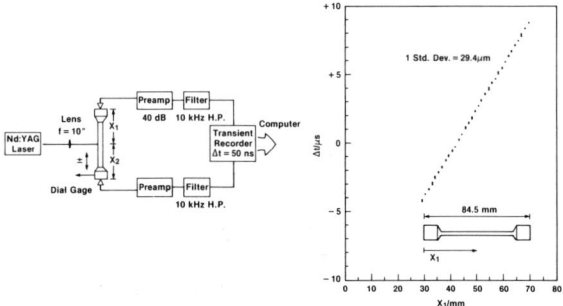

Fig.1 Source location calibration using pulsed Nd:YAG laser.

Tensile Testing. The specimens were loaded in tension in a screw-driven, constant crosshead velocity machine, on which load drops on fiber fracture are detectable [1,2]. Acoustic emission was monitored during testing using a technique and system described previously [1]. Conical piezoelectric transducers were mounted at each end of the sample to record the wave arrival times and thus source position. A pulsed Nd:YAG laser was used to calibrate source position versus difference in arrival times (Fig. 1) to within ± 30 μm [3]. In principle, with this accuracy, it is possible to determine the

---

[1] Use of a brand name in this paper is made solely for the purpose of accurately specifying materials and does not imply recommendation or endorsement of the product by NBS.

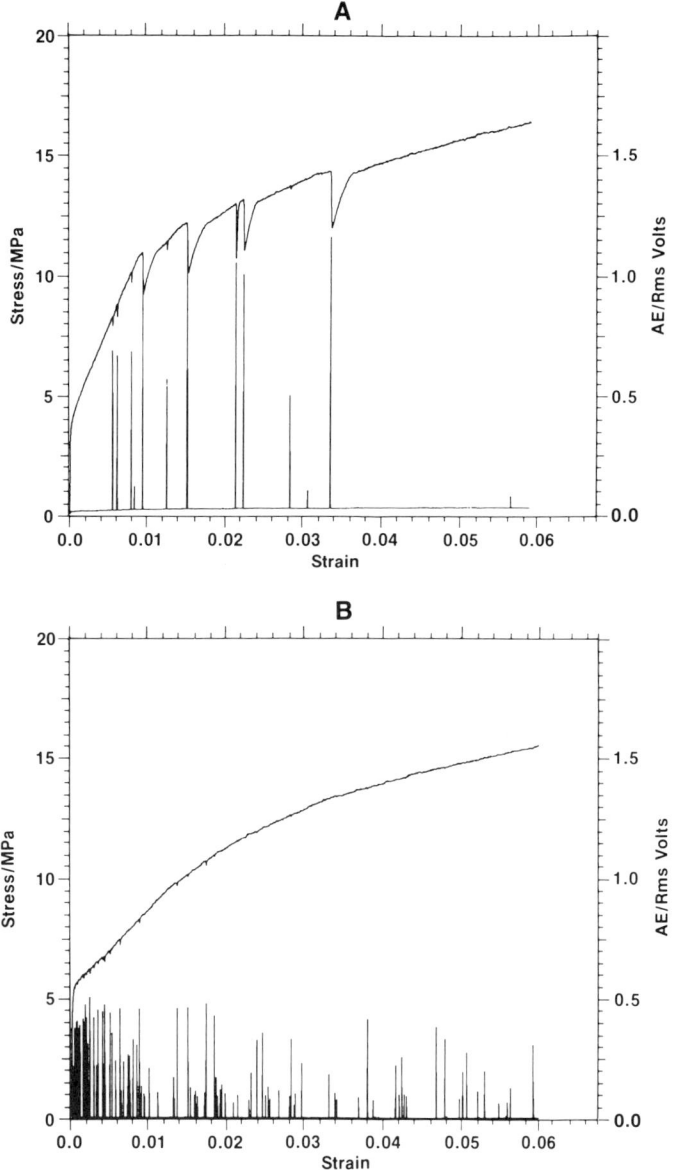

Fig.2. Stress-strain and AE-strain curves for model Al/SiC composites with SCS-2 fibers (C-rich surface). (a) More rapid solidification (140 s), (b) Less rapid solidification (6840 s).

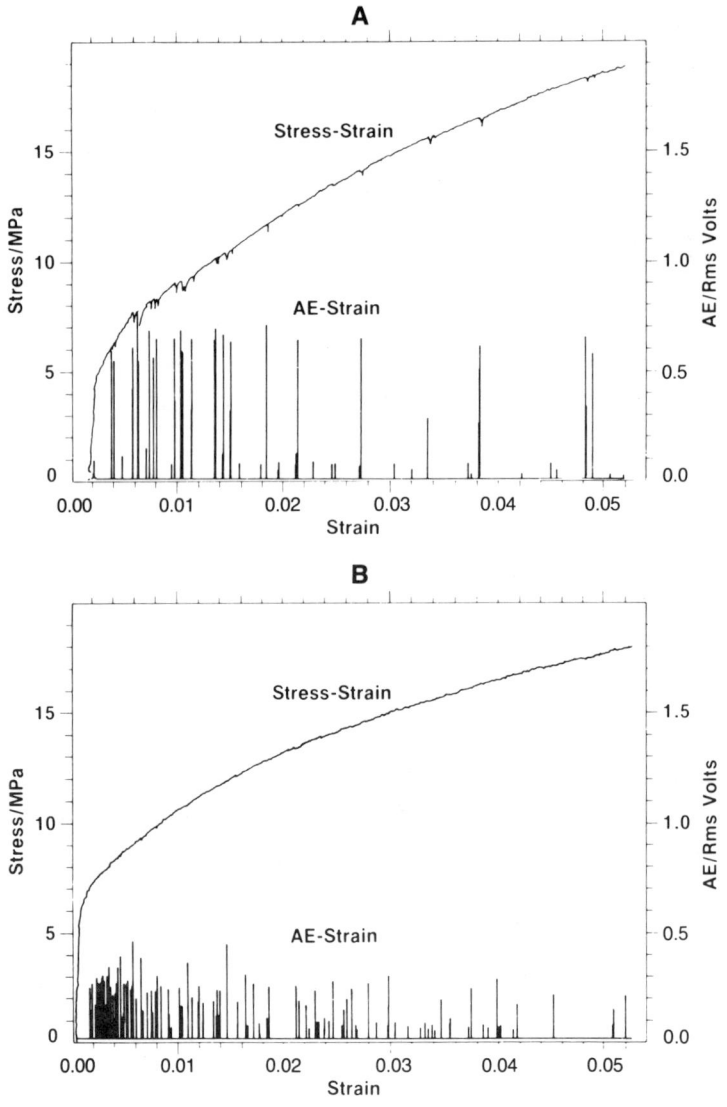

Fig.3 Stress-strain and AE curves for model Al/SiC composites using single SiC fiber with unmodified surface. (a) More rapid solidification; (b) Less rapid solidification (6840 s).

length of each fiber segment during loading. However, there are practical difficulties introduced during actual testing, chiefly by mechanical/electrical noise, as well as multiple fractures during specimen ring-down, which can reduce the accuracy to a value several times this. We are at present examining the use of lower strain rates, specimen damping, and high-pass filtering (since the lowest frequencies reverberate the longest) to minimize these sources of error.

RESULTS AND DISCUSSION

Figures 2-3 are representative tensile and acoustic emission test curves obtained with these single crystal composite samples. To determine if there were pre-existing fractures in the fiber, a more rapidly grown sample with an SCS-2 SiC fiber, that was not strained in the tensile machine, was sectioned longitudinally along the fiber, polished, and examined optically. There were no fractures in the fiber. Single crystal Al specimens grown without fibers were also strained [1,2]. These produced neither load drops nor acoustic emission bursts, indicating that the load drops are associated with the fibers, and the acoustic emission with fiber or interface failure. Both fiber surface composition and solidification rate produce significant effects on acoustic emission and load drop magnitude. The solidification time had a larger effect than variation in fiber surface treatment. Regardless of surface composition, the more slowly grown composites always exhibited smaller load drops and lower energy acoustic emission signals than the more rapidly grown material.

Fig. 4 is a scanning electron microscope micrograph showing fiber fractures in a slowly grown composite with an untreated fiber after a plastic strain of 6%. There is a close correlation between the number of acoustic emission events, for the more rapidly grown material, and the number of fiber fractures. These were counted by removing the matrix with a sat. sol.of NaOH at $50°C$. This indicates that fiber fractures were the only significant source of acoustic emission for the more rapidly grown material. On the other hand, for the more slowly grown material, fiber fractures could account for only about one half of the acoustic emission events. The other half consisted of lower amplitude acoustic emission bursts. Since plasticity sources are undetectable at the sensitivity levels used here, and the number of fiber fractures have already been accounted for, these extra signals must be associated with interface cracking. Additional optical and scanning electron microscopy [2] show numerous microfractures between the SiC fiber and Al_4C_3, near the fiber ends, where the shear stress is largest.

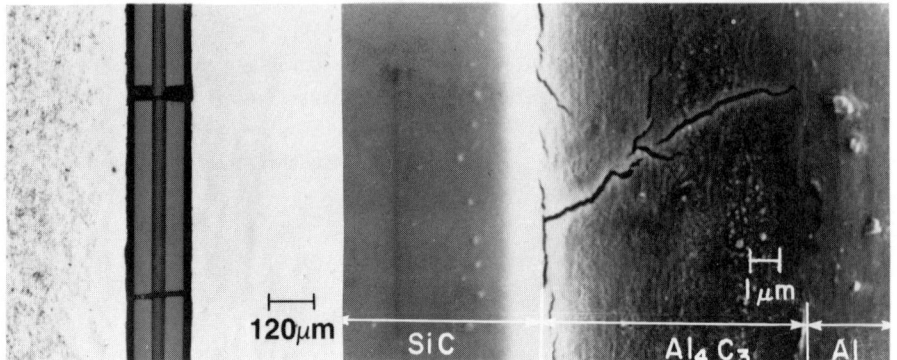

Fig. 4. SEM micrograph. Longitudinal section of slowly grown composite (unmodified SiC surfiber fractures.

Fig. 5. Interfacial and interzonal shear fractures in the interzonal region near end of fiber fracture in Fig. 4.

Fig. 5 shows an example of such interfacial cracks appearing in the sample shown in Fig. 4. These cracks appear at the interface between the SiC and the interzone, near the end of the fiber, and have branched into the carbide interzone and arrested near the Al matrix interface.

These results allow us to estimate the fiber and interfacial shear strengths as follows. First, the tensile test is used to determine the fiber strength σ_f from the size of the load drops. Post-test examination of the broken fibers and acoustic source location measurements provide critical lengths ℓ_c and the fiber diameter d. Finally, the interface shear stress is calculated from the shear lag relationship [4]:

$$2\tau_i = \sigma_f d/\ell_c \qquad (1)$$

Conversely, for a given interface strength, this relationship allows one to compute the fiber strength or ℓ_c. The relationship is obtained by equating the "gripping" forces due to interfacial shear stresses, which exist only at the ends of the fiber, with the tensile forces in the fiber, much like a tensile test of a bare fiber. Failure can occur either by "slipping" in the grips (i.e. interfacial failure) by shear failure of the grips themselves (i.e. matrix flow), or by fracture of the fiber, depending on which component is weakest. If the fiber fractures, the gripping stresses are reestablished at the fresh ends (making multiple fiber fractures during a load drop possible), and the process is continued until the length of fiber between the grips is zero. At that point ($\ell = \ell_c$), the tensile stresses are always below the fracture stress, and fiber fracture ceases.

Consider a composite tensile specimen, with a single fiber oriented along the tensile axis, elongating under load in a constant crosshead velocity tensile test machine. If the fiber breaks, there will be a sudden localized plastic strain at the break, since the matrix cross-section is no longer reinforced by the fiber. This sudden strain increment produces a load drop of magnitude ΔP due to contraction of the pull rods. The fiber strength is given by [2]:

$$\sigma_f = \frac{\Delta P N (A_s/A_f) \theta_L}{k L \Delta N} \tag{2}$$

where (A_s/A_f) is the ratio of cross-sectional areas of the sample and fiber, respectively, θ_L is the macroscopic rate of work hardening over the length of sample L with reduced cross-sectional area, and N and ΔN are the total number of fiber fractures and the number of fractures during that load drop, respectively.

The Weibull statistical model [5] was used to extract the maximum fiber strength σ_{max} from the data. Normally the model is used by neglecting the end effects and considering only the fully stressed length ($\ell - \ell_c$); here we have generalized the model to include the tensile stress gradients at the fiber ends. This effect predominates at the end of the break-up process, where the strength is most significant. This gives the fiber strength σ_f as a function of fiber length ℓ [2]:

$$(\sigma_f/\sigma_{max}) = (\ell_c(1-\alpha)/(\ell-\alpha\ell_c))^{(1-\alpha)/\alpha} \tag{3}$$

The factor $\alpha \equiv n/(n+1)$ is a material constant related to the Weibull exponent n [5]. The effects of fiber end stress gradients dominate precisely where strength is most significant: at the shortest fiber lengths. Fig. 6 shows the fiber fracture strengths versus average fiber length, $\ell = L/(N+1)$. The fiber strengths in Fig.6d were not measurable because the load drops were less than the noise limitation of the load cell; this situation can be remedied by the use of higher sensitivity load cells and additional signal processing. On each curve, which is the least squares fit of the statistical strength function given above, the two sets of symbols represent separate tensile tests.

Table I summarizes the fiber and interface strengths obtained from this analysis; the matrix stresses were obtained from the tensile results. The initial fiber strength was ≈3500 MPa. It can be seen that, invariably, the fiber and interface strengths decrease with exposure time in the melt, and that this effect is more pronounced with the untreated fiber. Both effects are indicative of chemical reactions between the fiber, any intervening layers, and the melt. The higher tendency of the SiC fibers with untreated surfaces to handling damage (surface defects), appears to make them more affected by reactions in the melt, as evidenced by the more rapid decrease in fiber strength with time in the melt, as compared to those fibers (SCS-2) with C coatings. In addition, recent TEM observations by Lee and co-workers [6] on specimens taken from the untested ends of samples tested here (untreated SiC fiber, slower solidification rate) show the presence of numerous notch-like intrusions of Al_4C_3 into the SiC fiber subboundaries. A similar effect is likely to occur at pre-existing surface defects. Both then are potential initiation sites for fracture of the fiber and are likely causes for the degradation in strength of the fibers in the melt. The interface strength, on the other hand, appears related to the *thickness* of the interzone, the growth rate of which can be affected by the presence of surface coatings. For the more rapidly solidified material, the more reactive C-rich surface of the SCS-2 fiber produced a thicker resultant layer of Al_4C_3 than resulted on the untreated fiber [2], and the strength of the Al_4C_3 interzone invariably decreased with increasing thickness [2].

TABLE I. Measured Fiber and Interface Strengths

Property	SCS-2 Fiber (C-Rich Surface)		Untreated SiC Fiber	
	Rap.Cooling	Slower Cooling	Rap.Cooling	Slower Cooling
Solidification Time (s)	140	6840	140	6840
Maximum Fiber Stress (MPa)	798 ±62	340 ±9	501 ±14	<35
Maximum Interface Shear Stress (MPa)	12 ±0.9	3.1 ±0.1	24 ±0.6	<3
Max. Matrix Shear Stress (MPa)	7.0 ±0.4	8.0 ±0.5	8.0 ±0.5	7.5 ± 0.5

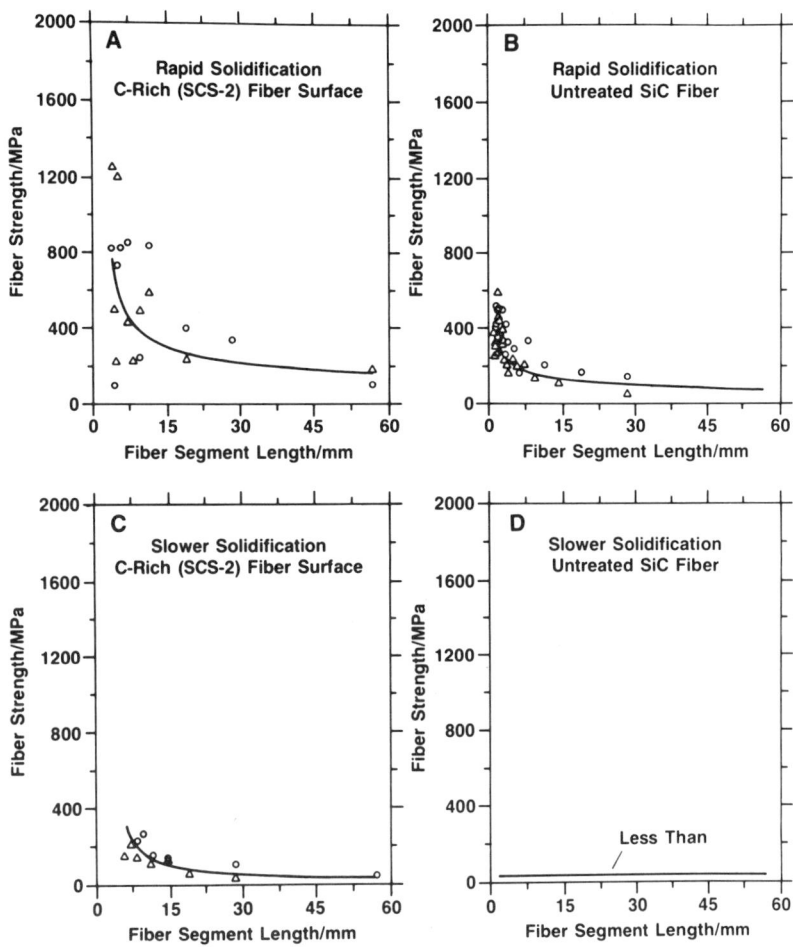

Fig.6. In situ fiber strength vs. average segment length for single crystal Al/SiC fiber composites. (a) SCS-2 fiber, more rapid solidification, (b) Untreated SiC fiber, more rapid solidification, (c) SCS-2 fiber, slower solidification, (d) Untreated SiC fiber, slower solidification.

CONCLUSIONS

The separate effects of processing and of fiber surface treatment on interface and fiber strength were studied by using model single fiber, single crystal composites. A micromechanical method for direct, simultaneous and separate measurement of the fiber and interface strengths was developed. Fiber strength as a function of length was also measured. The results were correlated with a statistical model of fiber break-up during specimen elongation which incorporates the shear lag model stress distribution into a Weibull failure probability distribution. With increased exposure to the melt, both fiber and interface strength degrade. The interface strength decreases as its thickness increases. All of these effects can be measured, and are ultimately related to the evolution of interfacial composition and microstructure during the sometimes complex growth sequence from the liquid to the solid.

ACKNOWLEDGMENTS

We wish to thank Larry Grumer of AVCO, who provided the SiC fibers, and R. Parke, J. Martinez, P. Quincoses, A. Manor and Dr. A. Shapiro, who assisted with various aspects of the work. This research was funded by the Strategic Defense Initiative Office and carried out under the auspices of the Office of Naval Research, Dr. Stephen Fishman, program manager.

REFERENCES

1. R. B. Clough, F. S. Biancaniello and H. N. G. Wadley, "Measurement of Fiber and Fiber-Matrix Interface Shear Strengths in Metal Matrix Composites", Proc. Conf. on Nondestructive Testing and Evaluation of Advanced Materials and Composites, H. Mindlin et al., eds. (Amer. Soc. for Metals, Metals Park, OH, 1987).
2. R. B. Clough, F. S. Biancaniello and H. N. G. Wadley, "Micromechanical Studies of Fiber and Interface Fracture in Single Crystal Aluminum/SiC Fiber Composites," to be submitted to Acta Met. (1988).
3. A. Manor, R. B. Clough, and H. N. G. Wadley, "High-Precision Location of Acoustic Emission Sources", in preparation.
4. A. Kelly, Strong Solids, (Clarenden Press, 1977), p. 131.
5. A. M. Freudenthal, "Statistical Approach to Brittle Fracture", Fracture, Vol. 2, Ed. by H. Liebowitz, (Academic Press, N.Y. 1968).
6. D-J Lee, M. Vaudin, C. Handwerker and U. Kattner, "Phase Stability and Interface Reactions in the Al-SiC System", MRS Symp. on High Temperature /High Performance Composites (1988).

EFFECT OF TEMPERATURE ON THE MECHANICAL PROPERTIES AND MICROSTRUCTURES
OF IN SITU FORMED Cu-Nb AND Cu-Ta COMPOSITES

W. A. SPITZIG*, P. D. KROTZ*, L. S. CHUMBLEY*, H. L. DOWNING**, and
J. D. VERHOEVEN*
*Ames Laboratory-USDOE, Iowa State University, Ames, IA 50011
**Drake University, Des Moines, IA 50311

ABSTRACT

The effect of temperature on the mechanical properties and microstructures has been evaluated for heavily cold drawn Cu-20% Nb and Cu-20% Ta composites. The strengths of the composites decrease with increasing temperature, with the decrease becoming most pronounced at temperatures above about 300°C and at larger draw ratios. Cu-20% Ta composites are stronger than Cu-20% Nb composites throughout the temperature range studied (22-600°C) with the improvement increasing with increasing temperature. Resistivity measurements and substructure analyses showed that at temperatures where softening accelerated, resistivity decreased indicating a substructural change which was observed to be coarsening of the Nb and Ta filaments in the composites.

INTRODUCTION

Deformation processing is a well established procedure for the fabrication of composites from ductile two-phase metal mixtures. During deformation processing these metal mixtures are transformed into a matrix containing filaments with very large aspect ratios. Recent emphasis has been concerned with the very high strengths developed in these types of composites at room temperature [1-6]. However, the increased strength of these composites is also of interest in high temperature applications. Some previous results on the effect of temperatures up to 495°C on the ultimate tensile stresses of Cu-Nb composites containing 14.8 vol% Nb showed that the strength decreased with increasing temperature and it decreased faster as the degree of deformation processing increased [7,8]. However, even at 495°C the composites retained a significant portion of their room temperature strength. In light of these previous results indicating appreciable strength retention in Cu-Nb composites at high temperatures, where the strength of Cu is very low, a study was undertaken to examine the high temperature (22-600°C) mechanical properties and microstructural evolution in Cu-Nb and Cu-Ta composites. The mechanical properties and microstructures of these composites at room temperature have been previously reported [4-6,9].

EXPERIMENTAL PROCEDURES

Ingots of Cu containing 20 vol% Nb or 20 vol% Ta were prepared by consumable arc melting electrodes containing Nb or Ta strips in a Cu cylinder [10]. The ingots were approximately 7.6 cm in diameter and 18 cm long and the average Nb and Ta dendrite sizes were 3.8 and 3.5 µm, respectively. The diameters were machined to 6.1 cm and the ingots were rod rolled to 1.3 cm in a series of reductions and subsequently drawn into wire using successively smaller dies at room temperature. For comparison purposes annealed Cu and Nb cylinders were also processed with the Cu-20% Nb and Cu-20% Ta ingots. Drawing reductions are given in terms of logarithmic strain by $\eta = \ln(A_o/A)$, where A_o and A are the initial and final cross sectional areas.

The Cu-20% Nb ingot was cold drawn to a diameter of 0.16 mm without difficulty which corresponds to $\eta = 11.9$ or a reduction of area of 99.9993%. Central bursting and fracture during drawing of the Cu-20% Ta ingot limited the final diameter obtainable to 0.7 mm which corresponds to $\eta = 8.1$ or a reduction of area of 99.99%. Three draw ratios were chosen for testing at temperatures of 150, 300, 450, and 600°C. The draw ratios were $\eta = 4.8$, 8.1, and 11.0 and were chosen on the basis that they were a good representation of the entire range of reductions. Cu-20% Ta could not be tested at $\eta = 11.0$ for the reasons discussed. Specimens were annealed for 24 h in vacuum at their respective testing temperature prior to testing. Tensile tests at the various temperatures were done in flowing argon at a nominal strain rate of $3.3 \times 10^{-4} s^{-1}$. Annealing at temperature for 24 h prior to testing was done to try to ensure that the microstructure was stable and would not change during testing. Preliminary experiments showed that measurable microstructural changes occurred only during the initial 6 h of annealing.

Filament sizes and spacings were measured on longitudinal sections of the wires using standard stereological intercept procedures [11]. Filaments were physically extracted from Cu-20% Nb and Cu-20% Ta wires for scanning electron microscopy analyses by dissolving the Cu matrix in an acid solution. Sample preparation procedures for TEM microstructural analyses of the composites after various reductions are quite involved and have been reported previously [4].

Electrical resistivity measurements were made on Cu-20% Nb and Cu-20% Ta wires deformed to various draw ratios over the temperature range 22 to 800°C. A standard four-probe dc technique [12] was used and samples were heated at a rate of 1°C/min up to a maximum temperature of 800°C followed by furnace cooling to room temperature.

EXPERIMENTAL RESULTS

Figure 1 compares the effect of draw ratio on the room temperature ultimate tensile stress of the Cu-20% Nb and the Cu-20% Ta composites and of Cu and Nb. The increased strength for Cu-20% Ta as compared to Cu-20% Nb at a given draw ratio is in agreement with the larger modulus of Ta [9]. The strengths of the composites show no signs of leveling off at the larger draw ratios and it is exponential in nature. The strengths of Cu and Nb level off at $\eta > 10.3$ and $\eta > 8.1$, respectively.

Figure 2 shows the effect of temperature on the ultimate tensile stress of Cu-20% Nb, Cu and Nb drawn to various draw ratios. The ultimate tensile stress of Cu-20% Nb decreases with increasing temperature with the effect being most pronounced at temperatures above 300°C and with increasing draw ratio. The high temperature strength of Nb parallels that of Cu-20% Nb, whereas, Cu shows a more pronounced decrease in strength that is independent of draw ratio.

Figure 3 shows the temperature dependence of the ultimate tensile stress for Cu-20% Ta drawn to $\eta = 4.8$ or 8.1 and compares it to that of Cu-20% Nb. Cu-20% Ta possesses greater strength at both draw ratios and the difference in strength increases with increasing temperature.

DISCUSSION OF RESULTS

It is apparent from Figs. 2 and 3 that at all draw ratios the composites have significantly superior high temperature strengths as compared to Cu. Comparison with data in the literature for high temperature Cu alloys and dispersion hardened Cu shows that the composites are considerably stronger even at 600°C [13,14]. Substitution of Ta for Nb as the filament material substantially increases the strength of the composite by about 20% at 22°C to about double at 600°C for $\eta = 8.1$.

To characterize microstructural changes occurring at the various temperatures Nb and Ta filaments were extracted from Cu-20% Nb and Cu-20% Ta wires drawn to different draw ratios and annealed for 24 h at different

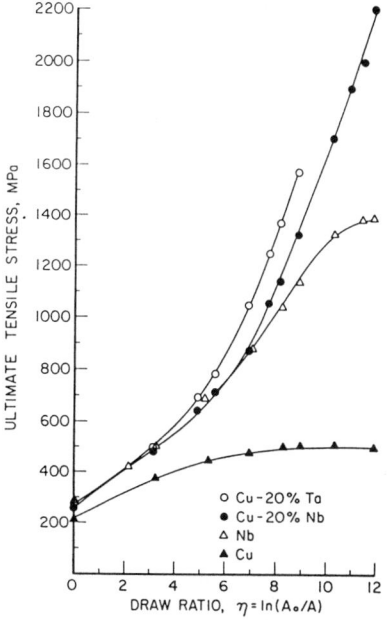

Fig. 1. Effect of draw ratio on the ultimate tensile stress of Cu-20% Nb, Cu-20% Ta, Cu and Nb.

Fig. 2. Effect of temperature and draw ratio on the ultimate tensile stress of Cu-20% Nb, Cu and Nb.

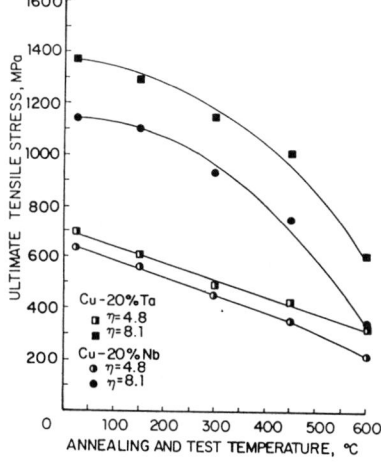

Fig. 3. Effect of temperature and draw ratio on the ultimate tensile stress of Cu-20% Nb and Cu-20% Ta.

temperatures. Figure 4 shows filaments of Nb extracted from Cu-20% Nb drawn to η = 8.1 and annealed at 450 and 600°C. Coarsening of the filaments is apparent especially after the 600°C anneal. After annealing at 150 or 300°C the filaments appear lamellar with sharp edges similar to their appearance prior to annealing. Filaments from Cu-20% Nb drawn to η = 11.0 showed more pronounced coarsening at 450°C and similar coarsening at 600°C as compared to Cu-20% Nb drawn to η = 8.1. However, after annealing at 300°C, filament coarsening was not apparent even in Cu-20% Nb drawn to η = 11.0.

Figure 5 shows extracted filaments from Cu-20% Ta drawn to η = 8.1 and annealed at 450 and 600°C. After annealing at 450°C there is only an indication of some surface roughening but no apparent indication of coarsening, but after annealing at 600°C coarsening is apparent. Filament

 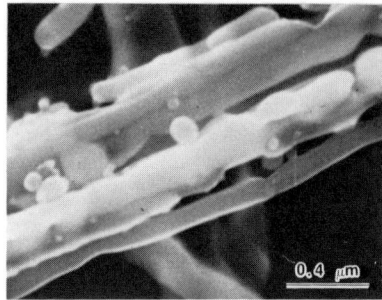

Fig. 4. SEM images of Nb filaments extracted from Cu-20% Nb that was drawn to η = 8.1 and annealed (24 h) at (a) 450°C and (b) 600°C.

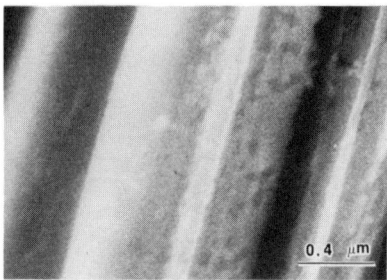 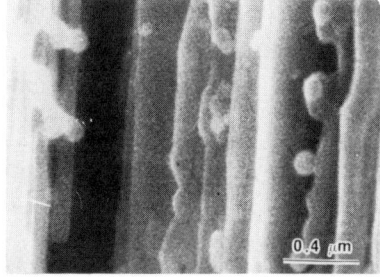

Fig. 5. SEM images of Ta filaments extracted from Cu-20% Ta that was drawn to η = 8.1 and annealed (24 h) at 450°C and 600°C.

coarsening in Cu-20% Ta is less advanced than that in Cu-20% Nb at 450 and 600°C in accord with the observed better high temperature strength of Cu-20% Ta. The reduced degree of coarsening of Ta as compared to Nb is most likely a result of its lower interface diffusion coefficient, predicted from its higher melting temperature.

Figure 6 shows TEM images from foils made from transverse sections of Cu-20% Nb wires drawn to η = 11.0 in the as drawn condition and after annealing for 24 h at 600°C. After drawing the filaments are ribbon-like in cross section. However, after annealing at 600°C the Nb filaments thicken and lose their ribbon-like morphology and become circular or elliptical in cross section. The observations in Figs. 4-6 appear to be typical of the mechanism of coarsening in lamellar structures [15]. In a lamellar structure terminations form by segments pinching off from the lamellar phase. These terminations subsequently spheroidize thereby changing the lamellar phase into a row of globular particles. Increased deformation is expected to accelerate this coarsening process by causing more terminations and enhancing diffusivity [15].

The results on filament coarsening in Cu-20% Nb at temperatures above 300°C are in contrast to earlier work on Cu-Nb composites [12] where the decrease in strength with temperature was attributed solely to dislocation annihilation and recovery in Cu. It was assumed that the Nb filaments were unaffected by temperatures up to 550°C, the maximum investigated, and electrical resistivity measurements were used to calculate dislocation densities.

 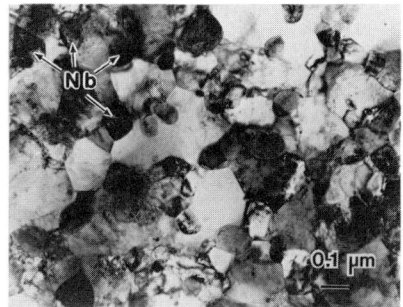

Fig. 6. TEM images of transverse sections of Cu-20% Nb that was drawn to η = 11.0; (a) as drawn condition, (b) after annealing (24 h) at 600°C.

The total decrease in resistivity with increasing temperature up to 550°C was attributed to dislocation annihilation resulting in calculated dislocation densities of 10^{12}-10^{13}/cm^2.

Figure 7 shows resistivity versus temperature curves for Cu-20% Nb and Cu-20% Ta drawn to η = 8.1. Because the conductivity of Nb and Ta is much lower than that of Cu they are treated as insulators in the analysis. The upper and lower curves for each composite correspond to the resistivity on heating from the initial as-drawn condition and cooling from the highest temperature, respectively. The difference in the heating and cooling curves is defined as Δρ and is taken to be a result of recovery and dislocation annihilation in the Cu and filament coarsening. The lower curve is retraced on subsequent reheating of a sample. The reasons for the upward shift and slightly larger Δρ for Cu-20% Ta have not yet been determined.

The temperatures where deviations from linearity occur on the heating curves for Cu-20% Nb and Cu-20% Ta in Fig. 7 are about 285 and 425°C, respectively. These temperatures are in good agreement with the temperatures where the strengths of the composites started to show pronounced softening (Fig. 3) and the filaments showed signs of coarsening (Figs. 4 and 5). The Δρ values increase with decreasing wire diameter or increasing draw ratio, as observed previously [12]. If the total decrease in resistivity (Δρ) is attributed to dislocation annihilation, dislocation densities of 10^{12} or 10^{13}/cm^2 are calculated for η = 8.1 or 11.0. These values are the same as those calculated earlier [12], but our preliminary results indicate that filament coarsening significantly affects the interface scattering component of resistivity, which was assumed to be unaffected by temperature in the earlier work. Therefore, dislocation densities calculated from resistivity changes without accounting for coarsening effects are unreliable and surely high. This is in agreement with TEM analyses on the Cu-20% Nb composites which showed maximum dislocation densities of about 10^{10}/cm^2 in the Cu matrix and the Nb filaments [4,5]. These observations also support a strengthening model for

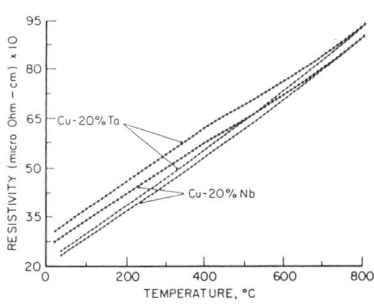

Fig. 7. Electrical resistivity as a function of temperature for Cu-20% Nb and Cu-20% Ta.

these composites based on the filaments acting as planar barriers to dislocation motion [5]. However, they are in contrast to models based on the accumulation of large dislocation densities as a result of strain incompatibility between the matrix and filaments [3].

CONCLUSIONS

Cu-20% Nb and Cu-20% Ta composites have much better high temperature strengths than current high temperature Cu alloys over the temperature range studied (22-600°C). The decreases in strength are relatively small up to about 300°C but become more pronounced at higher temperatures and at larger draw ratios. Cu-20% Ta composites have greater high temperature strengths than Cu-20% Nb composites and the improvement in strength increases with increasing temperature. Resistivity measurements exhibited pronounced decreases at temperatures where softening became pronounced. Substructural analyses showed that at the temperatures where softening accelerated, the Nb and Ta filaments had appreciably coarsened. The improved strength of Cu-20% Ta over that of Cu-20% Nb is attributed to the higher modulus and melting temperature of Ta as compared to Nb.

ACKNOWLEDGEMENTS

The authors express their appreciation to E. D. Gibson, L. K. Reed, F. A. Schmidt, and C. L. Trybus for discussions and for experimental assistance, H. H. Baker for metallographic work and F. C. Laabs for SEM and TEM assistance. Ames Laboratory is operated for the U.S. Department of Energy by Iowa State University under contract no. W-7405-ENG-82. This work was supported by the Director for Energy Research, Office of Basic Energy Sciences.

REFERENCES

1. G. Frommeyer and G. Wassermann, Acta Met. 23, 1353 (1975).
2. J. Bevk, W. A. Sunder, G. Dublon, and E. Cohen, in In Situ Composites IV, edited by F. D. Lemkey, H. E. Cline, and M. McLean (Elsevier Science Publishers, New York, 1982), p. 121.
3. P. D. Funkenbusch and T. E. Courtney, Acta Met. 33, 913 (1985).
4. A. R. Pelton, F. C. Laabs, W. A. Spitzig, and C. C. Cheng, Ultramicroscopy 22, 251 (1987).
5. W. A. Spitzig, A. R. Pelton, and F. C. Laabs, Acta Met. 35, 2427 (1987).
6. W. A. Spitzig and P. D. Krotz, Scripta Met. 21, 1143 (1987).
7. K. R. Karasek and J. Bevk, Scripta Met. 13, 259 (1979).
8. J. Bevk and K. R. Karasek, in New Developments and Applications in Composites, edited by D. Kuhlmann-Wilsdorf and W. C. Harrigan, Jr. (AIME, Warrendale, PA, 1979), p. 101.
9. W. A. Spitzig and P. D. Krotz, to be published in Acta Met.
10. J. D. Verhoeven, F. A. Schmidt, E. D. Gibson, and W. A. Spitzig, J. Metals 38, (9), 20 (1986).
11. E. E. Underwood, Quantitative Stereology (Addison-Wesley, Reading, MA, 1970), p. 80.
12. K. R. Karasek and J. Bevk, J. Appl. Phys. 52, 1370 (1981); Scripta Met. 14, 431 (1980).
13. H. H. Wawra, Metall. 32, 346 (1978).
14. J. J. Cronin, Met. Eng. Quart. 16, 1 (1976).
15. T. H. Courtney, in New Developments and Applications in Composites, edited by D. Kuhlmann-Wilsdorf and W. C. Harrigan, Jr. (AIME, Warrendale, PA, 1979), p. 1.

PRELIMINARY INVESTIGATIONS ON ALUMINA-FIBER
REINFORCED Ni_3Al MATRIX COMPOSITES

B. MOORE, A. BOSE, R.M. GERMAN AND N.S. STOLOFF
Materials Engineering Department, Rensselaer Polytechnic Institute,
Troy, NY 12180-3590

ABSTRACT

Composites containing alumina fibers in Ni_3Al+B or an Ni_3Al,Cr+B alloy have been prepared by several powder metallurgical processes. Tensile tests reveal some ductility and a mixed fracture mode in both monolithic and composite alloys. Previous reports of inadequate bonding of fibers in Ni_3Al+B have been confirmed.

INTRODUCTION

The need for high temperature structural materials with improved strength and oxidation resistance has led to consideration of intermetallic-based composites. Particularly interesting as a potential composite system is Ni_3Al reinforced with Al_2O_3. Ni_3Al possesses a lower density than Ni-base superalloys, and is very oxidation resistant. However, the alloy is inherently brittle in polycrystalline form. Recently, it has been shown that doping of hypostoichiometric alloys with a small quantity of boron provides extensive room temperature plasticity [1,2]. Unfortunately, boron does not prevent high temperature embrittlement in oxygen-containing environments. Chromium additions have proven effective in supressing embrittlement and several Cr-containing alloys have been developed at Oak Ridge National Laboratory [3,4].

This paper is concerned with the preparation of Ni_3Al-base alloys containing Al_2O_3 fibers, processed by reactive sintering[5] or reactive HIPing [6] at moderate temperatures, setting off an exothermic reaction to produce rapid densification. A deficiency of both these processes is the inability to align fibers prior to or during the reaction. Therefore this paper also describes preliminary attempts to injection mold prealloyed Ni_3Al with chopped Al_2O_3 fibers.

EXPERIMENTAL PROCEDURE

Elemental carbonyl nickel (type 123) was obtained from International Nickel Co. Helium atomized aluminum powder (type H-15) was obtained from Valimet. Particle sizes were 3 and 15μm respectively. Mixtures of 87w%Ni, 13%Al, 0.06B were prepared.

Prealloyed powder containing 81.2w%Ni, 7.5Cr, 8.5Al, 0.83Zr and 0.02B, labelled IC-218, was produced by gas atomization at Homogeneous Metals; average particle size was 70 μm.

Al_2O_3 fibers, 20μm dia x 3 to 5mm long, were obtained from Dupont. Density of these fibers is 3.95 g/cc. Elastic modulus and tensile strength are reported to be 380 GPa and 1380 MPa, respectively.

Three methods have been employed to produce Ni_3Al-base composites:
 a) reactive HIPing of Ni_3Al and B powders with 3v% Al_2O_3 fibers
 b) reactive HIPing of prealloyed IC-218 with 5% fibers
 c) injection molding of IC-218 with 5v% Al_2O_3 fibers

Tensile samples were prepared from Ni_3Al+B and IC-218 base composites; no tensile properties of fiber-reinforced IC-218 injection molded material have yet been determined.

Reactive HIPing of the Ni_3Al+B with alumina fibers has been carried out at 800°C for 30 min at 104MPa pressure. Reactive HIPing of monolithic Ni$_3$Al+B was done at 1100°C for 60 min at 172MPa and also under the same conditions as that of the Ni$_3$Al+B with alumina fibers.

A micrograph of the cross section of as-HIPed Ni$_3$Al+B is shown in Fig. 1a); a composite sample is shown in Fig. 1b). Note the large particles in Fig. 1a) and the randomness of the fibers and the apparent lack of attack of the fibers by the matrix in Fig. 1b).

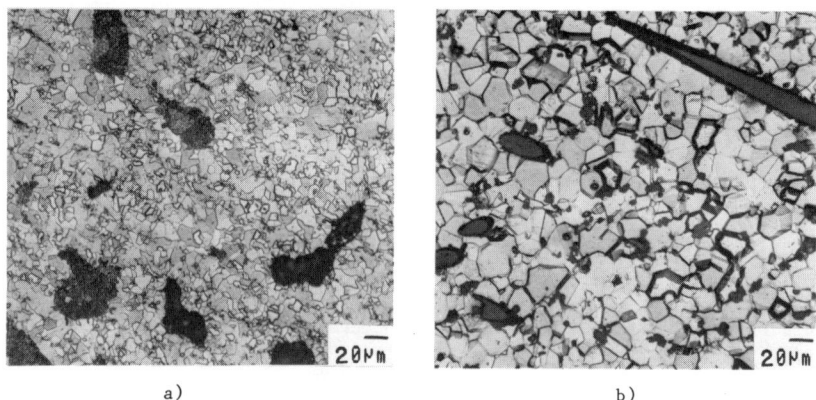

a) b)
Fig. 1 Micrograph of as HIPed Ni$_3$Al+B a) matrix only, HIPed at 1100°C
b) 5v%Al$_2$O$_3$, 800°C HIP.

Prealloyed IC-218/Al$_2$O$_3$ mixtures were HIPed at 1100°C or 1150°C for 1 hr at a pressure of 172MPa. Some samples were then heated to 1050°C for 1 hr, furnace cooled to 800°C, and held for 24 hr to remove second phase particles.

Typical microstructures of as HIPed and heat treated IC-218/Al$_2$O$_3$ are shown in Figs. 2a) and b), respectively. A second phase is known to exist in the matrix alloys[4]. Note the lack of fiber alignment in these samples. A precipitate was noted at fiber/matrix interfaces.

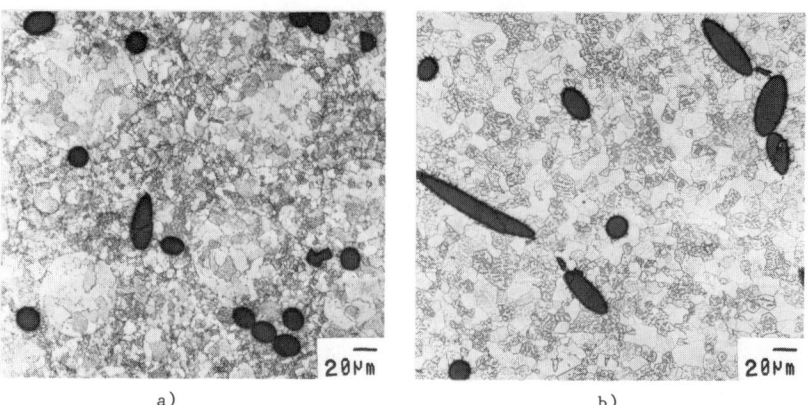

a) b)
Fig. 2 Micrographs of IC-218/Al$_2$O$_3$ composites, a) as HIPed at 1150°C
    b) heat treated

Injection molding was carried out with a low density polyethylene wax with a softening temperature of 90°C. The wax was first poured into the basket of a double planetary mixer maintained at a temperature between 125 and 135°C. The wax was liquified and the desired amount of IC-218 powder was added. The chopped fibers were added slowly to give 5v% fiber with respect to the original powder volume.

Small pellets were made from the (IC-218/Al$_2$O$_3$)/wax mixture, (66v% solid) by compacting in a preheated die. At 90°C the mixture becomes fluid and can be shaped at low pressures within the injection molding pressure range. The height and diameter of the green specimens were approximately 5 and 12 mm, respectively.

Debinding of this material proved to be extremely difficult; hence, various debinding conditions were studied. Initial debinding was carried out on samples of the IC-218/wax mixture. There was considerable slumping and flowing of the material when the binder was removed without any support, even with slow heating rates. Subsequently, it was determined that embedding the compact in loose alumina powder (1 μm particle size) greatly increased the rate of binder removal by wicking from the compact into the alumina by capillary action; also, the sample holds its basic cylindrical shape. As an example of the rapid debinding cycle, the pellet was embedded in alumina powder and heated in dry hydrogen with the following cycle:

1 K/min to 150°C, hold for 60 min, 2K/min to 400°C, hold for 120 min, 10K/min to 1200°C, hold for 30 min then furnace cool.

For the mixture without fibers, some problems with slumping of the top surface were encountered. However, for the composite mixture this was significantly reduced.

1000g of the IC-218/Al$_2$O$_3$/wax mix was introduced through the hopper of a reciprocating screw type injection molding machine. Molding was carried out at a pressure of 7.5 MPa at 125°C. The mold produced two different types of tensile bars. Final densification is to be obtained by HIPing the bars.

Tensile tests on reactively HIPed or HIPed prealloyed powders were carried out at room temperature or at 600°C. Cylindrical tensile samples had a gage diameter of 2.85mm and a gage length of 11.7mm. Crosshead rate was 8.5×10^{-3} mm^{-1}, corresponding to a strain rate of 7.25×10^{-4} s^{-1}. Room temperature tests were conducted in air, and tests at 600°C in 10^{-5} torr vacuum.

Fractured samples were examined optically and by scanning electron microscopy (SEM).

RESULTS AND DISCUSSION

Tensile data for reactively sintered Ni$_3$Al-based composites and matrix alloys are listed in Table I. Appreciable ductility was obtained at 25°C in Ni$_3$Al+B (HIPed at 800°C) and IC-218 (HIPed at 1100 or 1150°C). The poor properties of Ni$_3$Al+B HIPed at 1100°C arises from the large second phase particules seen in Fig. 1a). However, these values are far less than are usually observed in cast and wrought or conventional P/M alloys [3]. Also, fibers produced no strengthening (based on yield stress) and actually caused a decrease in tensile strength due to the sharply lowered ductility of the composites. However, heat treated IC-218/Al$_2$O$_3$, tested at 25°C, did show 6% elongation. The ductility problem undoubtedly arises from the randomness of the fibers and relatively poor bonding, as well as their relatively low strength. It is hoped to solve the alignment problem by use of injection molding.

Fractographs from fractured IC-218 and IC-218/Al$_2$O$_3$ tested at room temperature in air are shown in Figs. 3-5. Note the predominance of transgranular facets in as HIP IC-218, see Fig. 3. Figs. 4a) and 4b) show that while matrix fracture is transgranular, bonding

TABLE I Tensile Properties of Ni$_3$Al-Base Alloys

Sample	Test Temp °C	σ_{ys} (MPa)	σ_{UTS} (MPa)	ε_F %	Remarks
Ni$_3$Al+B	25	286	759	14.8	As HIP, 800°C
Ni$_3$Al+B	25	494	677	2.1	As HIP 1100°C
IC-218	25	638	1380	19.8	As HIP 1100°C
IC-218	25	663	1400	23.5	As HIP 1150°C
IC-218/Al$_2$O$_3$	25	663	890	3.5	As HIP 1150°C
Ni$_3$Al/Al$_2$O$_3$	25	474	548	1.0	As HIP 800°C
Ni$_3$Al+B	25	591	828	5.2	Heat Treat; 1100°C-HIP
IC-218	25	535	1428	23.1	Heat Treat; 1100°C-HIP
IC-218	25	518	1421	21.5	Heat Treat; 1150°C-HIP
IC-218/Al$_2$O$_3$	25	499	756	6.0	Heat Treat; 1150°C-HIP
IC-218	600	787	1070	15.9	As HIP 1100°C
IC-218	600	766	1049	15.2	As HIP 1500°C
IC-218/Al$_2$O$_3$	600	814	869	1.0	As HIP 1150°C

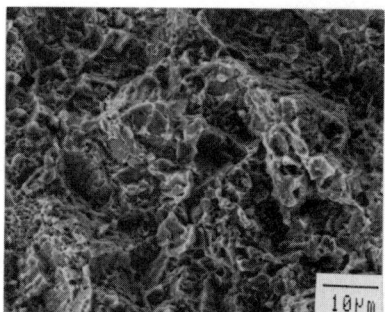

Fig. 3 SEM fractograph of as HIPed (1150°C) IC-218 tested at 25°C.

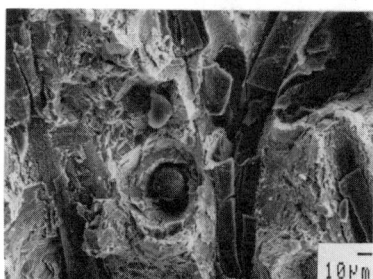 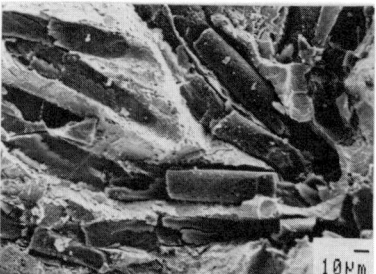
a)   b)

Fig. 4 SEM fractographs of IC-218/Al$_2$O$_3$, HIPed at 1150°C, tested at 25°C
a) as HIP, b) heat treated.

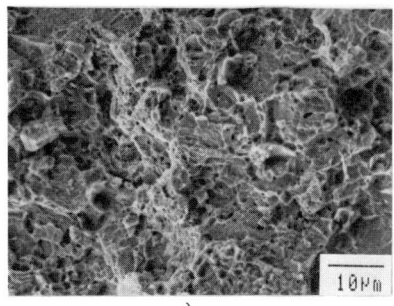 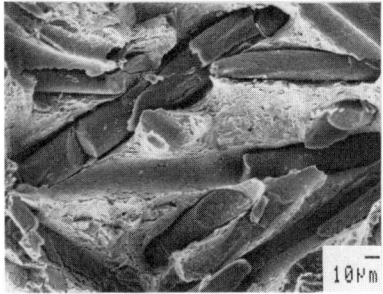

Fig. 5 SEM fractographs of alloys, HIPed at 1150°C, tested at 600°C
a) IC-218, b) IC-218/Al$_2$O$_3$.

is poor between fibers and matrix for as HIP and HIP + heat treated samples, respectively. At 600°C in vacuum the as HIPed IC-218 again showed predominantly transgranular fracture, see Fig. 5a). While IC-218/Al$_2$O$_3$ again displayed lack of bonding at the interface and the breakup of fibers, Fig. 5b).

The results reported in Table II may be compared with fragmentary recent observations by Povirk et al[7] on hot pressed IC-15 (Ni-24a%Al-0.24%B) and IC-218 alloys, Table II. IC-15 was hot pressed at 1300-1350°C and IC-218 at 1250-1300°C. All alloys were then annealed for 2 hr at 1000°C and 24 hr at 800°C and bend or tensile tested at 25°C. It appears that ductilities of Ni$_3$Al/Al$_2$O$_3$ are lower, and those of IC-218/Al$_2$O$_3$ are higher in the present work, as shown in Table I. However, hot pressing produces stronger unreinforced IC-218, with a yield stress of 900 MPa and substantial ductility. ZrO$_2$ particles observed at IC-218-Al$_2$O$_3$ interfaces were given as the cause of low ductility in the earlier work[7]. In the present work similar particles have been seen, see Fig. 6.

TABLE II Tensile Properties of Hot Pressed Alloys [7]

Alloy	g.s. μm	σ_{ys} (MPa)	σ_{UTS} (MPa)	ε_F %	Remarks
IC-15	50	---	---	0	Tensile
IC-15/Al$_2$O$_3$	--	500	---	>5	Bend
IC-218	10	900	---	>5	Bend
IC-218/Al$_2$O$_3$	10	---	170-200	<1	Bend

SUMMARY

Reactive HIPing of elemental and normal HIPing of prealloyed powders with randomly oriented short Al$_2$O$_3$ fibers has successfully produced fully dense alloys. Although the yield strength of Ni$_3$Al+B, HIPed at 800°C, is increased by the presence of fibers, no such beneficial effect is found with IC-218. In all cases ductility is reduced relative to monolithic material. However, IC-218/Al$_2$O$_3$ HIPed at 1150°C and then heat treated did exhibit 6% elongation at 25°C. Injection molding techniques have been developed to orient Al$_2$O$_3$ fibers, but tensile data are not yet available.

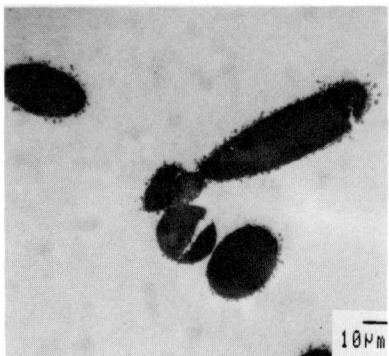

Fig. 6 Particles at Al_2O_3-IC-218 interfaces, HIPed at 1150°C

ACKNOWLEDGMENT

This investigation was sponsored by the Defense Advanced Research Projects Agency (DARPA) and Office of Naval Research under Contract N00014-86-K-0770.

REFERENCES

1. K. Aoki and O. Izumi, <u>J. Japan Inst. Met.</u>, vol. 43, p. 1190, 1979.
2. C.T.Liu, C.L. White and J.A. Horton, <u>Acta Met</u>, vol. 33, p. 213, 1985.
3. C.T. Liu and C.L. White, <u>High Temperature Ordered Intermetallic Alloys</u>, Mat. Res. Soc. Symposium Proceedings, vol. 39, Mat. Res. Soc., Pittsburgh, PA, p. 355, 1985.
4. C.T. Liu and V.K. Sikka, <u>J. Metals</u>, vol. 38, (5) p. 19, 1986.
5. D.M. Sims, A. Bose and R.M. German, <u>Prog. Powder Met.</u>, vol. 43, p. 575, 1987.
6. A. Bose, B. Moore, N.S. Stoloff and R.M. German, Proc. Conf. on P/M Aerospace Materials, Luzern, Switzerland, Nov. 1987, in press.
7. G.L. Povirk, C.G. McKamey, T.N. Tiegs and S.R. Nutt, Oak Ridge National Lab, 1988, unpublished.

HIGH TEMPERATURE INTERMETALLIC COMPOSITES

D.L. Anton
United Technologies Research Center, East Hartford, CT 06108

ABSTRACT

Many intermetallic compounds possess properties which make them excellent candidates for high temperature use in advanced gas turbine and aerospace applications. One method proposed for increasing damage tolerance in these brittle materials is to artificially composite them with high temperature fibers as utilized in both ceramic and glass composites. Fabrication of these composites is a formidable problem. One method of fabricating these structures, termed here Transient Liquid Phase Consolidation, TLPC, is demonstrated for a number of intermetallic/reinforcing fiber combinations. Thermal stability of the fibers in the intermetallic matrices was observed with FP Alumina being the most stable. Ambient temperature tensile property evaluations were made on monolithic, chopped and aligned FP fiber reinforced Al_3Ta with the aligned structure having the highest ultimate strength and the chopped fiber composite the greatest pseudo plastic response.

INTRODUCTION

Intermetallic compounds have long been candidates for high temperature materials in aerospace propulsion systems. They possess many of the same positive attributes found in ceramic materials; e.g., low density, very high melting points and superb oxidation resistance. Although brittle at ambient temperatures like ceramics, these compounds are of particular interest because they gain extended ductility at elevated temperatures. Given the low temperature similarities between these two materials types, it may be expected that the compositing approach to damage tolerance, which has been shown to be successful in ceramic matrix composites [1-3], can be utilized with intermetallic compounds.

The useful elevated temperature stress-strain envelope for intermetallics can be expanded with respect to ceramics as a result of their high temperature ductility, as shown in (see Fig. 1 [4]). At low temperatures and at strains below the reinforcing fiber failure strain, ceramic and intermetallic matrix composites are expected to behave linear elasticly to failure. At elevated temperatures, as shown in Fig. 1 a&b for ceramic and intermetallic matrix composites, respectively, the enhanced matrix failure strain of the intermetallic is expected to lead to greater strengths before matrix microcracking and thus an enlarged cyclic stress-strain envelope.

In order to maintain low temperature damage tolerance, fiber pull-out during failure has been shown to be necessary in ceramic composites [1-3]. To achieve this failure mode, the fiber-matrix interface must be controlled to not allow a matrix crack to penetrate the fibers. Thus the chemical compatibility of the matrix and fiber must be maintained either through judicious selection of composite components or through various fiber coating approaches [5].

Preparation of the intermetallic composites, as with any artificial structure containing chopped or continuous reinforcement, is of initial primary importance. Since these intermetallics are hard at high temperature with limited flow capabilities, simple compaction will result in severe damage. At the temperatures necessary for easy flow, T>1400°C, fiber degradation has been reported in those fibers of primary consideration, such as FP Alumina, PRD-166 and Nextel 480[*] [6].

   * FP Alumina and PRD-166 is a trademark of I.E. DuPont de Nourous Corp.
      Nextel 440 is a trade mark of 3-M Corp.

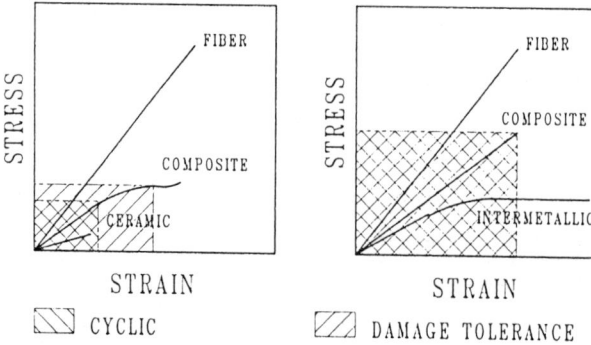

Fig. 1 Schematic diagram of intermetallic and ceramic matrix composite strength at elevated temperatures. The shaded regions indicate useful stress-strain envelope in which the composite can perform. fiber

Additionally, intermetallic compounds are a class of materials not fully understood with respect to physical or mechanical properties and manufacturing processes. The role of intersticial elements such as carbon and oxygen on strength is not known. In order to minimize their effects, material cleanliness is of concern.

This paper will report on a new means of fabricating chopped and continuous fiber reinforced intermetallic composites at intermediate temperatures, relative to the compound melting point, by a transient liquid phase technique. This has resulted in a homogeneous, fully dense composite structures.

Such composites have been used to study both fiber-matrix interactions in a number of intermetallic/reinforcement combinations as well as the elevated temperature mechanical properties in one specific system, Al_3Ta/FP, to verify the feasibility of producing these structures and to gain initial ambient temperature properties.

EXPERIMENTAL PROCEDURES

TLP Consolidation

The Transient Liquid Phase Consolidation, TLPC, technique was used to fabricate both compatibility test coupons and 4-pt. bend specimens. This technique consists of blending the elemental powders in the desired stoichiometry, mixing with a volatile binder, infiltrating the continuous fiber lay-up or chopped fibers, completely driving off the binder at low temperatures and hot pressing in a floating closed alumina die; although HIPing would work just as effectively. The procedure for laying up continuous fibers is explained elsewhere [3], while the chopped fibers were simply mixed with the elemental powders before consolidation in a v-blender. The TLPC technique requires one of the elemental constituents to be of low melting point, preferably below 1000°C to minimize fiber degradation. Thus, this technique is especially well suited to the fabrication of aluminide compounds or those intermetallics in which a low melting point eutectic exists. In the case of aluminides, pressure is applied when the powder mixture has reached the melting point of aluminum (660°C) and maintained for one hour. Further homogenization at temperatures to 1200°C was performed to obtain chemical homogeneity.

Compatibility Testing

Fiber-matrix compatibility test coupons were coin shaped having dimensions 1.9cm in diameter and approximately 3mm thick. Fiber-matrix pairs evaluated were Al_3Ta/FP, $Al_3Ta/440$, $TiAl/440$, $Nb_3Al/440$ and Nb_3Al/FP; where FP represents FP Alumina and 440 for Nextel 440. The fibers were chopped to 10mm lengths and blended with the appropriate powders. Fiber volume fractions were limited to 5% since only reactivity was of interest. Subsequent heat treatments for 4 hours at 1093, 1204, 1315 and 1426°C consecutively, were conducted followed by a metallographic examination after each heat treatment.

4-pt. Flex Testing

Four-point bend specimens of Al_3Ta plus both chopped and aligned FP Alumina fibers 32 x 6.4 x 2mm were cut from a larger 60 x 32 x 2mm plate containing 35% by volume fiber. In addition, specimens of the monolithic intermetallic, fabricated identically to the composites, were prepared for comparison. After compaction, the composite and monolithic specimens were given a homogenization heat treatment of 1000°C for 4 hours in Ar and furnace cooled.

Four point bend tests were conducted in a 10^{-6} torr vacuum at ambient temperatures using a 2:1 lower to upper span ratio and a deflection rate of 0.25mm/min. Center point deflection was measured as a function of load. ASTM D-790 was followed for testing procedures and stress-strain calculations. Fractography was performed in order to determine mode of fracture as functions of temperature and compositing.

RESULTS AND DISCUSSION

TLPC Microstructures

The microstructures resulting from TLPC of monolithic intermetallic compounds are fully dense compacts. They are approximately 95% transformed into the intermetallic compound, with evenly distributed small islands and ligaments of untransformed or partially transformed material. A post consolidation heat treatment of 1200°C for 4 hours was sufficient to chemically homogenize the compounds into nearly single phase materials with only trace precipitates of remnant elemental or intermetallic material.

The chopped fiber reinforced composite had a similar compound microstructure as found in the monolithic intermetallic with the addition of alumina fibers as shown in Fig. 2a. The fibers were evenly distributed throughout the specimen with minimal fiber damage observed.

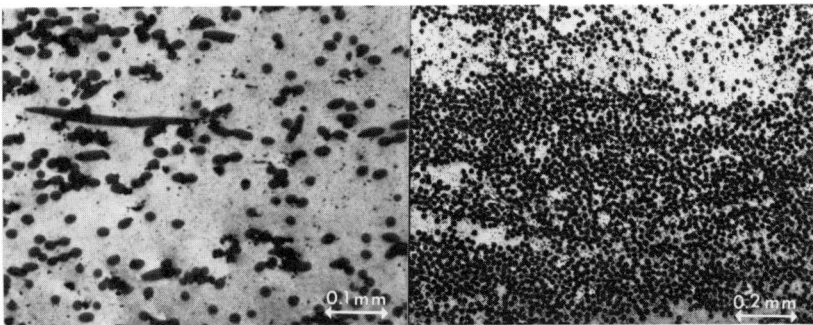

Fig. 2 Composite structures after hot pressing for both chopped and continuously aligned FP fibers in an Al_3Ta matrix.

Fig. 2b illustrates the transverse structure of the aligned composite. One notes nearly total infiltration of the dense fiber toes with the matrix material. Regions of fiber poor material were observed and are attributed to inter-ply regions.

Under SEM observation (see Fig. 3), the higher fiber volume fraction of the aligned composites hindered diffusional homogenization and more elemental segregation was found. The lower fiber content of the chopped fiber specimen did not inhibit homogenization and the microstucture resembled that of the monolithic compound. Subsequent homogenization at 1200°C in purified Ar is shown in Fig. 3b to minimize segregation and yield essentially phase pure intermetallic matrix.

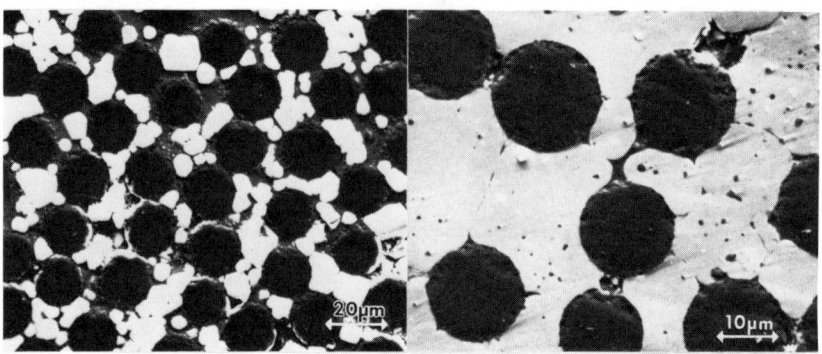

Fig. 3   Microstructure of (a) as hot pressed and (b) after 1200°C/4hr heat treatment. In (a) a relatively high degree of segregation after hot pressing and (b) most of the segregation has been eliminated by subsequent heat treatment.

Compatibility Testing

The chemical compatibility of fiber/matrix combinations can in many instances be deduced from simple thermodynamic calculations. In many cases however, unexpected reactions can occur with additive elements or the fibers may simply transform upon exposure to elevated temperatures as reviewed in [4]. In the final analysis it is always more technically sound to make the simple tests to assure compatibility.

The results of the fiber-matrix compatibility tests are given in Table I. The legend at the bottom of the table explains the abbreviations. Generally, the FP fibers were inert in the matrices studied. The Nextel 440 fibers were less stable and internally transformed to a crystalline state and reacted with the matrices to a larger extent than the FP fibers. Typical of the reacted microstructures containing 440 fibers is that of $Al_3Ta/440$ heated to 1093°C for 4 hours given in Fig. 4a. The core of the fibers are still present while a reacted zone 3um thick has developed. These fibers became fully reacted after exposure at 1204°C. The other two composite structures containing 440 fibers, $Nb_3Al/440$ and TiAl/440 reacted after similar heat exposure with the TiAl composite displaying less reactivity at the two lower temperatures.

The FP fibers appeared microstructurally stable under optical microscopy and only a slight interface reaction was noted in the Nb_3Al matrix composite shown in Fig. 4b after exposure to 1204°C. The dark speckles in Figs. 4a and b were identified as alumina which was found in all of the compatibility specimens and traced to a surface oxide layer on the aluminum powder used in these early studies and subsequently discarded prior to 4-pt. flex specimen fabrication.

Table I

Fiber-Matrix Compatibility Test Results

System\Temp.(°C)	1093	1204	1315	1426
Al_3Ta/FP	*	*	*	*
Al_3Ta/440	**	**	**	**
Nb_3Al/FP	*	*	*	*
Nb_3Al/440	**	**	**	**
TiAl/440	o	*	**	**

o = No Reaction
* = Interface Reaction
** = Severe Reaction

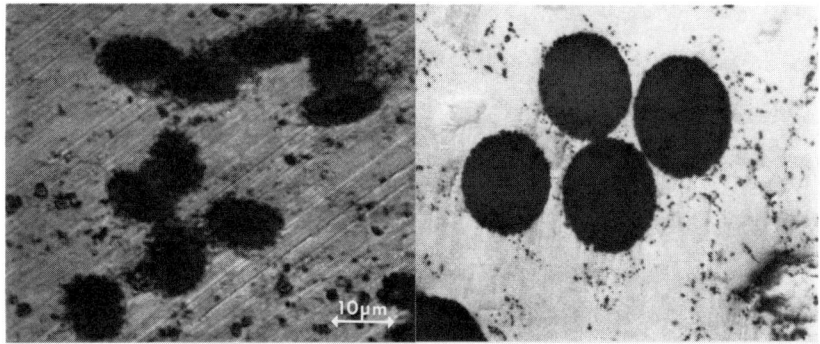

Fig. 4 Fiber-matrix compatibility trials showing (a) heavily reacted Nextel 440 fibers and (b) interface reaction on FP alumina fibers.

4-pt. Flex Testing

The tensile stress-strain curves were calculated from the load-deflection curves recorded during testing. Thus, it should be remembered that the data reported here is an equivalent tensile strain over the outside tensile ligament of the bend specimens. After a crack has initiated, the tensile stresses and strains reported are apparent stress and strain at the outer surface as if the specimen had not cracked.

Fig. 5 gives a typical tensile ligament stress-strain curve for the three microstructure types tested here and Table II contains the test results; i.e., ultimate strength, modulus, strain to maximum stress and strain to total specimen separation. This last parameter gives an indication of pseudo plasticity that results from fiber pull-out.

The base line monolithic intermetallic displayed a relatively low average ultimate strength of 96.8 MPa (14.1 ksi) and a modulus of 167 GPa (24.2x10^6 psi), similar to that of Ta metal. A featureless fracture surface resulted as shown in Fig. 6a. Evidence of both intergranular and transgranular cleavage is observed. In general, however, failure occurred via brittle transgranular fracture. The stress-strain curve displays totally brittle behavior with purely elastic loading followed by failure.

The addition of chopped alumina fibers lowered both the ultimate strength of the compound and reduced the modulus significantly, while the strain to maximum load remained constant at less than 0.2%. A higher degree of variability was observed in the chopped fiber results than for the monolithic compound. This and the lower strengths are due to inconsistencies in the chopped fiber distributions. Shorter fiber lengths and more careful

Table II

Four Point Bend Test Data

Composite Structure	Ultimate Strength (MPa)	Young's Modulus (GPa)	Strain to Max. Load (%)	Strain to Failure(%)
Monolithic	99.3	151.7	0.110	0.110
	94.4	182.6	0.137	0.136
Chopped Fiber	41.2	66.9	0.090	0.342*
	11.3	42.0	0.077	0.221
	11.8	80.6	0.052	0.197*
Aligned Fiber	174.2	188.6	0.097	0.157
	167.0	168.2	0.132	0.179
	164.1	148.9	0.106	0.203

*-failure strain exceeded test fixture capabilities specimen separation not obtained

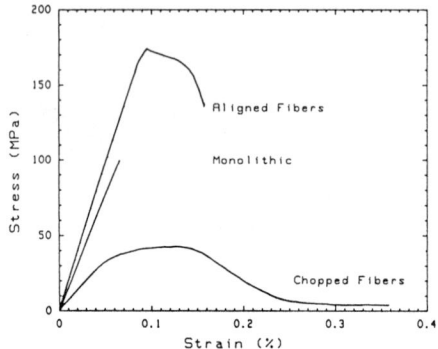

Fig. 5 Stress-strain relation reduced from four point bending tests of monolithic, chopped fiber and aligned fiber Al_3Ta at ambient temperature.

attention to mixing would be expected to yield a uniform fiber distribution and thus enhanced properties.

Fig. 5 shows a marked change in the stress-strain response with the addition of the chopped fibers. It is readily apparent that the fibers have added pseudo plasticity to the compound. Table II illustrates this clearly by comparing strain to failure for the monolithic and chopped fiber reinforced intermetallics. Where the monolithic compound displayed failure strains equal to ultimate load strains, the reinforced compound maintained engineering failure strains exceeding strain to ultimate stress by approximately 250%. Fractography revealed that extensive fiber pull-out has occurred as shown in Fig. 6b. The matrix has failed in an intergranular mode with many of the chopped fibers clearly visible. The fiber surfaces appeared clean in these fracture surfaces with no evidence of matrix interaction as previously shown in the compatibility studies.

The aligned fiber reinforced composites inherit the benefits of both the monolithic compound ultimate strength and the pseudo plasticity of the chopped fiber composite. The ultimate strength of this material is 74% stronger than the monolithic case with 51% enhancement in strain to failure capacity, as shown in Table II.

Fig. 5 shows that the absorbed strain energy is somewhat reduced with respect to the chopped fiber composite, but a pseudo plastic response is still evident. The high degree of fiber pull-out associated with the chopped

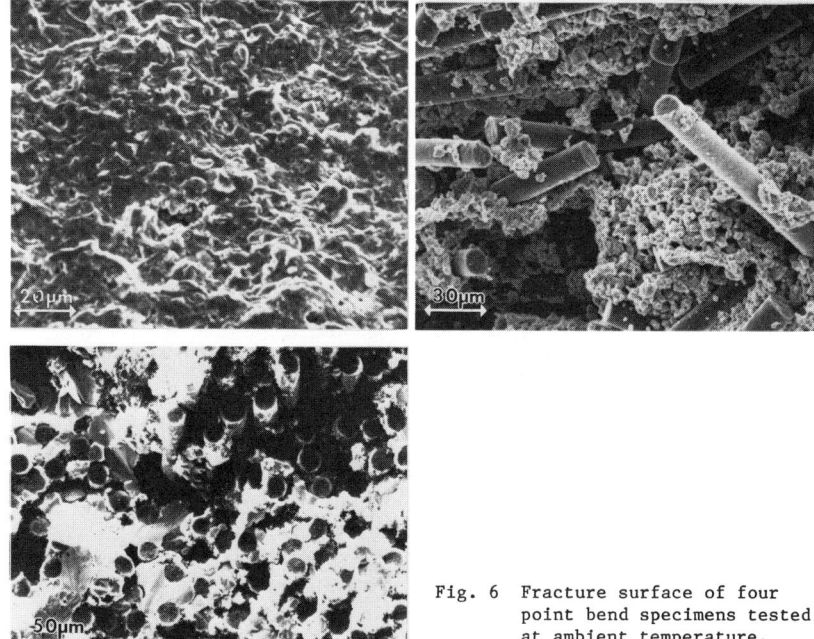

Fig. 6 Fracture surface of four point bend specimens tested at ambient temperature.

fiber composite is not observed in Fig. 6c for the aligned fiber composite. Most of the fibers appear to have fractured along with the matrix. Some areas of the fracture do reveal extensive fiber pull-out though. Remembering from Fig. 2 that the fiber distribution was not completely uniform, it is suspected that regions where the fiber density was lower resulted in more pull-out while those regions with higher fiber densities resulted in fiber fracture. It is clear, however, that better control of fiber density along with a lower overall fiber density is required.

CONCLUSIONS

The Transient Liquid Phase Consolidation (TLPC) technique for the fabrication of intermetallic composites was demonstrated. It involves the mixing of elemental powders in a volatile binder followed by infiltration of the fibers, hot consolidation and homogenization. The resulting microstructure is a fully dense nearly single phase intermetallic. Fiber/matrix compatibility studies were carried out on a number of matrix/fiber combinations. FP fibers showed minimal interface reaction while Nextel 440 reacted heavily at intermediate temperatures. Monolithic, chopped FP and uniaxial FP fiber reinforced Al_3Ta were fabricated and tested in four point bending. The monolithic compound failed by brittle transgranular fracture. The composite material exhibited showed extensive fiber pull-out and pseudo plastic behavior with strain to failure increases of greater than 200% over the monolithic compound.

ACKNOWLEDGEMENTS

The author would like to recognize the many helpful discussions with Drs. D.M. Shah, D.N. Duhl and A.F. Giamei. Many of their ideas and suggestions have been incorporated. The exceptional help of Messrs. R. Brown and L.H. Favrow are also greatfully acknowledged for their help in fabrication of the composites and mechanical evaluation respectively.

REFERENCES

1. T. Mah, M.G. Mendiratta, A.P. Katz, R. Ruh and K.S. Mazdiyasni, J. Am. Ceram. Soc. 68 , C27 (1985).
2. K.M. Prewo, J.J. Brennan and G.K. Layden, Ceram. Bull. 65 ,305 (1986).
3. E.Y. Luh and A.G. Evans, J. Am. Ceram. Soc. 70 ,466 (1987).
4. J.R. Strife, Personal Communication, United Tech. Res. Ctr., 1987.
5. D. Lewis and R.W. Rice in Metal Matrix, Carbon, and Ceramic Matrix Composites 1985 , edited by J.D. Buckley (NASA Conference Publication 2406 1985) pp. 13-26.
6. T. Mah, M.G. Mendiratta, A.P. Katz and K.S. Mazdiyasni, Ceram. Bull. 66 , 304 (1987).

PART II

Deformation Mechanisms in Metal Matrix Composites

MECHANISMS AND MODELS OF HIGH TEMPERATURE
DEFORMATION OF COMPOSITES

MALCOLM McLEAN
Division of Materials Applications, National Physical Laboratory, Teddington, Middlesex TW11 0LW, UK

ABSTRACT

The deformation mechanisms that can occur in advanced composites at high temperature are reviewed and the implications of different deformation modes in the constituent phases for the creep behaviour of the composite are considered. A generalised description of creep of composites in terms of internal state variables that have clear physical significance is presented.

1   INTRODUCTION

The resurgence in interest in metal matrix composites over the last five years or so has been associated with the increasing availability of suitable relatively cheap reinforcements and by the development of several reliable processing routes resulting in reproducible composite microstructures[1]. Potential structural applications of these materials depend on their high specific stiffnesses and increased temperature capabilities relative to the matrix alloys; for high temperature applications the latter characteristic is probably of more importance. Although short term creep and tensile data are sometimes available, there is very little information on the long time-scales relevant to likely service requirements. Moreover, existing extrapolation techniques developed for more conventional materials are inappropriate to modern composite materials which can have quite different deformation and fracture modes.

Metal matrix composites are available with a wide range of forms of reinforcement - eg particulate, randomly-oriented or aligned whiskers, continuous or chopped filaments, continuous eutectic phases. Also, the dimensions of the reinforcing phases can vary from the sub-micron whiskers of in-situ composites to fibres of ~ 100 μm diameter in some synthetic composites. The creep performances of this range of materials are quite different and any useful treatment of creep of metal matrix composites must be capable of addressing these important microstructural and behavioural dependences.

The limited available data, in conjunction with previous extensive studies on model systems and on in-situ composites, provide guidance on the mechanisms that operate in materials of current interest. This paper attempts to interpret this information in terms of the interacting deformations of the constituent phases with particular regard to the principal microstructural features, fibre volume fraction and aspect ratio. The possibilities of both direct strengthening through load sharing between the phases and indirect strengthening resulting from synergisms that modify the deformation mechanisms are considered as are the effects of degradation of the microstructure.

A generalised model of creep of composites, developed in detail elsewhere[2], that incorporates the knowledge of physical mechanisms and is expressed in terms of internal state variables, appears to be capable of describing a wide range of creep data.

2   CREEP BEHAVIOUR

Figure 1 shows a selection of creep curves for metal matrix composites with quite different microstructures. None have the simple form associated with pure metals and some single-phase alloys where most of the life is dominated by a steady state creep rate $\dot{\epsilon}_{ss}$ which can be expressed as a simple power-law function of stress σ and exponential function of temperature T.

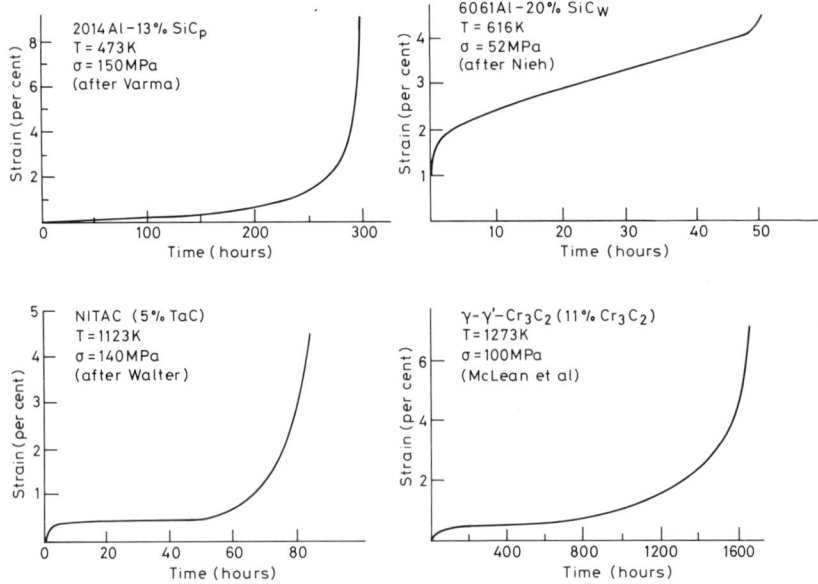

Figure 1 Creep curves for metal matrix composites with various microstructures
(a) 2014 Al − 13% SiC particulate (Varma[3])
(b) 6061 Al − 20% SiC whiskers (Neih[4])
(c) NITAC (5% TaC) (Jackson and Walter[6])
(d) $\gamma-\gamma'$-Cr$_3$C$_2$ in-situ composite (Bullock et al [5])

$$\dot{\epsilon}_{ss} = \dot{\epsilon}_o \left[\frac{\sigma}{\sigma_o} \right]^n \exp\left[-\frac{Q}{RT} \right] \quad (1)$$

where $\dot{\epsilon}_o$, σ_o are normalising constants, R is the gas constant and the parameters n ∼ 3 to 4 and Q ∼ self-diffusion activation energy are consistent with concepts of recovery controlled creep. It is useful to establish the extent of the differences between the creep behaviours of composites and simpler materials since these must be incorporated into any realistic model.

a) The particulate composite shows virtually no primary or secondary creep behaviour but is characterised by a progressively increasing creep rate over most of the creep life. This is very similar in form to the creep curves of a wide range of commerical precipitation strengthened alloys (eg nickel-base superalloys, Al-Li alloys) where the particles act as obstacles to dislocation motion and radically alter the deformation mechanisms in the matrix[7]. There is also evidence of indirect matrix strengthening in some in-situ composites[5] which have very long aligned fibres, where microstructural refinement produced by

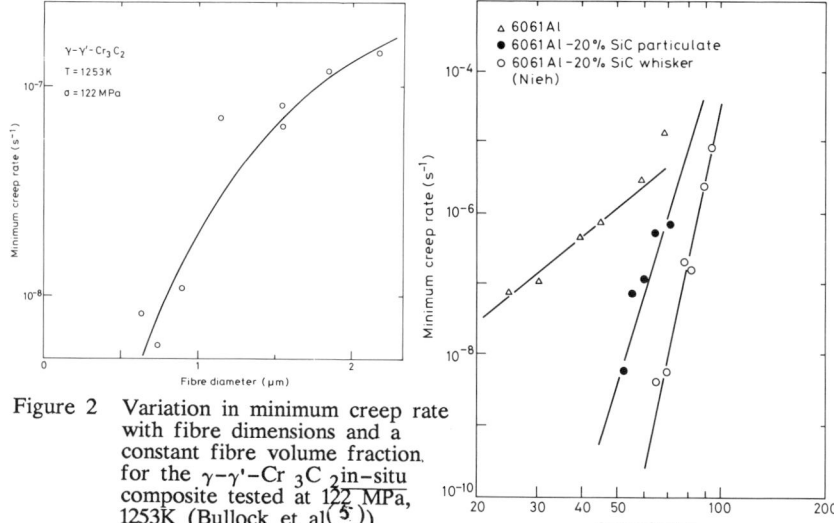

Figure 2 Variation in minimum creep rate with fibre dimensions and a constant fibre volume fraction, for the $\gamma-\gamma'-Cr_3C_2$ in-situ composite tested at 122 MPa, 1253K (Bullock et al[5])

Figure 3 Comparison of minimum creep rate as a function of stress for 6061 Al matrix alloy and for 6061 Al reinforced by 20% vol SiC whiskers and 30% volume SiC particles (Nieh[4])

increased solidification rates leads to very substantial decreases in minimum creep rate and increases in rupture life (Figure 2).

b) In aligned fibre reinforced materials the primary creep regime of decreasing creep rate is more apparent, and indeed is dominant in the short fibre reinforced Al-SiC system[4] (Figure 1b). Primary creep also extends over much of the life of in-situ composites and, in this case, the creep strain is almost entirely anelastic, ie it can be fully recovered by heat treatment in the absence of load[8]

c) When the minimum creep rates for composites are represented by the Bailey Norton power-law equation (Equation 1), physically unrealistic values of n (>> 4) and Q (~ $3 Q_{SD}$) are required to fit the data. However, in both particulate and aligned fibre systems the creep strengths and rupture lives are substantially superior to those of the matrix alloys[4] (Figure 3).

d) Whereas creep fracture in most conventional metallic alloys is associated with the development of cavities at grain boundaries, the initiation and growth of defects in the reinforcement is generally the life limiting factor for composites. However, the nature of this damage can differ significantly both within and between systems. For example, Walter[9] shows that variation in the directional solidification rate of an advanced Ni-TaC in-situ composite, which alters both the volume fraction and dimensions of the reinforcing fibres, leads to changes in (i) the nature of the creep fracture surface, (ii) the extent of fibre fragmentation with distance from the fracture surface and (iii) the mode of deformation in the matrix. (Figure 4)

e) The creep performance of simple metals is generally improved by plastic strain accumulation since the strength largely derives from the dislocation substructure. Metal matrix composites, however, degrade when strained, as for example by thermal cycling, due to either deterioration of the reinforcing phase or to modification of the matrix structure[10]. This latter characteristic is similar to precipiation hardened materials such as nickel-base superalloys.

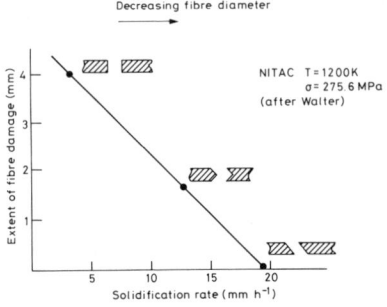

Figure 4

Width of zone of fibre damage measured from the fracture surface as a function of solidification rate for a NITAC in-situ composite tested at 275.6 MPa, 1200K (Walter[9])

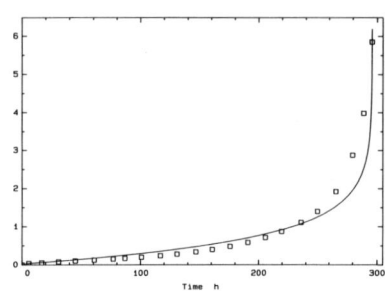

Figure 5

Comparison of the fit of the optimum curve described by Equation Set 2 to creep data for an Al-SiC particulate composite

3 MECHANISMS AND MODELS

3.1 Particulate reinforcement

The potential causes of tertiary creep in engineering alloys have recently been reviewed by Ashby and Dyson[9] who point to the usefulness of the Monkman-Grant parameter C_m and a creep damage factor λ in identifying the type of "damage" that causes the accelerating creep rate. Analysis of the creep curves for the particulate Al-SiC composites indicate values of $C_m \sim 10^{-2}$ and $\lambda \sim 10$ which are far outside the range of values associated with the damage mechanisms that lead to tertiary creep and fracture in more conventional metallic alloys - eg loss of external section ($C_m \sim 0.2$), development of creep cavities ($\lambda \sim 1$ to 3). The values of the measured parameters are much more appropriate to mechanisms dependent on degradation of the microstructure by, for example, thermal coarsening of the particles or the development of a dislocation substructure. The former can be discounted here because of the relatively low temperatures and high stability of the particle morphologies. However, the data are similar to those of nickel-base superalloys and Al-Li alloys where tertiary creep has been shown to be a strain-softening phenomenon resulting from the progressive accumulation of mobile dislocations[7]. Constitutive equations, based on this physical interpretation, have been developed and expressed, using the formalism of continuum damage mechanics, in terms of two state variables S and ω by Ion et al [11] and Maldini et al [12].

$$\dot{\varepsilon} = \dot{\varepsilon}_{min} \frac{(1 - S)}{(1 - S_{ss})} \exp(\omega)$$

$$\dot{S} = H \dot{\varepsilon}_{min} \left\{ \frac{(1 - S)}{(1 - S_{ss})} - \frac{S}{S_{ss}} \right\} \quad (2)$$

$$\dot{\omega} = H \dot{\varepsilon}$$

S is the ratio of an internal stress caused by local stress redistributions and the applied stress that leads to an element of primary creep; ω is a measure of increase in the density of mobile dislocations $(\rho - \rho_{min}/\rho)$, $\dot{\epsilon}_{min}$ is the minimum creep rate and S_{ss} is the steady state value of S. The four parameter set ($\dot{\epsilon}_{min}$, S_{ss}, H, C) completely defines the shape of the creep curve. Figure 5 shows the fit of Equations 2 to a creep curve for Al–SiC; there is close agreement but further work is required before the pertinence of strain softening to particulate composites can be fully established. However, it is quite clear that the strengthening associated with equiaxed particles is due to the suppression of the natural creep mechanism that can occur in the matrix material in isolation; for example, dislocation glide may be prevented by the particles thus requiring deformation to occur by the much slower climb around the obstacles. The same factors are likely to be present when the reinforcing phase is in the form of fibres. However, in this case additional strengthening occurs by direct load transfer to the fibres; this will be considered in the following section but it must be borne in mind that the matrix behaviour used in the following models may be quite different from that of the matrix alloy devoid of reinforcement.

3.2 Continuous fibres

When the load is directly shared between matrix and fibres, the creep behaviour of the composite is clearly determined by the combined deformations of the constituent phases. Even in the simplest case of fibre and matrix individually deforming with steady state creep rates that are simple power law functions of stress, the composite will exhibit:

(i) a transient creep behaviour during which stress is redistributed to give a new steady state creep rate

(ii) a complex dependence of creep rate on stress which approaches the matrix stress dependence at high stresses and the reinforcement stress dependence at low stresses.

This combination, which was first treated by McDanels et al[13], is inappropriate to modern advanced composites where there is generally a very large difference in melting points, and consequently of creep rates, of the matrix and reinforcing phases. It is more realistic to consider the fibres to deform elastically constrained by creep of the matrix.

McLean[2,14,15] has previously modelled the creep deformation of an axially stressed composite with infinitely long elastic fibres, of volume fraction φ, and a matrix extending of power law creep. The conditions for mechanical equilibrium when the applied stress σ is redistributed to maintain equal strains in the two phases are:

$$\left. \begin{array}{l} \dot{\epsilon} = \dfrac{1}{E_f} \dot{\sigma}_f \\[1em] \dot{\epsilon} = \dfrac{1}{E_m} \dot{\sigma}_m + \dot{\epsilon}_{mo} \left[\dfrac{\sigma}{\sigma_{mo}} \right]^n \\[1em] \sigma = \sigma_f \varphi + \sigma_m (1 - \varphi) \end{array} \right\} \quad (3)$$

where the subscripts f, m represent fibre and matrix respectively.

Re-arrangement of Equations 3 lead to the following coupled differential equations in which the creep rate of the composite is expressed as a power law modified by a parameter S_1, which is essentially a dimensionless internal stress which increases with strain

$$\dot{\epsilon} = \alpha \dot{\epsilon}_m (1 - S_1)^n \quad (a)$$

$$\dot{S}_1 = H \dot{\epsilon} \quad (b) \qquad (4)$$

where $\dot{\epsilon}_m = \dot{\epsilon}_{mo} \left[\dfrac{\sigma}{\sigma_{mo}} \right]^n$ = creep rate of the matrix with the full applied stress

$S_1 = \dfrac{\varphi \sigma_f}{\sigma}$, σ_f = stress carried by the fibres

$H = \dfrac{\varphi E_f}{\sigma}$, $\alpha = (1 - \varphi)^{-n} \cdot \dfrac{(1-\varphi) E_m}{(1-\varphi) E_m + \varphi E_f}$

Although Equations 4a and b can be integrated with appropriate boundary conditions it is convenient to retain them in their present form to be consistent with the treatments of short and fracturing fibres described below. Moreover, numerical integration of Equations 4a and b with different boundary conditions can elucidate the effects of internal stresses developed during processing on the mechanical performance of the composites. Figure 6 shows computed creep curves for such a continuous composite using values of the parameters appropriate to the γ-γ'-Cr_3C_2 in-situ composite and assuming different initial values of S_1. This model predicts a continuously decreasing creep rate that approaches an asymptotic strain $\epsilon_c = \sigma/\varphi E_f$ when the fibres carry the full load. Of course, in practice, other factors such as fibre fragmentation which is discussed below intervene to cause failure. The minimum creep rate in γ-γ'-Cr_3C_2 usually occurs at a strain of about 1%; Figure 7 shows the computed creep rate at 1% strain as a function of applied stress for this material plotted as log $\dot{\epsilon}$ as a function of log stress. This clearly cannot be described by a simple power law.

3.3 Short fibres

When the fibres are of finite aspect ratio, matrix flow around the fibre ends modifies Equations 4 to give a finite asymptotic steady state creep rate, rather than the zero creep rate at equilibrium required for continuous fibres. A full discussion of how to combine these end effects with the contributions due to fibre deformation is given in Reference 2. Two extreme averaging procedures are identified.

a) When the components of strain from the fibre end effects and the visco-elastic central zone act in series, then the Reuss averaging procedure which assumes the full stress to be acting on each sub-element leads to the expression:

$$\dot{\varepsilon} = \alpha_1 \dot{\varepsilon}_m (1 - S_1)^n + \alpha_2 \dot{\varepsilon}_m (1 - S_2)^n \quad (a)$$

$$\dot{S}_1 = H \alpha_1 \dot{\varepsilon}_m (1 - S_1)^n \quad (b)$$

(5)

(b) For the Voigt average, where the end effects and fibre elastic strain act in parallel to the deformation of the matrix, so that strain, rather than stress, is equal in each sub-element of microstructure, then the following expressions apply.

$$\dot{\varepsilon} = \alpha_1 \dot{\varepsilon}_m (1 - S_1)^n \quad (a)$$

$$\dot{S}_1 = H \dot{\varepsilon} - R \dot{\varepsilon}_m (1 - S_2)^n \quad (b)$$

(6)

where $\alpha_1 = (1 - \varphi')^{-n} \dfrac{(1-\varphi')E_m}{(1-\varphi')E_m + \varphi' E_f}$, $\alpha_2 = (1 - \varphi)^{-n}$

$\varphi' = \dfrac{n+1}{2n+1} \varphi$ = reduced volume fraction of fibres carrying the full tensile stress

$R = (1 - \varphi)^{-n} \dfrac{\varphi E_f}{\sigma}$, $S_2 = \left[1 + \dfrac{1}{b\lambda^{\frac{n+1}{n}}} \right]^{-1}$

$b = \dfrac{\varphi}{1 + \varphi} \cdot (0.667)^{1/n} \left[(0.95 \varphi)^{-\frac{1}{2}} - 1 \right]^{-1/n} \left[\dfrac{n}{2n+1} \right]$

λ = fibre aspect ratio.

The final term in Equation 6b is essentially a dynamic recovery factor that leads to a steady state value of S_2 and, consequently of $\dot{\varepsilon}$. The end effects also restrict the length of fibres, and therefore the volume fractions that can be fully stressed.

Figures 8a and b shows creep curves for composites of the same fibre volume fraction but different aspect ratios calculated using Equations 6. Values of the parameters appropriate to Al-SiC and $\gamma-\gamma'-Cr_3C_2$ composites for which the matrices have quite different creep stress exponents have been selected. It is quite clear that the effect of fibre aspect ratio on the creep behaviour is greatest for the matrix with the low stress exponent for creep. Thus, long fibres are required to reinforce metals with a low stress sensitivity to creep while shorter fibres are adequate for materials with high stress exponents. However, it must be borne in mind that the appropriate matrix behaviour to be considered is that of the indirectly strengthened material, modified by the presence of the fibres, rather than of the isolated matrix.

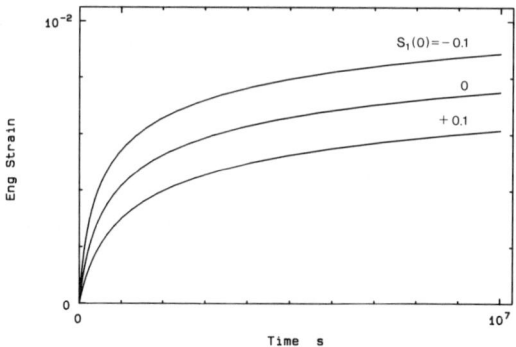

Figure 6 Creep curves calculated using Equations 4 and parameters relevant of the $\gamma-\gamma'-Cr_3C_2$ in-situ composite at 150 MPa, 1273K and showing the effect of different initial fibre stresses

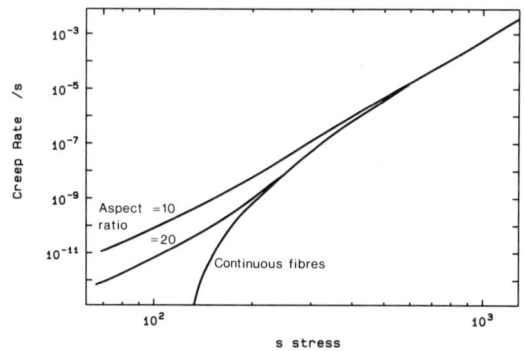

Figure 7 Calculated variation in creep rate at 1% strain with stress for the $\gamma-\gamma'-Cr_3C_2$ in-situ composite at 1273K for continuous fibres and fibres with aspect ratios of 20 and 10

The general shapes of the curves shown in Figure 8 are similar to those of the short aligned fibre material - Al-SiC, shown in Figure 1b. However, in that case fracture occurs at relatively low strains without any significant tertiary creep. For in-situ composites a more complete description of damage initiation and growth, and its influence on deformation is required.

Equation sets 5 and 6 lead to very similar creep curves[2], but the stresses accumulated in the fibres are quite different for the two averaging procedures. This difference is important in determining both the nature of damage that leads to fracture and the anelastic contribution when stresses are changed. The effect of fibre aspect ratio on the creep rate at 1% strain is shown in Figure 7.

3.4 Fibre fragmentation and fracture

The treatment to this point has taken no consideration of the development of damage to the reinforcing microstructure and its effect on the final stages of creep deformation and fracture. Because of the key role of the fibres in

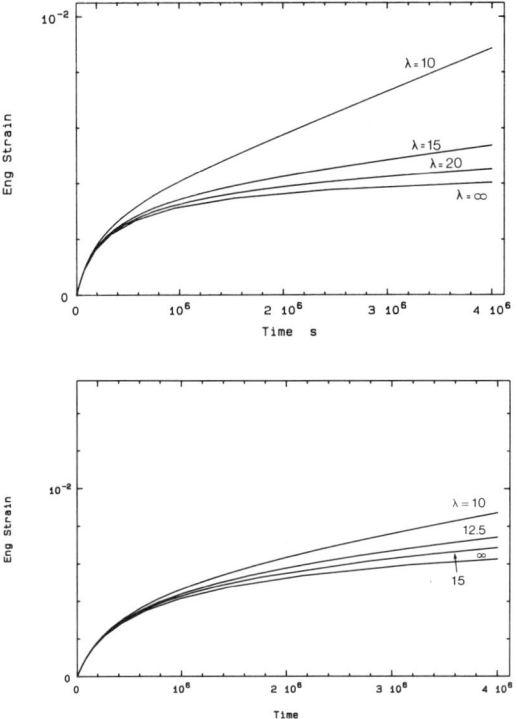

Figure 8   Calculated creep curves for short fibre reinforced composites with different fibre aspect ratios showing the importance of stress exponent in:
a) Al–15% SiC at 150 MPa/473K   n = 3
b) $\gamma-\gamma'-Cr_3C_2$ at 150 MPa/1273K   n = 6

supporting a considerable proportion of the load, it is self evident that changes to the morphology of the reinforcing phase will have an important effect. In the following discussion thermodynamic and chemical instabilities due to, for example, spheroidisation and fibre–matrix interactions, will not be considered although in some circumstances these may be important effects. Rather, the consequences of fibre fragmentation will be described. McLean[2] has considered the effects of three types of damage development following fibre breakage on the shape of the creep curve of a composite. These are shown schematically in Figure 9 and have the following characteristics.

a) Crack propagation. When the fibre diameter exceeds the critical length for crack propagation, then a single dominant crack progresses rapidly to cause fracture without significantly affecting the shape of the creep curve. Consequently Equations 5 still describe the shape of the creep curve for a short fibre reinforced composite except for the final short instability. Figure 1b would appear to fall into this category. The strain at which failure occurs depends on when the fibres support a critical breaking stress. For continuous fibres, where fibre stress is proportional to total strain (ie elastic and creep), the fracture strain should be independent of applied stress. However, for short fibre

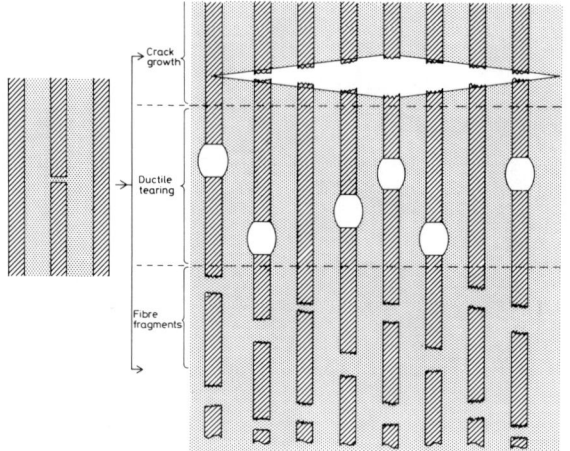

Figure 9 Schematic illustration of different modes of damage development

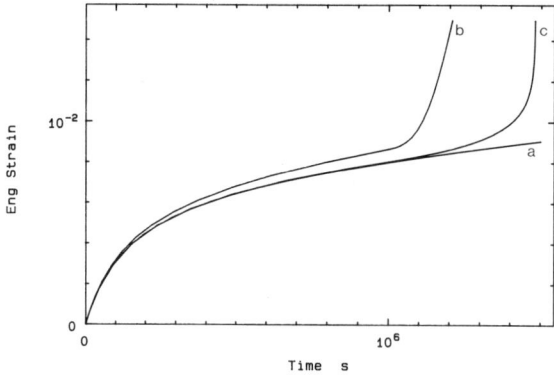

Figure 10 Calculated creep curves incorporating three different types of damage propagation:
a) crack growth   b) reducing fibre aspect ratio   c) void growth

composites the end effects provide variations in the strain before the fibres are loaded to a critical stress with both fibre aspect ratio and applied stress.
fracture strain – increases with decreasing aspect ratio for a given stress
  – decreases with increasing stress for a given aspect ratio.

b) Reducing aspect ratio. If fibre fragmentation occurs throughout the composite on a statistical basis then the load bearing capacity of the composite can be treated using continuum damage mechanics[2]. Here we consider the case where the matrix flows to fill the gap between the broken fibre ends to leave a composite with a lower mean fibre aspect ratio. McLean has considered the case where fibre fracture is compatible with Weibul statistics. Equations 5 can be extended to describe the evolution of fibre aspect ratio λ when Reuss averaging is appropriate. This mechanism does not apply to Voigt averaging since in this case a broken fibre reduces the likelihood of further fibre breaks.

$$\dot{\epsilon} = \alpha_1 \dot{\epsilon}_m (1 - S_1)^n + \alpha_2 \dot{\epsilon}_m (1 - S_2)^n \quad (a)$$

$$\dot{S}_1 = H \alpha_1 \dot{\epsilon}_m (1 - S_1)^n \quad (b)$$

$$\dot{S}_2 = 0 \qquad \text{for } S_1 < \frac{\varphi \sigma_f^{th}}{\sigma} \quad (c) \quad (7)$$

$$= -\left[\frac{n+1}{n}\right] \Gamma S_2 (1 - S_2) \dot{S}_1 \quad \text{for } S_1 > \frac{\varphi \sigma_f^{th}}{\sigma} \quad (d)$$

where σ_f^{th}, σ_f^f are the fibre stresses at which fibres begin to fragment and when final fracture occurs;

$$\Gamma = \frac{\sigma}{\varphi} \cdot \frac{\ln(\lambda_o/\lambda_f)}{\sigma_f^f - \sigma_f^{th}} \; ; \quad \lambda_o \text{ and } \lambda_f \text{ are the initial and final fibre aspect ratios.}$$

Figure 10 shows a creep curve computed using Equations 7. This is similar in form to those for the in-situ composites shown in Figures 1c and d. The differences in the extent of the tertiary creep regime can be explained qualitatively in terms of the uniformity of microstructure and the range of fibre breaking stresses in each material. Thus the COTAC alloy (Figure 1c) has a regular, evenly sized distribution of fibres which leads to a short, sharp tertiary; the γ-γ'-Cr_3C_2 alloy has an irregular and variably sized reinforcement which is associated with a long tertiary zone. In both cases, the final fracture must occur by another mechanism.

c) Ductile tearing

The final fracture is likely to occur when the stress is sufficiently high to cause the cavities associated the fractured fibres to extend by a ductile tearing at a rate constrained by creep of the surrounding sound material. Cocks and Ashby[15] have considered this effect for metals assuming the growth of a pre-existing population of voids. This is unlikely to be appropriate to voids generated by fracturing fibres. McLean has suggested that such voids will be continuously generated with increasing creep strain. Here Equation 6 can be modified as follows

$$\dot{\epsilon} = \alpha_1 \dot{\epsilon}_m (1 - S_1)^n \exp(\omega) \quad (a)$$

$$\dot{S}_1 = H \alpha_1 \dot{\epsilon}_m (1 - S_1)^n - R\dot{\epsilon}_m(1-S_2)^n \quad (b)$$

$$\dot{\omega} = 0 \quad \text{for } \epsilon < \epsilon_{th} \quad (c)$$

$$\dot{\omega} = \frac{n \varphi \dot{\epsilon}}{(\epsilon_f - \epsilon_{th})} \quad \text{for } \epsilon > \epsilon_{th} \quad (d)$$

(8)

Equations 7 has all of the characteristics required to account for the creep behaviour of the ductile composites such as the NITAC alloys and $\gamma-\gamma'$-Cr_3C_2 shown in Figure 1c and d.

4 DISCUSSION

The equation sets 7 and 8 each have the forms that can describe creep curves exhibiting primary, steady-state and tertiary creep behaviour in constant stress creep conditions and the two models make very similar quantitative predictions of the creep curves. However, there are major differences in the likely nature of damage and the consequences of complex loading conditions that are associated with the maximum stress experienced by the fibres in the two averaging procedures. Thus, Equations 4 require that S_1 increases asymptotically to a value of 1 as the creep rate approaches a steady state value, while Equations 6 lead to a steady state value of $S_1 < 1$ which is associated with the same steady state creep rate. Of more importance, when fibres break and the fibre aspect ratio decreases the steady state value of fibre stress decreases according to Equations 6 making further fibre fracture less likely whereas Equations 5 require S_1 to progressively increase. Thus, although examination of simple creep curves does not discriminate between the various mechanisms, it should be possible to devise appropriate diagnostic tests that will do so.

The modification of matrix creep behaviour by the presence of the fibres, as indicated by the strengthening by particulate reinforcement, makes it impossible to quantitatively validate the models in terms of the behaviours of the constituent phases. However, the equations can be generalised quite simply to provide empirical constitutive laws, inspired by physical reasoning as advocated by Ashby[17], that can account for important microstructural and processing variables, such as fibre volume fraction and internal stresses.

5 ACKNOWLEDGEMENT

The author wishes to thank Dr Ana Barbosa for writing the software used to calculate and display the theoretical creep curves.

REFERENCES

1. T.W. Chou, A. Kelly and A. Okura, Composites 16, 187 (1985).

2. M. McLean, proceeding of conference "Materials 88", May 1988, The Insitute of Metals, London (to be published).

3. R.K. Varma, private communication.

4. T.G. Nieh, Metall. Trans. 15A, 139 (1984).

5. E. Bullock, P.N. Quested and M. Mclean, Acta Met. 5, 333 (1977).

6. M.R. Jackson and J.L. Walter, in "In-situ Composites IV", edited by F.D. Lemkey et al, Elsevier North Holland, 1984.

7. B.F. Dyson and M. McLean, Acta Met. 17, 17 (1983).

8. T. Khan, J.F. Stohr and H. Bibring in "Superalloys 1980", edited by J.K. Tien et al, p531, ASM, Metals Park, Ohio, 1980.

9. J.L. Walter, in "In-situ Composites IV", edited by F.D. Lemkey et al, Elsevier North Holland, 1984.

10. F.M. Dunlevey and J.F. Wallace, Metall. Trans. 5, 1351 (1974).

11. J.C. Ion, A. Barbosa, M.F. Ashby, B.F. Dyson and M. McLean, NPL Report No DMA(A)115, National Physical Laboratory, April 1986.

12. M. Maldini, A. Barbosa, M.F. Ashby, B.F. Dyson and M. McLean, NPL Report No DMA(A)126, National Physical Laboratory, January 1987.

13. D. McDanels, R.A. Signorelli and J.W. Weeton, NASA Report No TND-4173, 1967.

14. M. McLean, "Directionally Solidified Materials for High Temperature Services", Book No 296, The Metals Society, London (1983).

15. M. McLean, Proceedings of 5th International Conference on Composite Materials, p37, edited by W.C. Harrigan et al, The Metallurgical Society Inc, Warrendale, PA, 1985.

16. A.C.F. Cocks and M.F. Ashby, Progress in Materials Science, volume 27, p189 (1982).

17. M.F. Ashby, Phil. Trans. Royal Society, A322, 307 (1987).

CRACK-TIP SHIELDING IN METAL-MATRIX COMPOSITES: MODELLING OF CRACK BRIDGING BY UNCRACKED LIGAMENTS

JIAN KU SHANG AND R. O. RITCHIE
Department of Materials Science and Mineral Engineering, University of California, Berkeley, CA 94720

ABSTRACT

As part of an investigation into the micro-mechanisms of crack-tip shielding associated with the growth of fatigue cracks in metal-matrix composites, simple models are developed for the role of crack bridging in high-strength aluminum alloys reinforced with SiC particulate (Al/SiC_p). Based on experimental observations of crack growth, crack-tip shielding and crack-path morphology in these alloys, the bridges are found to be associated with uncracked ligaments in the wake of the crack tip, and are modelled in terms of approaches based on a critical crack-opening displacement or critical tensile strain in the ligament.

INTRODUCTION

Over the past ten years, much work in metals, ceramics, rocks and composites has focused on the role of crack-tip shielding in enhancing toughness, or more generally in impeding crack advance, by locally reducing the "crack driving force" actually experienced at the crack tip; notable examples are transformation toughening in ceramics and fatigue-crack closure in metals, as reviewed in [1,2]. In certain composite and monolithic materials, however, a prominent shielding mechanism occurs from bridging between the crack faces by strong fibers or unbroken ligaments in the wake of the crack tip [3-13]. In brittle fiber-reinforced ceramic-matrix composites where the fibers are sufficiently strong and the fiber/matrix interface sufficiently weak, preferential failure in the matrix can leave intact fibers spanning the crack for some distance behind the crack tip [1,3-8]. The fibers act as a series of springs which restrain crack opening and thereby shield the crack tip from the applied far-field loading, resulting in lower, yet crack-size dependent, growth-rate behavior [3-8].

In metallic materials, similar effects have been reported for aluminum laminates reinforced with epoxy-resin sheets impregnated with aramid fibers (ARALL Laminates®), where the fiber/epoxy interfaces now are weak enough to permit controlled delamination and thus bridging of unbroken fibers across the crack [13,14]. However, in most metal-matrix composites, such as aluminum alloys discontinuously reinforced with SiC fibers (or whiskers or particles), the reinforcement phase invariably fractures due to its strong interface with the matrix, with the result that fiber-bridging is essentially insignificant [15-17].

Recently, however, studies on fatigue-crack growth in aluminum alloy/SiC-particulate composites (Al/SiC_p) have revealed a different mechanism of bridging, induced by the presence of uncracked ligaments behind the crack tip [15,16]. Such ligaments, although not continuous in three dimensions, act in any one two-dimensional section to inhibit crack opening. This mechanism, which has also been observed in monolithic materials [10-12], appears to result from fracture events triggered ahead of the crack tip or from general non-uniform or discontinuous advance of the crack front; in Al/SiC_p composites it predominates at intermediate fatigue-crack growth rates ($\sim 10^{-9}$ to 10^{-6} m/cycle) where cleavage of SiC particles ahead of the crack tip becomes significant [15].

It is the objective of the present note briefly to characterize such uncracked-ligament bridging in Al/SiC$_p$ composites, and to develop simple models to quantify the magnitude of the induced shielding.

FATIGUE CRACK BRIDGING IN Al/SiC$_p$ COMPOSITES

Two types of ligament bridging have been observed during fatigue-crack propagation at intermediate stress intensities in Al/SiC$_p$ composites, as described in [15,16] for SiC-particulate reinforced P/M Al-Zn-Mg-Cu alloys. In alloys with higher SiC volume fractions and small interparticle spacings, the uncracked ligaments are predominantly co-planar with the crack, and directly associated with fracture of carbides ahead of the crack tip (Fig. 1a); however, by comparison to behavior in the unreinforced alloy, the resulting effect on crack-growth rates is small (Fig. 1b). In alloys with lower volume fractions of more dispersed SiC particles, conversely, the ligaments are principally formed by overlapping cracks on different planes (Fig. 1c); the effect on growth rates is now considerably larger (Fig. 1d).

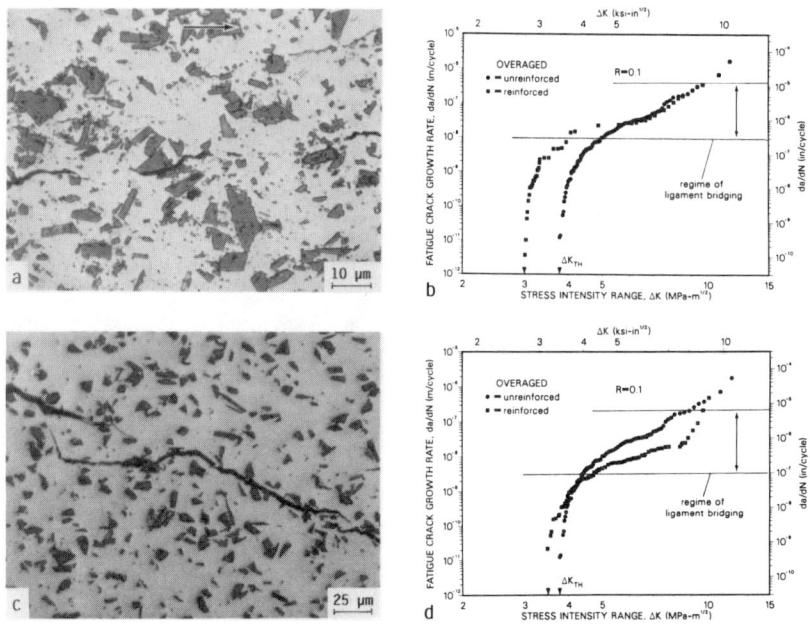

Fig. 1: Uncracked-ligament bridging in Al/SiCp composites, showing a) co-planar ligaments in 20 vol% (6-μm-sized) SiC alloy, and b) corresponding effect on fatigue-crack growth rates (by comparison of behavior in the reinforced and unreinforced alloy over the range ($\sim 10^{-8}$ - 10^{-6} m/cycle); c) "overlapping" ligaments in 15 vol% (11-μm-sized) SiC alloy, and d) corresponding effect on growth rates. Note: Growth-rate differences at low ΔK levels are associated with primarily crack closure; no bridging is observed in this regime. Arrow indicates general direction of crack growth.

CRACK-BRIDGING MODELS

There have been several previous models to evaluate the role of bridging in metals, ceramics and composites [3-13], although only one (for ARALL Laminates) is specific to cyclic loading [13]. In essence, the key problem lies in defining the force in the bridges as a function of the crack-opening displacement or distance from the crack tip. A listing of the force-separation functions utilized in five prominent models [5,9-12] is given in Table I. The fiber-bridging model of Marshall et al. [5] and the rubber-particle toughening model of Kunz-Douglass et al. [9] compute the strain in the bridges from the strain compatibility between the fiber and matrix during fiber pull-out or from the shape change of rubber particles, and are thus mechanism-specific. Mai and Lawn [10] in their ligament-bridging model, conversely, simply adopt a trial exponential force-separation function, with parameters set by the particular mechanism. The equilibrium-crack models of Gerberich [11] and Rosenfield and Majumdar [12], on the other hand, are more general, but assume simply that the stress in the bridges is equal to the yield or fracture stress, respectively.

In the current work, two approaches are taken to model the forms of ligament bridging observed during fatigue-crack growth in Al/SiC$_p$, i.e., based on a limiting crack-opening displacement or a limiting strain in the uncracked ligaments; analyses are described below.

General Principles

The effect of an area fraction, f, of uncracked ligaments on the crack plane, existing over a distance $x = \ell$ (the bridging zone) behind the crack tip (Fig. 2), is represented by a distributed force, p(x), given in terms of the stress $\sigma(x)$ in the ligaments by:

$$dp(x) = f\,\sigma(x)\,dx \qquad (1)$$

Table I. Force-Separation Functions for Various Models

	Marshall Cox Evans	Mai Lawn	Kunz-Douglass Beaumont Ashby	Gerberich	Rosenfield Majumdar
Separation-Function	$\sigma(x) = 2 \cdot [u \cdot \tau \cdot E_f \cdot (1+n)/R]^{\frac{1}{2}}$ τ - interfacial stress E_f - fiber modulus R - fiber radius $n \sim E_f V_f / E_m (1 - V_f)$	$p(u) = p^* (1 - \frac{u}{u^*})^m$ u^*, p^* - limiting values m - mechanism-dependent $p = f \cdot \sigma(x)$	$\sigma = G \cdot (\epsilon - \frac{1}{\epsilon^2})$ G - modulus ϵ - true strain	$\sigma = \sigma_0$	$\sigma = Y_0$
Method	pull-out mechanics (strain compatibility)	trial function	constitutive eqn. for rubber	yield strength	fracture stress
Applications	fiber bridging in ceramic matrix composites	interfacial bridging	rubber-toughened plastics	general	

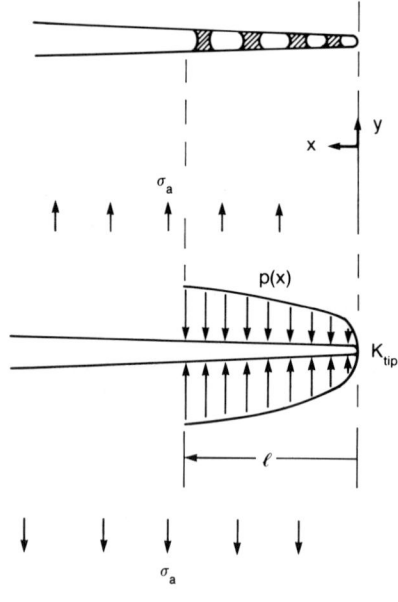

Fig. 2: Idealization of the bridging zone over distance ℓ behind the tip of a crack subjected to an externally applied stress σ_a.

For a semi-infinite crack, this distributed force induces a stress intensity given by [18]:

$$K_s = \frac{\sqrt{2}}{\pi} \int_0^\ell \frac{dp(x)}{\sqrt{x}} \quad (2)$$

If the crack is subjected to an applied stress field, superposition of the shielding stress intensity, K_s, due to bridging with the globally applied (far-field) stress intensity, K_a, yields an expression for the effective (near-tip) stress intensity, K_{tip}, experienced locally at the crack tip (Fig. 2):

$$K_{tip} = K_a + K_s \quad (3)$$

This is related to a crack-opening displacement, δ_{tip}, in terms of Young's modulus, E, and yield strength, σ_y, of the ligament, by:

$$K_{tip} = \sqrt{\frac{E' \sigma_y \delta_{tip}}{d}} \quad (4)$$

where $E' = E$ in plane stress and $E/(1 - \nu^2)$ in plane strain, ν is Poisson's ratio, and d is a constant varying between 0.3 and 1.0 depending upon the yield strain and work-hardening exponent and whether plane-strain or plane-stress conditions apply [19]. Solutions to Eqs. 2-4 are given below.

Limiting Crack-Opening Displacement Approach

The basis of this approach is that the stress in any ligament behind the crack tip is related to the crack opening at that point; specifically the displacement in the last intact ligament at the end of the bridging zone must approach the limiting crack-opening displacement, δ_c, for fracture of that ligament. By assuming for simplicity that an idealized fatigue crack can be taken as trapezoidal (Fig. 3), the crack-opening displacement, δ_x, at any distance x along the crack length can be determined in terms of the crack-tip opening displacement, δ_{tip}, and specimen ligament, b:

$$\delta_x = \delta_{tip} \cdot \left(\frac{x + rb}{rb}\right) \quad (5)$$

assuming that the crack opens about some rotational axis at a distance, rb, ahead of the crack tip; r is the rotational factor and takes values between 0.195 for elastic deformation and 0.470 for plastic deformation [20].

To maintain equilibrium such that the crack may extend without breaking ligaments along the bridging zone, the crack-opening displacement at any point within the zone, δ_x, must satisfy:

$$\delta_x \leq \delta_c, \qquad 0 \leq x \leq l \quad (6a)$$

whereas at the end of the bridging zone:

$$\delta_x = \delta_c, \qquad \text{at } x = l \quad (6b)$$

where δ_c, the maximum displacement in the ligament corresponding to its failure, is independent of the size of bridging zone but varies with the area fraction f of ligaments. Thus, assuming that a partially-bridged crack, with $f < 1$, is analogous to a fully-bridged crack with an effective thickness f times the full specimen thickness, Eq. 2 becomes:

$$\sqrt{\frac{\delta_c \, rb \, E' \, \sigma_y}{d(l + rb)}} = K_a + \frac{\sqrt{2}}{\pi} \int_0^l \frac{\sigma(x) \, dx}{\sqrt{x}} \quad (7)$$

yielding an expression for the stress, $\sigma(x)$ in the ligaments:

$$\sigma(x) = -\frac{\pi}{2\sqrt{2}} \sqrt{\frac{\delta_c \, rb \, E' \, \sigma_y}{d}} \left[\frac{\sqrt{x}}{(x + rb)^{3/2}}\right] \quad (8)$$

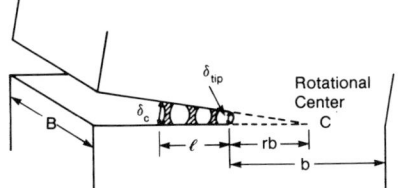

Fig. 3: Schematic illustration of idealized fatigue crack with bridging zone, showing definitions of the rotational axis and crack-opening displacements.

With substitution, Eq. 8 provides an expression for the degree of crack-tip shielding due to uncracked-ligament bridging, in terms of the area fraction of ligaments, the applied stress intensity and the ratio ℓ/rb:

$$K_s = -f\, K_a[(1 + \ell/rb)^{\frac{1}{2}} - 1]/[1 - f + f(1 + \ell/rb)^{\frac{1}{2}}] \qquad (9)$$

Limiting Strain Approach

An alternative, first-order solution to Eqs. 2-4 can be obtained by representing the bridges as tensile ligaments, where the stress in a ligament is proportional to the strain. With reference to Fig. 3, the strain, $\varepsilon(x)$, in any ligament within the bridging zone can be estimated by assuming it to be equivalent to the strain in a bent beam, with rotational center at point C and neutral plane at the crack tip, such that:

$$\varepsilon(x) = x/rb \qquad (10)$$

Converting to a true strain and substituting into a constitutive law for the uncracked-ligament material, of the form:

$$\sigma(x) = \sigma_y + k \cdot \varepsilon(x) \qquad (11)$$

where σ_y is the (initial) yield stress and k is a constant, provides a second expression for the degree of crack-tip shielding due to uncracked-ligament bridging, in terms of the area fraction of ligaments, the ratio ℓ/rb and the (constrained) flow properties of the ligament:

$$K_s = -f\, \sigma_y \cdot \frac{2\sqrt{2\ell}}{\pi} \left[1 + \frac{k}{\sigma_y}(1 + \frac{\ell}{rb}) - 2\left(\sqrt{\frac{rb}{\ell}} \cdot \tan^{-1}\sqrt{\frac{\ell}{rb}}\right) \right] \qquad (12)$$

RESULTS AND DISCUSSION

As ligament bridging in the Al/SiC$_p$ alloys predominates between $\sim 10^{-9}$ and 10^{-6} m/cycle, estimates of the degree of shielding were made for a stress-intensity range ΔK of 8 MPa$\sqrt{m}$ (K_a = 9 MPa$\sqrt{m}$). Metallographic studies of the crack-path morphology using serial sectioning indicated an area fraction of bridges of ~ 27 to 31% along a bridging zone of approximately 400 μm behind the crack tip; rb is typically 1 mm for the C(T) geometry and the values of σ_y and k are 380 and 1600 MPa for the overaged alloy.

To apply the proposed models to the cyclic crack-growth behavior in Fig. 1, we note that the stretching of co-planar uncracked ligaments is controlled by the crack opening; the degree of crack-tip shielding is thus more appropriately described by the limiting crack-opening displacement model (Eq. 9). Using the measured values of f, ℓ and rb defined above, the stress intensity K_s due to bridging is predicted to be approximately 0.5 MPa$\sqrt{m}$ at an applied K_a of 9 MPa$\sqrt{m}$. This form of bridging thus induces minimal shielding (~ 6% in this case), consistent with the minimal difference in growth rates between the reinforced and unreinforced 20 vol% SiC alloys at $\Delta K = 8$ MPa$\sqrt{m}$ (Fig. 1b). Conversely, the deformation of uncracked ligaments resulting from overlapping cracks is less a function of the crack opening but rather is limited by the strength of the ligament; the limiting-strain model (Eq. 12) is therefore more appropriate. Here using measured values of f, ℓ, rb, σ_y and k at K_a = 9 MPa$\sqrt{m}$, the stress intensity K_s due to bridging is predicted to be 3.2 MPa$\sqrt{m}$. This clearly represents a more substantial degree of shielding (~30%) and is consistent with the larger shift in growth-rate curves between the reinforced and unreinforced 15 vol% SiC alloys at $\Delta K = 8$ MPa$\sqrt{m}$ (Fig. 1d).

Finally, it might be noted that co-planar uncracked ligaments can be considered either as a bridging zone behind the crack tip or a damage zone ahead of it, the only difference being the definition of the crack tip. Recent work by Thouless, however, has shown that the two approaches are equivalent and that identical crack-extension rates are predicted [21].

CONCLUSIONS

Based on a study of crack bridging via uncracked ligaments in Al/SiC$_p$ composites, simple models are developed to predict the magnitude of crack-tip shielding during fatigue-crack growth. It is found that bridging models based upon a limiting crack-opening displacement in the bridge predict only minimal shielding but are most appropriate to co-planar ligaments. Conversely, models based on a limiting strain are appropriate to ligaments formed by overlapping cracks, and predict larger levels of shielding.

Acknowledgments

This work was supported by the Air Force Office of Scientific Research under University Research Initiative No. F49620-87-C-0017 to Carnegie Mellon University. Thanks are due to Dr. Alan Rosenstein for his continued support and to Warren Hunt and Dr. Bob Bucci of Alcoa for supplying the alloys.

References

1. A. G. Evans, in *Fracture Mechanics, 20th Symp.*, ASTM STP, edited by R.P. Wei and R.P. Gangloff (ASTM, Philadelphia, PA, 1988).
2. R. O. Ritchie, in *Mechanical Behaviour of Materials - V*, edited by M.G. Yan, S.H. Zhang and Z.M. Zheng (Pergamon, Oxford, U.K., 1988), vol. III.
3. J. Aveston, G. Cooper and A. Kelly, in *Properties of Fiber Composites*, NPL Conf. Proc. (IPC Sci. & Tech. Press, Surrey, U.K., 1971), pp. 15-26.
4. B. Budiansky, J. W. Hutchinson and A. G. Evans, J. Mech. Phys. Solids, 34, 167 (1986).
5. D. B. Marshall, B. N. Cox and A. G. Evans, Acta Met., 33, 2013 (1985).
6. L. N. McCartney, Proc. Roy. Soc., A409, 329 (1987).
7. L. R. F. Rose, J. Mech. Phys. Solids, 35, 383 (1987).
8. B. Budiansky, in *Proc. 10th U.S. Cong. Appl. Mech.*, (Austin, TX, 1986).
9. S. Kunz-Douglass, P. W. R. Beaumont and M. F. Ashby, J. Mater. Sci., 15, 1109 (1980).
10. Y. Mai and B. R. Lawn, J. Am. Ceram. Soc., 70, 289 (1987).
11. W. W. Gerberich, in *Fracture: Interactions of Microstructure, Mechanisms and Mechanics*, edited by J.M. Wells and J.D. Landes (TMS-AIME, Warrendale, PA, 1984), pp. 49-74..
12. A. R. Rosenfield and B. S. Majumdar, Metall. Trans. A, 18A, 1053 (1987).
13. R. Marissen, in *Fatigue 87*, Proc. Third Intl. Conf. on Fatigue, edited by R.O. Ritchie and E.A. Starke (EMAS Ltd., 1988), vol. 3, pp. 1271-79.
14. R. O. Ritchie, W. Yu and R. J. Bucci, Eng. Fract. Mech, (1988) in press.
15. J.-K. Shang, W. Yu and R. O. Ritchie, Mater. Sci. Eng., (1988) in press.
16. J.-K. Shang and R. O. Ritchie, Metall. Trans. A, 19A (1988) in review.
17. T. Christman and S. Suresh, Mater. Sci. Eng., (1988) in press.
18. G. C. Sih, *Handbook of Stress Intensity Factors* (Lehigh University Press, Bethlehem, PA, 1972).
19. C. F. Shih, J. Mech. Phys. Solids, 29, 305 (1981).
20. C. C. Veerman and T. Muller, Eng. Fract. Mech., 4, 25 (1972).
21. M. D. Thouless, J. Am. Ceram. Soc., 71 (1988) in press.

ELEVATED TEMPERATURE SLOW PLASTIC DEFORMATION
OF NiAl/TiB$_2$ PARTICULATE COMPOSITES

R. K. Viswanadham[1], J. Daniel Whittenberger[2], S. K. Mannan[3] and B. Sprissler[3]
[1] Formerly with Martin Marietta Laboratories, 1450 South Rolling Rd., Baltimore, MD 21227-3898; currently at Multi-Metals 715 Gray St., Louisville, KY 40202.
[2] NASA-Lewis Research Center, Cleveland, OH 44135.
[3] Martin Marietta Laboratories, 1450 South Rolling Rd., Baltimore, MD 21227-3898.

ABSTRACT

To enhance the high temperature strength of aluminides, NiAl-TiB$_2$ composites with particulate contents up to 30 vol. pct. were made by XD^{TM} synthesis and hot pressed to full density. Microstructures of these composites were characterized by optical, scanning and transmission electron microscopy (TEM). The average size of the TiB$_2$ particles was about 1 μm, and the average grain size of the NiAl matrix was on the order of 10 μm. Elevated temperature compression testing was conducted on these composites in air at 1200 and 1300 K with strain rates varying from $^-10^{-4}$ to $^-10^{-7}$ s^{-1}. Flow strengths were found to increase with increasing TiB$_2$ content; for example, the 20 vol. pct. TiB$_2$ composite was three times stronger than unreinforced NiAl. Post test TEM analysis showed that the primary feature of the dislocation substructure of deformed NiAl was well defined subgrain boundaries, whereas the structure of the higher volume fraction composites consisted of a very high density of tangled dislocations, loops and subgrain boundaries connecting particles. These observations suggest that TiB$_2$ particles can stabilize a completely different dislocation structure than that normally found in NiAl.

INTRODUCTION

A new method to form discontinuously reinforced metal matrix composites, the XD^{TM} process, has been under development at Martin Marietta Laboratories for the past several years. Under a program (Contract N0014-85-C-0639) sponsored by the Office of Naval Research, an attempt [1] is being made to apply this technology to produce composites based on the B2 crystal structure nickel aluminide NiAl with enhanced high temperature strength and low temperature ductility.

Of the various aluminides, NiAl has many attractive features: (1) a high melting point of 1912 K [2], (2) a low density of about 5.9 Mg/m^3 [3] and (3) excellent isothermal oxidation resistance [4, 5]. Its major liabilities are low strength at high temperature [6] and inadequate low temperature ductility [7]. Although grain refinement might partially alleviate the ambient temperature ductility problem [8], the lack of creep strength will continue to be a disadvantage. If XD^{TM} synthesis can be utilized to produce a dispersion of fine, discrete particles of a refractory phase in NiAl, then creep resistance might be improved, and, if the concept of dispersion toughening [9] is applicable, some enhancement in toughness could also result. Although an alternative approach to increase the elevated temperature strength of NiAl through precipitation of ternary compounds is possible [10], fine dispersions of the second phases, necessary for strength, would probably not be thermodynamically stable at high temperatures. The present method does not

---

XD^{TM} is a registered trademark of Martin Marietta Corporation.

suffer from this limitation since the strengthening phase can be preselected to be thermally stable, and the XD^{TM} process has the added advantage that the matrix and particulate can be formed in a single step.

EXPERIMENTAL PROCEDURES

Ni-50Al (atom pct.) composites containing 0 to 30 vol. pct. of TiB$_2$ particles were produced by XD^{TM} synthesis, and these materials were densified by vacuum hot pressing at 28 MPa at 1675 to 1775 K in graphite tooling. Cylindrical compression specimens, whose length was parallel to the hot pressing direction, were electro-discharge machined from each compact and ground to final size: ~5.5 mm in diameter by about 12.5 mm in length. Constant velocity compression tests at crosshead speeds ranging from 2.12×10^{-3} to 2.12×10^{-6} mm/s were conducted in a universal test machine to ~10 percent strain at 1200 and 1300 K in air. The autographically recorded load - time charts were converted to true compressive stresses, strains, and strain rates via the offset method [11 & 12] and the assumption of conservation of volume. Microstructural characterization of both as fabricated and compression tested materials was conducted using light optical, scanning electron and transmission electron (TEM) microscopy techniques and X-ray diffraction.

RESULTS AND DISCUSSION

Examination of the hot pressed NiAl-TiB$_2$ composites revealed that they were polycrystalline and fully dense with few visible signs of porosity or cracking (Fig. 1(a)). The TiB$_2$ particles (the light gray phase) tended to be evenly dispersed in the low volume fraction materials but not at the higher loadings. Because the XD^{TM} process utilizes metal powders, the materials also contained a low volume fraction (about 1 vol. pct.) of Al$_2$O$_3$ particles [1] which appear as the dark phase in the microstructure. Although grain sizes in the higher volume fraction composites could not be determined; they were on the order of 10 μm for the lower volume fractions.

TEM of as fabricated composites indicated that the average size of the TiB$_2$ particles was about 1 μm. While the low volume fraction materials were relatively dislocation-free, the higher particulate loadings (Fig. 1(b)) exhibited both subgrains and dislocations in NiAl.

Typical 1300 K true stress - strain curves for the NiAl-TiB$_2$ composites as functions of particulate loading are presented in Fig. 2. The plots demonstrate the two types of behavior observed in these materials: For those materials with less than 10 vol. pct. TiB$_2$, deformation took place at a constant flow stress after a small amount of work hardening, while the 20 & 30 vol. pct. composites exhibited diffuse yielding followed by strain softening.

Clearly from Fig. 2 it is evident that composite strength increases with TiB$_2$ content, and this is also the case in Fig. 3 where true compressive stress - strain rate behavior at 1300 K is shown. The flow stress σ and strain rate $\dot{\varepsilon}$ for both test temperatures were fitted to the standard power law and temperature compensated power law rate expressions

$$\dot{\varepsilon} = A\sigma^n, \text{ and} \quad (1)$$
$$\dot{\varepsilon} = B\sigma^n \exp(-Q/RT) \quad (2)$$

where A and B are constants, n is the stress exponent, Q is the activation energy for creep, R is the gas constant and T is the absolute temperature. Linear regression fits of the data indicated that the stress exponent and activation energy for the unstrengthened NiAl are about 6 and 350 kJ/mol respectively; whereas $8 \leq n \leq 12$ and $Q \approx 400$ kJ/mol for the composites.

Light optical examination of compression tested specimens indicated that elevated temperature deformation to nominally 10 pct. strain had little effect on the microstructure. Grain growth was not found, nor was grain boundary cracking or particle/matrix separation observed (Fig.4). Transmission electron microscopy of tested materials demonstrated that the TiB_2 particles strongly affected deformation behavior. Whereas relatively dislocation-free subgrains were formed in the unreinforced NiAl (Fig. 5(a), the $NiAl-TiB_2$ composite structure (Fig. 5(b)) was generally composed of a high density of tangled dislocation loops in the matrix with subgrain boundaries connecting adjacent second phase particles. As the volume fraction of TiB_2 increased, the structure was similar to that in Fig. 5(b) except the density of dislocations in the matrix usually increased.

From the above data (Figs. 2 & 3) it appears that NiAl can be effectively strengthened by a particulate dispersion of TiB_2. Additionally the 1300 K compressive flow strength - strain rate properties in Fig. 6 indicate that $NiAl-10TiB_2$ is as deformation resistant as various forms of precipitation hardened NiAl + 5 at. pct. Ta [13] for strain rates less than 10^{-5} s^{-1}, and based on these data, $NiAl-TiB_2$ composites have the potential to be much stronger than Ta modified NiAl at very slow strain rates ($< 10^{-7}$ s^{-1}).

The strength of $NiAl-TiB_2$ composites is due in part to (1) the excellent interface between particles and matrix which is unaffected by fabrication processes (Fig. 1) or elevated temperature straining (Figs. 4 & 5(b)) and (2) the dislocation structure in the matrix. As opposed to unreinforced NiAl which forms relatively dislocation-free subgrains during creep (Fig. 5(a) and [7 & 14]), the presence of small TiB_2 particles leads to a much higher dislocation density between subgrains (Fig. 5(b)). Additionally the particles in the composite appear to be able to anchor subgrains which probably yields a

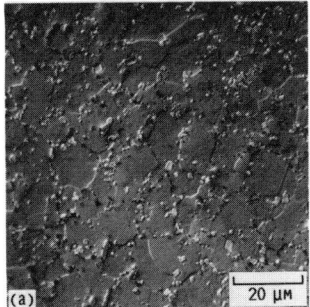
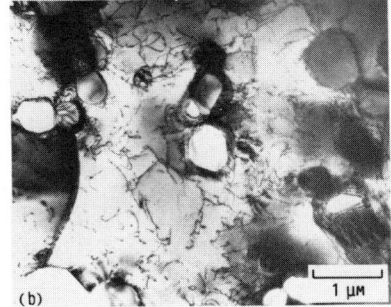

Figure 1.

Typical photomicrographs of as fabricated composites (a) etched structure of $NiAl-7.5TiB_2$ under light optical differential interference contrast conditions; hot pressing direction is horizontal, and (b) $NiAl-20TiB_2$ as revealed by TEM.

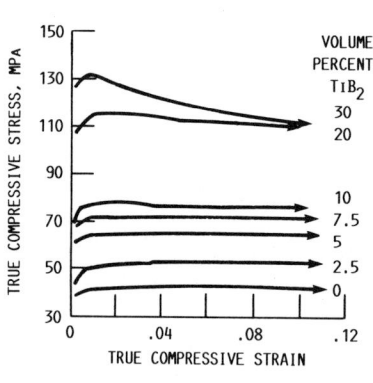
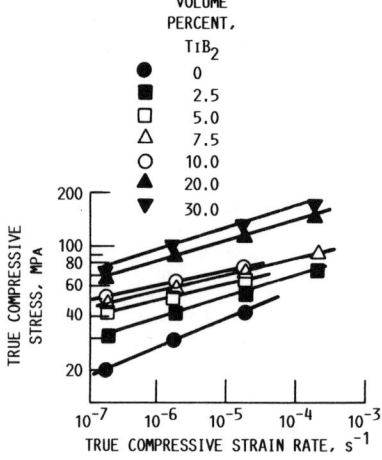

Figure 2.

True compressive stress - strain diagrams for NiAl - TiB$_2$ composites tested at 1300 K and a nominal strain rate of 2×10^{-5} s^{-1}.

Figure 3.

True compressive stress - strain rate behavior for NiAl - TiB$_2$ composites tested at 1300 K.

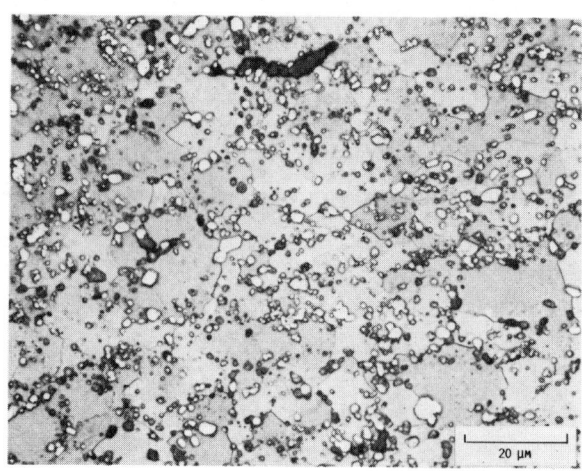

Figure 4.

Typical light optical microstructure of etched NiAl-10TiB$_2$ tested at 1300 K and an approximate strain rate of 2×10^{-7} s^{-1} to 9.3 pct. strain; the hot pressing and testing direction is horizontal.

 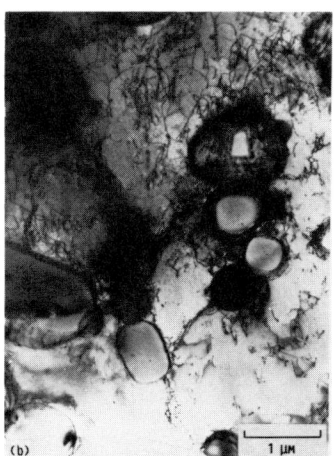

Figure 5.

TEM photomicrographs of NiAl - TiB_2 composites tested at 1300 K.

	Volume Fraction, Pct.	Flow Stress, MPa	Strain Rate, s^{-1}	Strain, Pct.
(a)	0	42.3	2×10^{-5}	10.6
(b)	7.5	91.5	2×10^{-4}	10.7

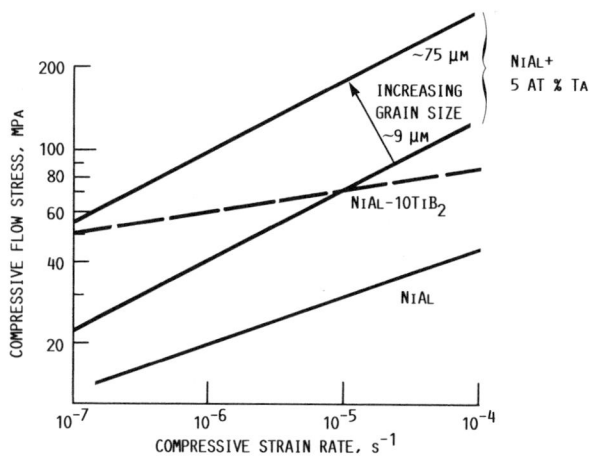

Figure 6.

Comparison of the 1300 K compressive flow strength - strain rate behavior of NiAl [6], NiAl-10TiB$_2$ and NiAl + 5 at. pct. Ta [13].

smaller than expected substructure size [15]. Both of these latter two
factors can yield increased creep resistance: a higher dislocation density
indicates significant work hardening, and stable, small subgrains can
strengthen via the Sherby, Klundt and Miller [16] model of dislocation climb
controlled creep.

SUMMARY

An initial study of the elevated temperature compressive properties of
NiAl-TiB$_2$ particulate composites fabricated by XD^{TM} synthesis and hot pressing
has shown that these materials possess a significant strength advantage over
the unreinforced matrix. This increased resistance to deformation appears to
be due to the TiB$_2$ particles which stabilize a different dislocation
substructure than that found in NiAl.

REFERENCES

1. R.K. Viswanadham, S.K. Mannan and B. Sprissler, "Nickel Aluminide /
   Titanium Diboride Composites" Martin Marietta Laboratories TR 87-66c,
   1987.

2. M. Hansen and K. Anderko, Constitution of Binary Alloys, (McGraw Hill,
   New York, 1958). Additional Supplements by R. P Elliot, 1965 and F. A.
   Shrunk, 1969.

3. A.J. Bradley and A. Taylor, Proc. R. Soc. Series A 159, 56 (1937).

4. E. Fetzer and P. Gerasimoff, Z. Metallk. 50, 187 (1959).

5. V. Imai and M. Kumazawa, Sci. Rept. Res. Inst. Tohoku Univ. 11 (1959)
   [referenced by E. A. Aitken in Intermetallic Compounds edited by J. H.
   Westbrook (John Wiley, New York 1967) p. 507.

6. J.D. Whittenberger, J. Mat. Sci. 22, 394 (1987).

7. E.M. Grala. in Mechanical Properties of Intermetallic Compounds,
   edited by J.H. Westbrook (John Wiley, New York, 1960) pp. 358-404.

8. E.M. Schulson and D.R. Barker, Scripta Metall. 17, 519 (1983)

9. J.C.M. Li and S.C. Sanday, Acta Metall. 34, 537 (1986).

10. M. Sherman and K. Vedula, J. Mater. Sci. 21, 1974 (1986).

11. J.D. Whittenberger, Mater. Sci. Eng. 57, 77 (1983).

12. J.D. Whittenberger, Mat. Sci. Eng., 73, 87 (1985).

13. V.M. Pathare, PhD Thesis, Case Western Reserve University, 1987.

14. W.J. Yang and R.A. Dodd, Met. Sci J. 7, 41 (1973).

15. S.V. Raj and G.M. Pharr, Mat. Sci. Eng. 81, 217 (1986).

16. O.D. Sherby, R.H. Klundt and A.K. Miller, Met. Trans. A 8A, 843 (1977).

TOUGHENING MECHANISMS IN INTERMETALLIC γ-TiAl ALLOYS CONTAINING DUCTILE PHASES

C. K. ELLIOTT, G. R. ODETTE, G.E. LUCAS and J.W. SHECKHERD
Department of Materials, University of California, Santa Barbara, CA 93106

ABSTRACT

This work is aimed at developing understanding of ductile phase toughening in powder-processed intermetallic γ-TiAl alloys. The nominally ductile phases studied include Nb and Ti6Al4V, in the form of pancake shaped particles. Toughening is primarily due to the formation of crack face bridging zones. Toughness increases in the composites with a Ti6Al4V phase are limited due to particle embrittlement during processing. In this case, toughening is associated with the formation of short, low ductility/high strength bridge zones. Larger toughness increases and significant resistance curves are observed in composites containing Nb particles. In this case, long, high ductility/low strength bridge zones are observed. The behavior of the Nb composites is consistent with current understanding of constrained deformation of embedded particles and bridge zone toughening models. Additional toughening was observed in the Nb composites due to crack deflection and blunting in orientations where the crack intersects the broad pancake face.

INTRODUCTION

Intermetallic γ-TiAl alloys offer substantial promise for high temperature applications where low density and high creep strength are desirable. However, due to the limited number of slip systems in the LI_O crystal structure, γ-TiAl has low ductility (<2%) and toughness (<12MPa$\sqrt{m}$) at temperatures less than about 600°C.
Ductile particles can be used to increase toughness, primarily by forming crack face bridging zones. Tractions produced by the bridges reduce crack tip stress intensities. Significant ductile phase toughening has been reported for glass/Al [1], Al_2O_3/Al [2] and WC/Co [3] composites. The theoretical framework for toughening by crack bridging has been developed by Budiansky [4], and others [2,5]. This work is aimed at investigating the application of this concept to γ-TiAl composites. The objectives are to: 1) quantify the effectiveness of various nominally ductile phases in toughening powder-processed γ-TiAl; and 2) test and improve micromechanical models of ductile phase toughening.

EXPERIMENTAL

This paper concentrates on two γ-TiAl composites containing either high purity Nb or Ti6Al4V. The alloys were supplied by McDonnell Douglas Corporation. The target volume fractions of ductile phases were 20%; actual volume fractions are lower, and vary from specimen to specimen. The alloys were forged following hot isostatic pressing to full consolidation. Forging deformed the ductile phases into irregularly shaped pancakes. The Nb pancakes average $\approx 25\mu$m thick (t) by $\approx 300\mu$m diameter (d) with an average spacing (s) of $\approx 100\mu$m. The corresponding Ti6Al4V particle averages are t $\approx 30\mu$m, d $\approx 550\mu$m and s ≈ 240 μm. The microstructures of both composites are highly heterogeneous.

Due to limits on the amount and geometry of available material, toughness testing was carried out on small 0.25 cm thick three point bend specimens, with spans of 2.0 cm and widths varying from 0.61 to 0.76 cm. The specimens were tested in the two orientations shown in Figure 1. Measurement of initiation fracture toughness and resistance curves in these alloys required development of new testing techniques which, due to space limitations, will be only briefly summarized here.

Precracking was carried out by stable crack growth out of a 90° electro-discharge-machined chevron notch at very slow rates under computer dislacement control. The cracks were subsequently stably extended in small increments (Δa) using a series of loading-unloading cycles, permitting resistance curve measurements (K_R as a function of Δa) on single specimens. Some specimens were tested after grinding a notch to within ≈ 0.02 cm of the crack front to remove most of the bridge formed in the chevron during precracking. Crack extension was defined as the distance from either the edge of the chevron notch or the bottom of the ground notch. Toughness values (K_R) are based on ASTM stress intensity formulas [6] using optically determined surface crack lengths coupled with loads measured at the initiation of crack extension. The initiation loads were detected by loading rate changes and direct (optical) observation. Uncertainites associated with this procedure are estimated to be ≈ 1.5 MPa$\sqrt{m}$.

Details of other mechanical, microstructural and fractographic measurements will be given in future publications. The measurements included: optical metallography; microhardness of individual phases; analytical scanning transmission and scanning electron microscopy (SEM) of matrix, reaction zone and nominally ductile phases; indentation-induced microcracking to observe crack-particle interactions; tensile tests of precracked samples with the remaining matrix ligament removed to characterize the properties of the intact bridges; SEM and optical fractography; SEM studies of crack-particle interactions and particle deformation and fracture in cracked specimens under load using a SEM fixture which provides sample deflection equal to that at final extension in the loading frame; tensile stress and elastic modulus.

RESULTS

The K_R-Δa resistance curves are shown in Figure 2. Data for specimens with the chevron removed after precracking are indicated by the larger symbols. There is little effect of crack length or orientation in the Ti6Al4V composites, which have an average toughness of $K_R \approx 12.4 \pm 1.1$ MPa$\sqrt{m}$. In contrast, a significant resistance curve is observed for the Nb composites. For the E orientation, where the crack front intersects the edge of the Nb pancakes, K_R increases from an apparent initiation value ($\Delta a \approx 0$) of ≈ 7-9 MPa$\sqrt{m}$ to ≈ 16 MPa$\sqrt{m}$ at $\Delta a > 1.5$ mm. The Nb composite tested in the F orientation, where the crack intersects the face of the pancakes, has higher toughness, with K_R values slightly in excess of 20 MPa$\sqrt{m}$ at $\Delta a > 2$ mm. The initiation toughness of the Nb alloy in the F orientation also appears to be higher, in the range of 9-12 MPa$\sqrt{m}$.

Other relevant observations on the Nb composites include:

1) Ductile Nb particle bridge zones, up to $\approx 2.20 \pm 0.4$ mm long at steady-state, are observed.

2) Nb surface area fractions are $\approx 9 \pm 4\%$. Density measurements indicate bulk Nb volume fractions of $\approx 12 \pm 3\%$. The average Nb volume fraction is taken as $f_p \approx 10 \pm 2\%$.

3) The Nb particles are surrounded by a thin (≈ 1.5 μm) brittle interface reaction zone. However, the particles remain soft with diamond pyramid hardness (DPH) values of $\approx 125 \pm 25$ DPH compared to $\approx 400 \pm 50$ DPH for the γ-TiAl matrix.

4) The Nb bridges normally fail in a highly ductile, knife edge manner as shown in Figure 3. Particle fracture occurs at crack face separations of $\approx 20 \mu$m. The Nb particles decohere below the fracture surface. Infrequently, particle bypass occurs by the cracks running around the brittle Nb-matrix interface. Note that in some other alloys, the Nb particles are harder, and fail by cleavage. This is attributed to impurities

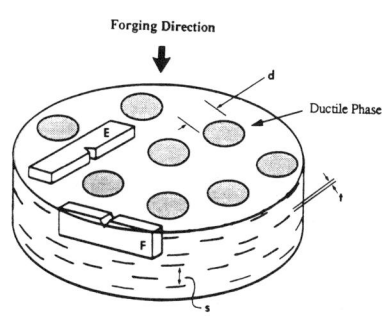

Figure 1. Fracture toughness specimen orientations.

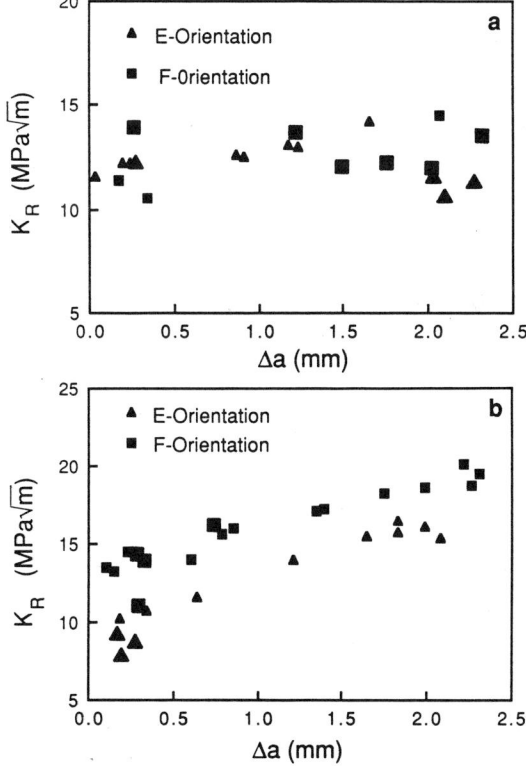

Figure 3. Transverse view of fracture surface, F orientation, showing deformed Nb particles.

Figure 2. K_R-Δa resistance curves for *a)* Ti6Al4V composites and *b)* Nb composites. Specimens with the chevron removed after precracking are indicated by larger symbols.

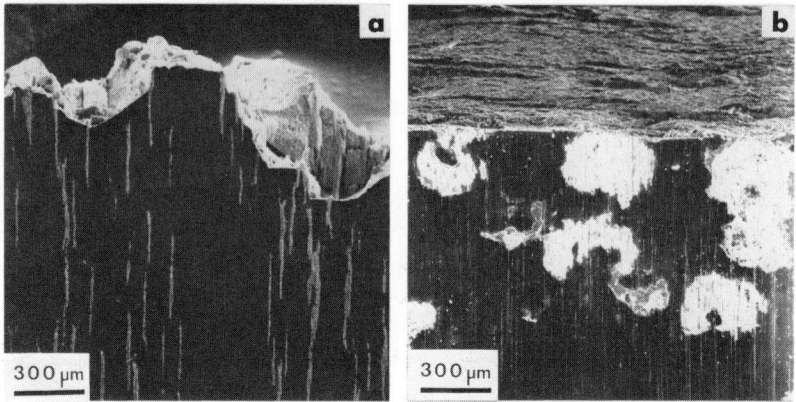

Figure 4. Transverse views of fracture surfaces in the Nb composite a) F orientation and b) E orientation showing particle orientation effects on crack morphology.

in the Nb powder, possibly introduced during processing.

5) The orientation dependence appears to be due to crack deflection. Figure 4 shows that cracks in the F orientation follow a highly tortuous path, in contrast to straight cracks observed in the E orientation. The total to projected crack extension ratio $\Delta a_{tot}/\Delta a$ is ≈ 1.4 for the F orientation and is ≈ 1 for the E orientation.

6) Deflection occurs when cracks intersect the Nb particle faces. After intersection, the crack opens to the point where it initiates on, or travels to, the other side. Decohesion occurs first on the front and subsequently on the back of the Nb in association with necking, preceding the knife-edge fracture. Cracks deflect an average of $\approx 45°$ with the most probable angle being $\approx 70°$. This is similar to the angle of the plastic zone lobes in the Nb.

7) The average strength of the Nb bridge particles, found by tensile testing of cracked specimens with the matrix ligament removed, was $\sigma_{av} \approx 460 \pm 120$ MPa. The work to fracture the Nb bridge particles was estimated to be $J_{br} \approx 1.25 \pm 0.65$ kJ/m^2. However, these estimates are subject to large experimental and analytical uncertainties.

The Ti6Al4V composites were not thoroughly characterized, but limited observations indicate:

1) Particles are irregularly shaped pancakes with a wide size range and a highly heterogeneous spatial distribution. Processing produces thick ($\approx 10 \mu$m) brittle reaction zones. The average particle volume fraction is $\approx 16 \pm 3$%.

2) The Ti6Al4V particles are hard (DPH $\approx 375 \pm 25$) and brittle. Limited toughening ($\approx 4 \pm 2$ MPa$\sqrt{m}$) occurs due to the formation of short bridges. Both plastic deformation in and crack bridging by large particles are observed. However, particle failure is by cracking.

Additional studies will be needed to fully characterize both composite systems.

ANALYSIS AND DISCUSSION

The reduction (shielding) of the crack tip stress intensity (K_{tip}) due to stresses ($\sigma(x)$) acting along the (bridged) face is [8]

$$K_{tip}(L) = K_{app} - C \int_0^L (\sigma(x)/\sqrt{x})dx \qquad (1)$$

where K_{app} is the nominal (applied) stress intensity, x is the distance from the crack tip, C is a constant that depends on the crack geometry and L is the bridge length. Equation 1 can be integrated to predict the steady-state toughness (K_{ss}). For an edge crack [8] this yields

$$K_{ss} \approx K_m + 1.9 f_p \sigma_{eff} \sqrt{L_{ss}} \qquad (2)$$

The nominal matrix toughness $K_m \approx 8 \pm 2$ MPa$\sqrt{m}$, σ_{eff} is the weighted engineering stress in the bridging particles

$$\sigma_{eff} = \frac{\left[\int_0^{L_{ss}} (\sigma(x)/\sqrt{x}) dx\right]}{2\sqrt{L_{ss}}} \qquad (3)$$

and f_p is the particle volume fraction.

The $\sigma(x)$ is governed by the combination of the crack opening at x and the engineering stress in the particles as a function of crack opening. The engineering stress is influenced by constraint arising from deformation of the ductile particles cohered, or partially cohered, to an elastic matrix. Increased triaxiality raises the flow stress and decreases the ductility of the embedded particles. Other factors which influence $\sigma(x)$ are strain-hardening and necking. The steady-state bridge length (L_{ss}) is governed by the combination of crack opening as a function of K_{app} and $\sigma(x)$, and the particle ductility.

Eberhard and Ashby [9] studied deformation of lead wires in cracked glass capillaries and found that the constrained stress decreases approximately linearly from a maximum near zero crack face displacement to zero at the fracture displacement. This behavior is reasonably consistent with mechanics calculations [2,9]. The σ_{eff} can be evaluated from the measured bridge strength σ_{av} assuming a linear stress versus displacement behavior coupled with the measured crack profile by equating the average stresses. This gives $\sigma_{eff} \approx 1.3\sigma_{av} \approx 600\pm155$ MPa for the Nb composite. Alternately, an average constraint-strain hardening-necking factor A can be used to multiply the yield stress (σ_y) as $\sigma_{eff} \approx 1.3A\sigma_y$. Both theory [2,9] and experiment [9] suggest a reasonable value for A is $\approx 2.25\pm.25$ (i.e. a maximum constraint factor of about 4.5$\pm$0.5). The uniaxial tensile strength (σ_t) and σ_y of the Nb can be estimated from hardness as $\sigma_t \approx 3$DPH [7] $\approx 375\pm75$ MPa, and $\sigma_y \approx 0.8\sigma_t \approx 300\pm60$ MPa. This gives $\sigma_{eff} \approx 875\pm125$ MPa for the nominal constraint factor.

Particle bridge toughening can also be evaluated from the extra work (ΔJ) required to create and, at steady-state, maintain a bridge zone. The ΔJ associated with a steady-state bridge is [9]

$$\Delta J_{ss} \approx Q f_p \sigma_y r_p \qquad (4)$$

where Q is a factor which depends on the particle decohesion behavior and r_p is the wire radius. We have found that Eberhard and Ashby's measurements [9] of Q correlate with the particle ductility. The Nb ductility, measured as the ratio of crack opening at particle fracture to the pancake thickness is $\approx 0.8\pm0.2$. The correlation gives $Q \approx 3.6$ if t/2 is equated to r_p. Finally, $\Delta J_{ss} \approx J_{br}$ as experimentally estimated. There is insufficient data to evaluate a similar set of properties for the embedded Ti6Al4V particles.

Table 1 lists the steady-state toughness predictions based on the various modeling approaches. Note, $K_R = [J_R E/(1-\nu^2)]^{\frac{1}{2}}$, where E is Young's modulus and ν is Poisson's ratio. The predictions are in reasonable agreement with the observed $K_{ss} \approx 16$ MPa$\sqrt{m}$ for the E orientation.

TABLE 1
Steady State Toughness Predictions

Model	Predicted K_{ss} (MPa$\sqrt{m}$)
Eq. 2, $\sigma_{eff} \approx 1.3\sigma_{av} \approx 600$ MPa	13.5
$\Delta J_{ss} = J_{br} \approx 1.25$ kJ/m^2	17.5
Eq. 2, $\sigma_{eff} \approx 1.3A\sigma_y \approx 875$ MPa	15.8
Eq. 4, $Q = 3.6$, $\sigma_y = 300$ MPa	18.0

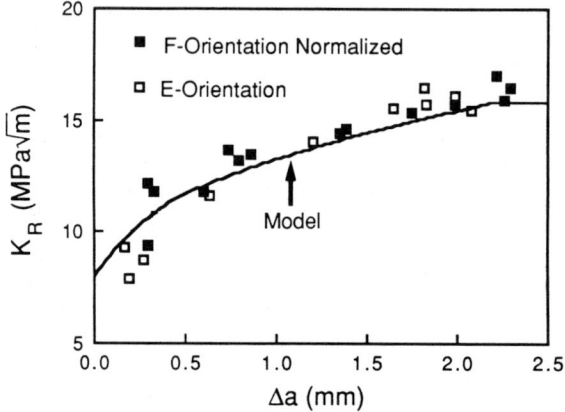

Figure 5. K_R-Δa resistance curve for the Nb composite predicted by the constraint factor model and data from the E and F (crack length normalized) orientations.

Evaluation of the transition regime toughness requires a self-consistent mechanics analysis of the evolving elastic-plastic bridge and elastic matrix. However, it can be crudely modeled simply by replacing L_{ss} in Equation 1 with Δa. Predictions based on the constraint factor model and this assumption track the transition regime toughness trends as shown in Figure 5.

Crack deflection in the F orientation results in an additional increase in toughness of $\approx 3.5\pm0.75$ MPa$\sqrt{m}$. The J_R-Δa curves can be normalized by dividing by the $\Delta a_{tot}/\Delta a$, given above. As shown in Figure 5, the normalized K_R data overlap except at small Δa. The differences at small Δa may indicate that crack blunting increases the initiation toughness in the F orientation.

SUMMARY AND CONCLUSIONS

The major conclusions of this study are as follows:
1) Brittle γ-TiAl alloys can be toughened by the introduction of Nb and Ti6Al4V particles.

2) Toughening in Ti6Al4V ($\approx$ 16 volume %) composites is limited by hardening and embrittlement of the nominally ductile phase and the formation of a relatively thick, brittle reaction zone during processing. These alloys do not have a significant resistance curve, but are modestly toughened ($\approx$ 12.4 MPa$\sqrt{m}$) by short, low ductility/high strength bridge zones.

3) The Nb composites ($\approx$ 10 volume %) have a strong resistance curve and a higher steady-state toughness $\approx$ 16 MPa$\sqrt{m}$ in the E orientation associated with the formation of a long, high ductility/low strength particle bridge zone.

4) The K_R behavior in the Nb composites in the E orientation is consistent with experimental and theoretical studies of constrained deformation and models of bridge zone toughening.

5) Pancake shaped particles introduce an orientation dependence to ductile particle toughening. Steady-state toughening is increased to $\approx$ 20 MPa$\sqrt{m}$ by crack deflection in the F orientation, where the crack intersects the the pancake face.

These initial results are very encouraging and provide useful guidance to both alloy development efforts and additional fundamental studies of ductile phase toughening.

ACKNOWLEDGMENTS

This work was supported by the DARPA University Research Initiative program at UCSB, ONR Prime Contract N00014-86-K-0753. The authors thank the McDonnell Douglas and Pratt and Whitney Corporations for supplying alloys used in this research and Professors A. Evans and B. Budiansky for helpful discussions.

REFERENCES

1. V. D. Krstic, P. S. Nicholson and R. G. Hoagland, *J. Am. Ceram. Soc.*, **64**, pp. 499-504 (1981).

2. L. S. Sigl, P. Mataga, B. J. Dalgleish, R. M. McMeeking and A. G. Evans, 1987 (in press).

3. L. S. Sigl and H. E. Exner, *Met Trans. A*, **18A**, pp. 1299-1308 (1987).

4. B. Budiansky, J. C. Amazigo and A. G. Evans, (AD-A183 208/8/GAR, Harvard University, 1987) p. 34.

5. L. R. F. Rose, *J. Mech. Phys. Solids*, **34**, (6), pp. 609-616 (1986).

6. ASTM E399-83 Plane-Strain Fracture Toughness of Metallic Materials, *1985 Annual Book of ASTM Standards* (ASTM, Philadelphia, 1983).

7. D. Tabor, *Hardness of Metal*, (Clarenden Press, London, 1951).

8. H. Tada, P. C. Paris, G. R. Irwin, *The Stress Analysis of Cracks Handbook*, (Del Research, St. Louis, MO, 1985).

9. J. Eberhart and M. F. Ashby, *Cambridge University Engineering Dept. Report*, (CUED/C-MATS/TR 140, Cambridge University, 1987).

MICROSTRUCTURAL EFFECTS ON DUCTILE PHASE TOUGHENING OF Nb-Nb SILICIDE COMPOSITES

J.J. LEWANDOWSKI#, D. DIMIDUK*, W.KERR*, and M.G. MENDIRATTA**
Dept. Matl's Sci. and Eng., Case Western Reserve University, Cleveland, OH 44106
* AFWAL/MLLM, Wright Patterson AFB, Dayton, OH 45433
**Universal Energy Systems, Inc., Dayton, OH 45432

ABSTRACT

In the Nb-Si system, the terminal Nb phase and the Nb_5Si_3 phase are virtually immiscible up to approximately 2033K. This system offers the potential of producing composites consisting of a ductile refractory metal phase and a strong intermetallic phase. In-situ composites containing different volume fractions of the ductile Nb phase were produced via vacuum arc-casting. Microhardness testing as well as smooth bend bar testing was conducted at temperatures ranging from 298K to 1673K in an attempt to determine microstructural effects on the yield strength and smooth bar fracture strength. Notched bend specimens were similarly tested to determine the effects of the ductile phase (i.e. Nb) on enhancing the notched bend toughness. It is shown that the Nb phase often behaves in a ductile manner during testing, thereby toughening the in-situ composite. The mechanism of toughening appears to be due to crack bridging.

INTRODUCTION

The typical requirements of high temperature strength and stability for high temperature structural materials necessitates the use of composites capable of withstanding extreme service temperatures. Issues such as reinforcement/matrix compatability, oxidation resistance, thermal stability, and ambient temperature properties are a few areas of particular concern. Unfortunately, many of the matrices capable of withstanding the extreme temperatures are brittle at ambient temperatures, while brittle reaction products and reinforcement degradation have been observed at elevated temperatures in a variety of composites. Attempts at toughening the brittle matrices include fiber or whisker toughening(1), as well as ductile phase toughening(2). In both cases (1,2) toughness increases have been realized by crack bridging, whereby the reinforcement (e.g. fiber, ductile phase) remains intact in the wake of a crack. Thus, the crack tip is effectively shielded from the applied stress intensity, as shown in the schematic in Figure 1(2).

As described above, high temperature microstructural stability is necessary due to the requirements for strength retention. In the Nb-Si system shown in Figure 2, the terminal Nb phase and the Nb_5Si_3 phase are virtually immiscible at temperatures up to approximately 2033K (i.e. 1760 C). This system offers the potential of producing composites consisting of a ductile refractory metal phase and a strong intermetallic phase, which should be stable at high temperatures. Systematically decreasing the amount of Si in the liquid alloy enables the production of in-situ composites containing progressively larger volume fractions of the ductile (i.e. Nb) phase. It is the purpose of this paper to summarize some of the preliminary microstructural observations and mechanical properties obtained on these ductile phase toughened composites. Other work(3) presents details of the microstructures and phase relationships in Nb-Nb silicide composites.

EXPERIMENTAL

The alloys tested in the present work were prepared by non-consummable electrode vacuum arc-casting and consisted of Nb containing the following amounts of Si (in at.%): 0.0, 0.25, 1.5, 6.0, 10.0, and 15.0. The amounts of dissolved interstitial elements were determined and are presented elsewhere(4). Microstructures were characterized using both optical and scanning electron microscopy. Metallographic specimens were polished through 0.25 um diamond polish and subsequently examined in a JEOL JSM-35 or 840 Scanning Electron Microscope(SEM) using either secondary electron or back-scatter electron imaging. Semi-quantitative chemical analyses were conducted via energy dispersive x-ray analysis(EDAX).

Microhardness tests were made on the various polished specimens using a Vickers indentor and loads ranging from 100-5000 grams, with indentation times of 15 seconds. Indentations were placed in the various constituents (i.e. pure Nb, intermetallic, Nb + Si solid solution). Subsequent examination at high magnification was utilized to reveal the presence of cracking.

Smooth bend bars of dimensions 2.54mm X 6.0mm X 32mm were EDM machined. Notched bend specimens of average dimensions 6.0mm X 12.7mm X 50mm were also EDM machined and were notched to half depth (i.e. a/w = 0.5) with a slow speed diamond saw. The root radius of the machined notch was approximately 0.25mm. Smooth bend bar testing was conducted in four point bending in vacuum (e.g. 1 x 10-5 torr) at temperatures of: 298K, 673K, 873K, 1073K, 1273K, 1373K, and 1673K on as-cast specimens as well as specimens which were held in vacuum at 1773K/3 hours and subsequently furnace cooled. Notched testing was via three point bending. Bend testing was conducted on an INSTRON Universal Testing Machine at cross-head speeds of 0.004 mm/sec and 0.02 mm/sec for the smooth and notched specimens, respectively. Duplicate specimens were tested in all cases. Analysis of the fractured specimens was accomplished using the SEM, while metallographic cross-sections were made perpendicular to the crack growth direction in an attempt to determine the effects of microstructure on the path of crack propagation. Matching surface fractography(5) and stereo imaging were utilized to determine additional details of the fracture surface, while EDAX analyses were utilized to determine the composition of phase(s) responsible for fracture initiation.

RESULTS AND DISCUSSION

Microstructures

SEM micrographs of as-cast microstructures of the Nb-1.5Si and the Nb-15Si in-situ composites are presented in Figure 3a and 3b, respectively. The light grey phases are the solid solution of Si in Nb, while the dark grey phase is the intermetallic silicide. Also evident is the eutectic/oid microstructure consisting of the solid solution of Si in Nb, and the intermetallic. EDAX dot maps showing regions of high Si content for the corresponding microstructures are included in Figure 3. X-ray analyses of the as-cast product revealed that greater than one intermetallic phase was present(3) although the predominant intermetallic present was Nb_5Si_3. Heat treatment at 1773K/3 hours did not significantly change the microstructures from that shown in Figure 3, although subsequent x-ray analyses indicated that only one intermetallic phase was detected(3). Additional microstructural details are provided elsewhere(3,4).

The microhardness measurements revealed that pure Nb exhibited a significantly lower microhardness (i.e. VHN 145) than did the single phase solid solution of Si in Nb (i.e. VHN 450) present in the as-cast composites. However, qualitative indications of the crack blunting effectiveness of the Nb-rich phase is illustrated by the SEM micrograph of the microhardness indent shown in Figure 4. Cracks in the brittle intermetallic phase are evident in Figure 4, while blunting of the cracks is observed to occur on two size scales. Both the large islands of Nb, and the smaller Nb particles present in the eutectic/oid structure appear to provide effective crack blunting. The high magnification photomicrograph in Figure 4 illustrates the scale of crack blunting occurring at both the large and small Nb particles.

The smooth bend bar data obtained on specimens held at 1773K/3 hours prior to testing is summarized in Table I for the various specimens. Consistent with the microhardness

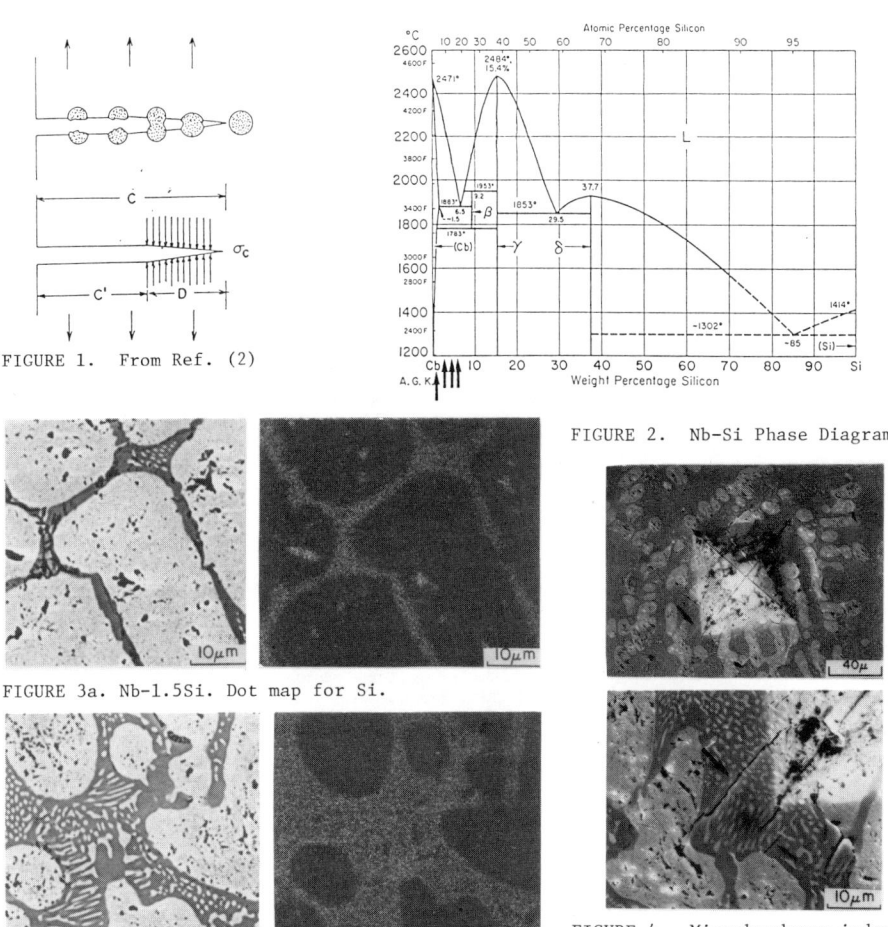

FIGURE 1. From Ref. (2)

FIGURE 2. Nb-Si Phase Diagram

FIGURE 3a. Nb-1.5Si. Dot map for Si.

FIGURE 3b. Nb-15 Si. Dot map for Si.

FIGURE 4. Microhardness indent

measurements, it is shown that the solid solution of Si in Nb(i.e. Nb-0.25Si) exhibits a significantly higher yield strength than does the pure Nb. Also evident is that the composite microstructures exhibit significantly higher yield and fracture strengths in comparison to the pure Nb. The pure Nb smooth bend specimens did not fail after large amounts of plastic offset, while the room temperature tests on the Si containing alloys exhibited fracture at higher loads and lower levels of displacement than that obtained in the pure Nb specimens.

Fracture in the smooth bend bars for the Nb-1.5Si was typically by transgranular cleavage. Fracture was initiated by the brittle intermetallic at temperatures lower than 1273K. However, areas of ductile fracture were observed as shown in Figure 5 for the Nb-1.5Si specimen tested at 673K. Metallographic cross-sections of failed specimens further revealed that at temperatures greater than 873K, the Nb phase blunted cracks which had initiated in the intermetallic, as Figure 6 shows. Fracture surfaces of the Nb-10Si smooth bend specimens tested at 1273K provided additional evidence of crack tip blunting. The presence of considerable local ductility on the surface of the Nb-10Si specimen tested at 1273K is revealed in the arrowed regions of Figure 7.

Table II summarizes the notched bend toughnesses obtained on as-cast specimens for the various compositions and test temperatures. As for the smooth bend bars, the pure Nb notched bend specimen tested at 298K exhibited significant ductility and did not fail. The remaining specimens failed at higher loads and lower total displacements than the pure Nb specimen. As shown in Table II, the toughness of the low Si alloys (i.e. larger proportion of toughening Nb phase) significantly exceeded that of the high Si alloys, while estimates of the toughness of the intermetallic alone did not exceed 1 MPa$\sqrt{}$m. Examination of the fracture surfaces again revealed that fracture was often initiated by the the brittle intermetallic phase(s). It thus appears that significant toughening (e.g. a factor of 10) is acheived by the dispersion of the ductile Nb phases. Although the Nb-rich phase often behaved in a ductile manner, as the arrowed regions in Figure 8 show, other Nb-rich regions on the fracture surface shown in Figure 8 exhibited transgranular cleavage fracture, in addition to fracture which appeared to propagate around the interface of the Nb-rich phase. Thus, in the present work, it appears that the 'ductile' phase does not always fracture in a ductile manner. Metallographic cross-sections confirmed that fracture often propagated along the Nb-Nb silicide interface, with other regions of the fracture surface again exhibiting fracture of the Nb by either ductile or brittle modes (3,4).

It is clear from the microhardness results in addition to the notched fracture data and corresponding fractography that the toughness of the Nb-Nb silicide in-situ composites increased with an increase in the amount of the 'ductile' phase. Cracking of the brittle intermetallic occurred at very low loads and these cracks were often effectively blunted by the Nb-rich phase. This is in qualitiative agreement with the theories of ductile phase toughening of brittle matrices(7), although quantifying the microstructural effects (e.g. size of ductile phase, distribution, length of bridged ligaments) on the magnitude of toughening is continuing on specimens where fracture is being monitored in-situ(8). Recent experiments have revealed that the ductile Nb phase provides effective bridging behind the crack tip, thereby shielding the crack tip and increasing the toughness(8). The reason(s) for the dual (i.e. brittle, ductile) behavior of the 'ductile' phase is under continuing investigation. It was noted earlier that the microhardness and the yield strength of the solid solution Nb-Si specimen (i.e. Nb-0.25Si) was significantly higher

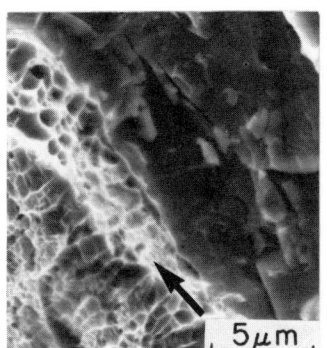

FIGURE 5. Arrow denotes ductile region in Nb-1.5Si specimen.

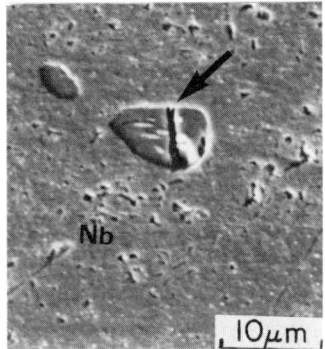

FIGURE 6. Arrow denotes crack in intermetallic blunted by Nb phase.

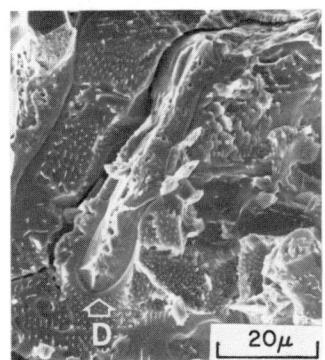

FIGURE 7. Arrow denotes ductile region on fracture surface.

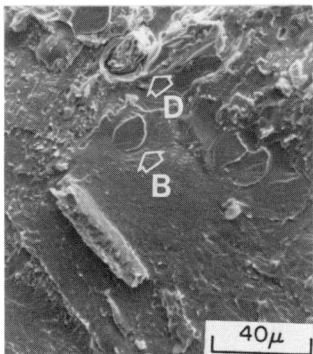

FIGURE 8. Arrows denote regions behaving in a ductile manner(D) and brittle manner(B).

TABLE I

SMOOTH BEND YIELD STRENGTH (MPa)

TEMPERATURE	PURE Nb	Nb-0.25Si	Nb-1.5Si	Nb-10Si
298K	262	540*	742*	-----
673K	-----	-----	610	-----
873K	70**	315	-----	-----
1073K	70**	280	448#	-----
1273K	56**	266	385#	546*
1373K	-----	266	390#	400*
1473K	42**	-----	-----	-----
1673K	-----	-----	98#	287

* Fracture before yield   # Did Not Fail   ** From Ref. (6)
ALL SMOOTH SPECIMENS WERE HELD 1773K/3 HOURS PRIOR TO TEST

TABLE II

NOTCHED BEND K_Q (MPa$\sqrt{m}$)

(AS CAST)

TEMP	PURE Nb	Nb-0.25Si	Nb-1.5Si	Nb-6Si	Nb-10Si	Nb-15Si
298K	DNF	10, 22*	10, 20**	8.4	7.6	6.2, 8.5#
873K	---	--------	11	---	---	-------
1273K	---	--------	10.1	9.6	6.7	5.1
1473K	---	--------	9.9	---	---	-------

  * Exposed 1773K/3.0 hours before testing
 ** Exposed 1773K/1.5 hours before testing
  # Exposed 1473K/3.0 hours before testing

than that of the pure Nb specimens. It is possible that the regions behaving in a locally brittle manner are due to Si enrichment of the Nb phase, while the locally ductile regions are depleted in Si. Another possibility is the effect of the orientation of the propagating crack with respect to the Nb phase. Finally, the observation of interface failure between the Nb phase and the silicide additionally suggests that reaction products at the interface may facilitate interface fracture near the Nb-rich phase, thereby preventing it from imparting additional toughening. Additional work is focussing on determining reason(s) for the dual behavior of the 'ductile' phase, in addition to quantifying microstructural effects on both the bridge lengths and the effectiveness of the bridged ligaments on enhancing the toughness of these materials.

CONCLUSIONS

Experiments have been conducted on a variety of Nb-Nb silicide composites produced via vacuum arc-casting in an attempt to verify the potential benefits to toughness acheived by a dispersion of a ductile phase (e.g. Nb solid solution). It was shown that the microstructures produced were stable at temperatures of 1773K/3 hours, while both microhardness testing and fracture testing revealed the potential toughening available in systems where a dispersion of ductile phase(s) inhibit crack growth via crack bridging. In general, it was observed that the notched bend toughness increased with an increase in the amount of the ductile phase. However, the ductile phase often exhibited dual fracture characteristics. Work is continuing in an attempt to quantify the magnitude of toughening acheivable in such microstructures, as well as optimize the microstructures with respect to toughness.

ACKNOWLEDGEMENTS

The support of a summer faculty fellowship at WPAFB under the support of Universal Energy Systems under Air Force Contract No. F33615-84-C-5071 for one of the authors (JJL) is appreciated. Additional work is being conducted at Case Western Reserve University, with experimental support supplied by DARPA-ONR N00014-86-K-0773.

REFERENCES

1. D.B. Marshall, B.N. Cox, and A.G. Evans, Acta Metall., 33, 2013, 1985.
2. V.V. Krstic, P.S. Nicholson, and R.G. Hoagland, Jnl. Amer. Cer. Soc., 64, 499, 1981.
3. M.G. Mendiratta, D. Dimiduk, and J.J. Lewandowski, Presented at MRS-1988 Spring Meeting, Reno, Nevada.
4. J.J. Lewandowski, D. Dimiduk, M.G. Mendiratta, and W. Kerr, manuscript in preparation.
5. J.J. Lewandowski, C. Liu, and W.H. Hunt, in Powder Metallurgy Composites, (eds. M. Kumar, K. Vedula, and A.M. Ritter) TMS-AIME, Warrendale, PA., 1988.
6. C. English, in Niobium, (ed. H. Stuart), TMS-AIME, Warrendale, PA., p. 263, 1984.
7. A.G. Evans and R.M. McMeeking, Acta Metall., 34, 2435, 1986.
8. J.J. Lewandowski, unpublished results, 1988.

Internal Friction of Cast Graphite-Magnesium Composites[†]

J. H. ARMSTRONG, S. P. RAWAL AND M. S. MISRA
Martin Marietta Astronautics Group, Materials and Structures, Denver, CO 80201

ABSTRACT

Internal friction behavior in cast 8-ply [0°] P55Gr/Mg-0.6%Zr alloy and P55Gr/Mg-1%Mn composites as a function of vibratory strain amplitude was measured at 80 kHz using a Marx-type piezoelectric composite oscillator. Both the matrix and composite exhibited strain amplitude independent internal friction below $\varepsilon \approx 10^{-6}$, while significant strain amplitude dependence was noted at higher strain levels. A maxima in damping was observed for most of the specimens tested. Heat treatment to enlarge grain size was found to increase both the strain amplitude independent and dependent internal friction of the composite. Strain amplitude dependence of the internal friction, including the existence of the maxima, was explained by the Granato-Lucke (G-L) dislocation internal friction model. Dislocation densities obtained from various TEM images from the fiber-matrix interface were compared to values predicted by G-L theory.

INTRODUCTION

With the advent of large precision space structures (LPSS), a new class of materials must be developed with high specific mechanical properties, high resistance to both environmental attack and outgassing, low coefficient of thermal expansion (CTE), and high thermal and electrical conductivities for improved performance. For these reasons, metal matrix composites, in particular graphite-fiber reinforced magnesium, are candidate structural materials for LPSS. Because of in-flight disturbances, enhanced damping of these structures can also improve their dynamic stability.
Typically, damping enhancement of space structures can be divided into active controls and passive measures. Active vibration controls increase design and operational complexity, and thus increase weight and decrease reliability. Passive measures such as placement of discrete dampers and/or constrained layer viscoelastic materials at strategic locations on the structure also increase weight. To reduce complexity and weight, while increasing reliability of LPSS, a logical compliment to active controls and passive damping measures is the development of structural materials with enhanced inherent damping, or internal friction.
Previous investigations into the graphite/aluminum system showed that the internal friction characteristics were significantly influenced by the fiber, matrix alloy and fiber-matrix interface[1,2]. Thus it was the intent of this investigation to study two alloys in the magnesium system, Mg-0.6%Zr (K1A) and Mg-1%Mn (M1A), both of which have been shown to possess high internal friction in cast condition[3]. Composite panels of pitch 55 graphite-reinforced Mg alloys were fabricated using vacuum investment casting technology. Strain amplitude dependence of the internal friction was studied to identify internal friction mechanisms in these materials. On the basis of these studies, microstructural optimization through heat treatments was conducted to enhance internal friction.

[†] Work performed under Office of Naval Research Contract N00014-85-C-0857, Dr. Steven G. Fishman, Technical Monitor.

EQUIPMENT

All internal friction measurements were made with a piezoelectric composite oscillator apparatus at a resonant frequency of 80 kHz in air shown schematically in Figure 1. A pair of quartz crystals were driven in a standing wave configuration by a closed-loop crystal driver manufactured by Solid State Equipment, New Zealand. Resonant frequency was measured to within 1 Hz. This apparatus was capable of measuring internal friction and resonant frequency of the specimen while being driven in a longitudinal standing wave configuration at preset strain amplitudes ranging from 10^{-9} to 10^{-4}.

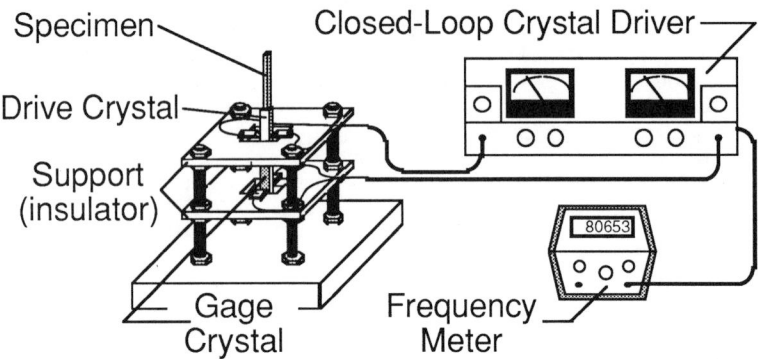

Figure 1
Schematic of Piezoelectric Composite Oscillator Damping Apparatus.

EXPERIMENTAL PROCEDURE

Unidirectional 8-ply composite panels were fabricated from pitch 55 graphite and Mg-0.6%Zr and Mg-1%Mn alloys by Materials Concepts, Inc., Columbus, OH, using vacuum investment casting technology. Each panel was subjected to X-radiography to verify fiber collimation and to identify the presence of defects. Specimens were fabricated from these panels in the [0°] fiber orientation, as well as from the K1A and M1A alloys for comparison. Physical properties of the specimens are given in Table 1. While sand-cast K1A material was used, M1A was only available in extruded billet form for this investigation. M1A specimens were heat treated at 427°C (800°F) for 4 hours to obtain equiaxed grain structure.

Table I — Material Properties of Matrix Alloys and Composites in this Investigation

Material	Density g/cm^3 (lb/in^3)	v/o %	Modulus GPa (msi)
[0°] P55Gr/Mg-0.6%Zr	1.91 (0.690)	46.8	164.8 (23.9)
[0°] P55Gr/Mg-1%Mn	1.89 (0.683)	46.1	166.2 (24.1)
Mg-1%Zr	1.74 (0.0629)	—	44.9 (6.51)
Mg-1%Mn	1.74 (0.0629)	—	44.2 (6.41)

Each specimen was carefully hand sanded to 1/2 wavelength as determined experimentally. Resonant frequency of the specimen was maintained to within 0.5% of the resonant frequency the quartz bars. Specimen surfaces were hand-lapped to provide a smooth, parallel surface for bonding with E910 adhesive onto the quartz bar assembly.

Internal friction of each specimen was determined for a series of strain amplitudes within the range of 10^{-8}—10^{-4}. Sufficient time was taken between strain amplitudes to stabilize the apparatus. All internal friction measurements were expressed in terms of the specific damping capacity ψ which is related to other damping terminology in the following fashion:

$$\psi = 2\delta = \Delta W/W = 2\pi Q^{-1} = 4\pi\zeta = 2\pi E''/E' = 2\pi\phi = 2\pi\eta \tag{1}$$

where:

$\delta \equiv$ Log Decrement
$\Delta W \equiv$ Energy Lost per Cycle
$W \equiv$ Total Vibrational Energy per Cycle
$Q \equiv$ Quality Factor
$\zeta \equiv$ Damping Ratio or Damping Factor
$E'' \equiv$ Loss Modulus
$E' \equiv$ Storage Modulus
$\phi \equiv$ Loss Angle
$\eta \equiv$ Loss Factor.

Calibration of vibratory strain amplitude and internal friction was accomplished with techniques found in the literature[4].

RESULTS

Significantly different internal friction behavior as a function of strain amplitude was observed in the matrix alloys Mg-0.6%Zr and Mg-1%Mn as shown in Figures 2 and 3 respectively. While both alloy exhibited strain amplitude independent internal friction behavior below $\varepsilon \approx 10^{-6}$, values for Mg-0.6%Zr were substantially lower ($\psi_i \approx 0.3\%$) compared to those of Mg-1%Mn ($\psi_i \approx 0.6\%$). However, a relatively high value of strain amplitude dependent internal friction for sand cast Mg-0.6%Zr was observed ($\psi \approx 14\%$), in contrast to a value of only $\approx 4\%$ found for the Mg-1%Mn alloy. These differences were attributed primarily to the condition of the material, i.e. sand cast versus extruded/heat treated. An interesting phenomenon was a peak in internal friction observed in Mg-1%Mn at higher strain amplitude ($\varepsilon \approx 1.6 \times 10^{-5}$). This type of peak was not observed in the Mg-0.6%Zr alloy, although presumably it might appear at higher strain amplitude levels not obtainable with this apparatus.

Internal friction behavior of the cast composites was similar in nature to the alloys as shown in Figure 4 for Gr/K1A and Gr/M1A. Internal friction of P55Gr/K1A did not exceed 3.5%, although the composite did exhibit a maxima in internal friction at high strain amplitudes. Strain amplitude independent internal friction was significantly higher in the composite than in the alloy, with $\psi_i \approx 0.8\%$. Internal friction behavior of the Gr/M1A, on the other hand, was nearly identical to that of its respective matrix, with the exception that the peak in internal friction appeared at a higher strain amplitude.

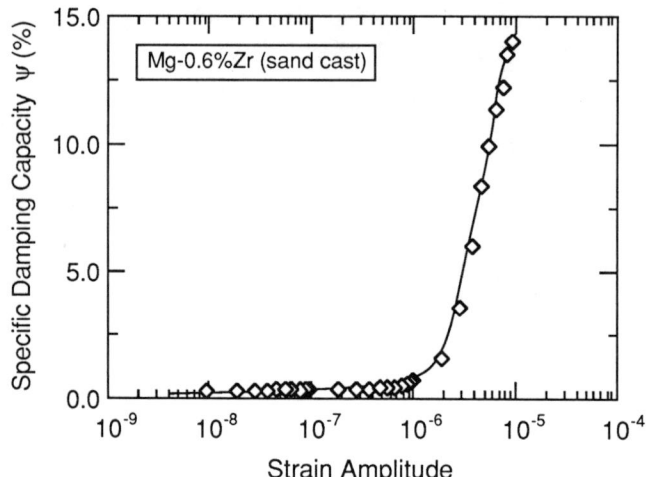

Figure 2
Damping Behavior of Sand Cast Mg-0.6%Zr Alloy Exhibiting High Strain Amplitude Dependence for $\varepsilon > 10^{-6}$.

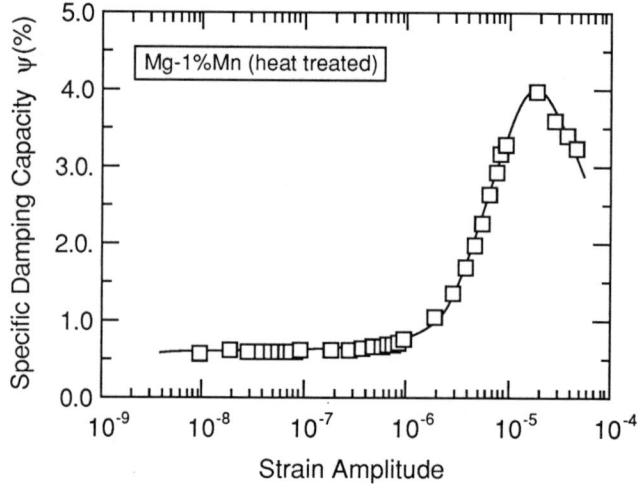

Figure 3
Damping Behavior of Extruded/Annealed Mg-1%Mn Alloy Exhibiting Moderate Strain Amplitude Dependence for $\varepsilon > 10^{-6}$.

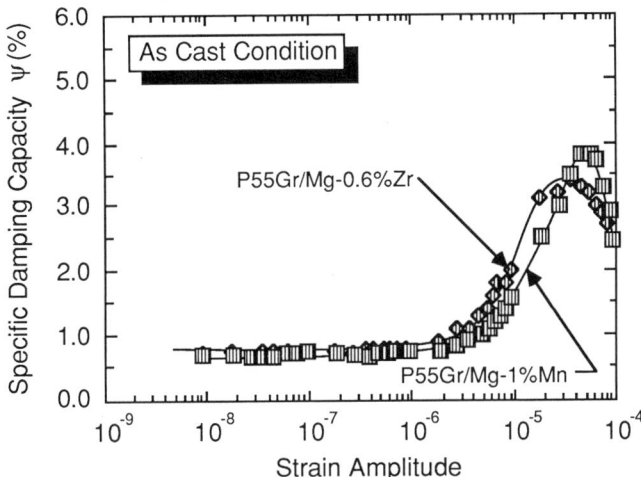

Figure 4
Damping Behavior of As-Cast P55Gr/Mg-0.6%Zr and P55Gr/Mg-1%Mn.

DISCUSSION

Internal friction of K1A and M1A alloys is highly dependent upon grain size. It has been shown in the past that optimum grain size is > $10 \mu m^8$. However, micrographs and TEM images of as-cast composites indicated that grain size was often less than 2 μm. Two heat treatments of these composites were attempted to increase the grain size in hopes of enhancing the internal friction of these composites. As shown in Figure 5, the internal friction of Gr/M1A increased systematically with heat treatments of 6 hours at 450°F and 4 hours at 800°F respectively. Both strain amplitude independent and dependent damping increased with respect to heat treatment. Strain amplitude independent internal friction values were $\psi_i \approx$ 0.55%, 0.62% and 1.0% for as-cast, 6 hours at 450°F, and 4 hours at 800°F respectively. Peak position appears to shift towards lower strain amplitudes. A micrograph of the heat treated specimen, shown in Figure 6, indicates that a number of grains have grown to an average of $\approx$ 10 μm, often encompassing two or more graphite fibers.

Strain amplitude independent internal friction can be attributed to anelastic mechanisms which can be separated into contributions from its constituent fiber and matrix. According to Hashin theory, internal friction of unidirectional composites (ψ_{11}) can be expressed as

$$\psi_{11} = V_f [E_f / E_{11}] \psi_f + V_m [E_m / E_{11}] \psi_m \tag{2}$$

where E_f, E_m, and E_{11} are the elastic moduli of the fiber, matrix and composite respectively and ψ_f and ψ_m are the damping capacity values of the fiber and matrix respectively. Equation 2 has been used successfully to predict strain amplitude independent damping values for Gr/Al[2] and B/Al[5,6] in agreement with measured values. Given the strain amplitude independent damping for Gr/M1A mentioned earlier, the contribution from the matrix can be calculated as $\psi_m \approx$ 3.10%, 3.59%, and 6.24% for the as-cast and heat treatments respectively. This increase of matrix contribution to damping from heat treatment corresponds well with the noted growth in grain size. While these values for the matrix internal friction are above those reported for the

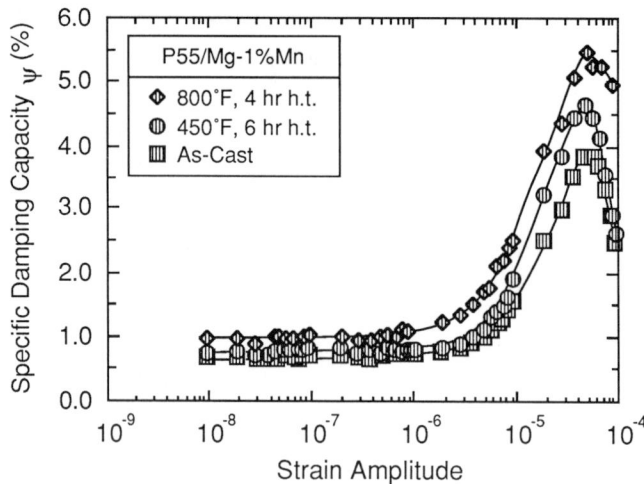

Figure 5
Internal Friction of Gr/Mg-Mn Composite in As-Cast and Heat Treated Conditions.

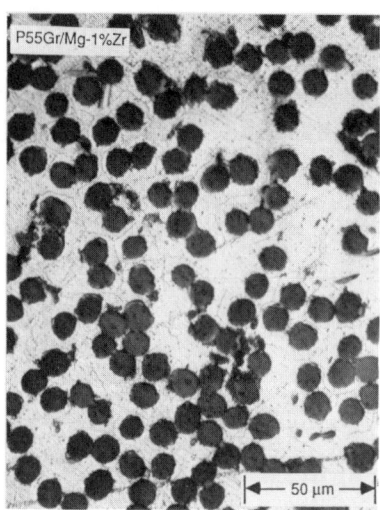

Figure 6
Micrograph of As-Cast P55Gr/Mg-0.6%Zr Showing a Number of Grains Larger than 10 μm.

M1A billet specimen in this paper, they are still significantly below those values for sand-cast M1A and K1A found in the literature[7,8].

In the case of graphite/aluminum mentioned earlier[2], strain amplitude dependent internal friction was attributed to dislocation motion as described by Granato and Lucke (G-L)[9,10]. According to G-L theory, the strain amplitude dependence of the internal friction (ψ_H) can be described as:

$$\psi_H = C_1/\varepsilon^{1/2} \exp\{-C_2/\varepsilon\} \qquad (3)$$

where ε, the vibratory strain amplitude, is sinusoidal with respect to specimen length (standing wave). For a standing-wave configuration, constants C_1 and C_2 are defined by the following:

$$C_1 \equiv 2\Omega\Lambda L_N^3 \{2K\eta^*a/\pi^5 L_c^3\}^{1/2}$$
$$C_2 \equiv K\eta^*a/L_c$$

where Ω and K are orientational constants, Λ is the mobile dislocation density, L_N is the dislocation network length, η^* is the Cottrell misfit parameter, a is the lattice spacing, and L_C is the average dislocation loop length. To investigate the possibility that internal friction from dislocation motion could be responsible for the strain amplitude dependence, a G-L plot ($\psi_H\varepsilon^{1/2}$ vs. $1/\varepsilon$) for the Mg-0.6%Zr alloy was performed as shown in Figure 7. These data show an excellent fit to the predicted G-L strain amplitude dependence. Least-squares fit of G-L plots for both Gr/K1A and Gr/M1A were performed, and in each case, very good agreement with G-L theory was found. By substituting material constants of a = 3.21 Å, η^*(K1A) = 0.141, η^*(M1A) = 0.219, K = 0.02, Ω = 0.04 and L_n/L_c = 100, values were obtained for L_c and Λ as shown in Table II. Dislocation loop length increased and density decreased with heat treatment, which is consistent with increasing grain size, and preliminary TEM measurements of the dislocation density[11].

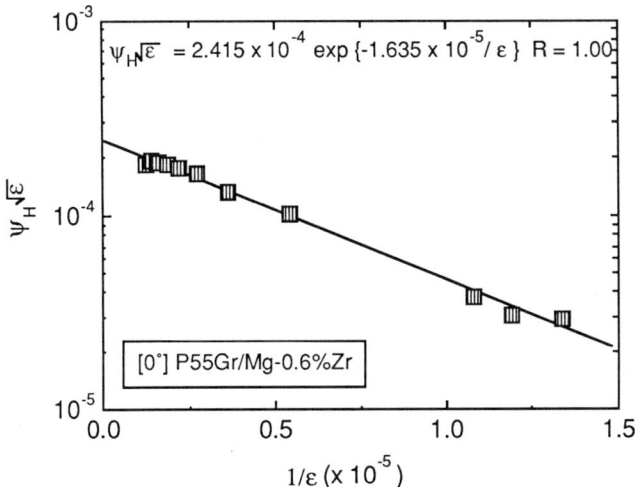

Figure 7
G-L Plot of Gr/Mg-Zr Composites Indicating an Excellent Fit.

Table II — Results of Granato-Lucke Analysis of Strain Amplitude Dependent Internal Friction of Gr/K1A and Gr/M1A

Material	Condition	$C_1 (\times 10^4)$	$C_2 (\times 10^5)$	$L_C (\times 10^8 m)$	$\Lambda (\times 10^{-9} 1/m^2)$
P55Gr/Mg-0.6%Zr	As Cast	2.415	1.635	5.53	3.01
	450°F, 6 hr.	2.567	0.631	14.3	0.768
	800°F, 4 hr.	2.249	0.532	17.0	0.520
P55Gr/Mg-1.0%Mn	As Cast	3.075	3.936	3.57	5.94
	450°F, 6 hr.	3.107	2.954	4.76	3.90
	800°F, 4 hr.	4.227	0.809	17.4	0.762

The internal friction maxima with respect to strain amplitude (Figs. 2,3,4 and 5) can also be explained in terms of G-L theory. By taking the first and second derivative of equation (2), a maxima in internal friction at strain ε_p is predicted by G-L theory where $\varepsilon_p = 2C_2$. Predicted values from the G-L slopes are $\varepsilon_p = 7.872 \times 10^{-5}$, 5.908×10^{-5}, and 1.618×10^{-5} for as-cast and heat treated Gr/M1A respectively. In reference to Figure 5, the actual peak position is close to the predicted values, with the exception of the 800°F, 4 hr. test which does not shift nearly as much as anticipated.

SUMMARY

Internal friction behavior of P55Gr/Mg-0.6%Zr and P55Gr/1.0%Mn composites was measured at 80 kHz. All materials exhibited strain amplitude independent and dependent damping behavior. Application of Hashin theory showed that the internal friction contribution from the matrix in the composite is far below its intrinsic values due to fine grain microstructure near the fiber-matrix interface. Dislocation motion as described by Granato and Lucke appears to be the operative mechanism in strain amplitude dependent internal friction.

ACKNOWLEDGEMENTS

The authors wish to thank Dr. Steve H. Carpenter, Department of Physics, University of Denver, Denver, CO for the generous use of his equipment for these measurements. This research was funded by the Office of Naval Research, Contract #N00014-85-C-0857, Dr. Steven G. Fishman, Technical Monitor.

REFERENCES

1. M.S. Misra and P.D. LaGreca: "Damping Behavior of Metal Matrix Composites," Vibration Damping 1984 Workshop Proceedings, AFWAL-TR-84-3064, Nov. 1984., p. U-1.
2. S.P. Rawal, J.H. Armstrong and M.S. Misra: "Interfaces and Damping in Metal Matrix Composites," Martin Marietta Report #MCR-86-684, prepared for the Office of Naval Research, December 1986.
3. D.F. Walsh, J.W. Jensen and J.A. Rowland: "Vibration Damping Capacity of Various Magnesium Alloys, "U.S. Bureau of Mines Report #6116, 1962.
4. W.H. Robinson and A. Edgar: PEL Technical Note No. 204, NZ, DSIR, 1970 (unpublished).

5. J.A. DiCarlo and J.E. Maisel: "High-Temperature Dynamic Modulus and Damping of Aluminum and Titanium Matrix Composites," Advanced Fibers and Composites for Elevated Temperatures, I. Ahmad and B.R. Noton, eds., Conference Proceedings, Metallurgical Society of AIME, p. 55, 1979.
6. J.A. DiCarlo and J.E. Maisel: "Measurement of the Time/Temperature-dependent Dynamic Mechanical Properties of Boron/Aluminum Composites," Composite Materials: Testing and Design (Fifth Conference), ASTM STP 674, S.W. Tsai, ed., American Society for Testing and Materials.
7. J.W. Fredrickson: "Damping Capacity of K1A," Dow Chemical Corporation Report #MT17753-Final, Dec. 1958.
8. K. Sugimoto, K. Niiya, T. Okamoto and K. Kishitake: "A Study of Damping Capacity in Magnesium Alloys," Trans JIM, 18, pp 277-288, 1977.
9. A. Granato and K. Lucke: "Theory of Mechanical Damping Due to Dislocations," J. Appl. Phys., 1956, vol. 27, p. 583.
10. A. Granato and K. Lucke: "Applications of Dislocation Theory to Internal Friction Phenomena at High Frequencies," J. Appl. Phys., 1956, vol. 27, p. 789.
11. S.P. Rawal, J.H. Armstrong and M.S. Misra: "Damping Characteristics of Cast Graphite-Magnesium Composites," presented at the American Ceramics Society Conference, Cocoa Beach, FL, January 1988.

THE ELEVATED TEMPERATURE RESPONSE OF SILICON CARBIDE AND BORON REINFORCED ALUMINUM AND TITANIUM METAL MATRIX COMPOSITES

M.S. MADHUKAR*, A. FAREED, J. AWERBUCH* & M.J. KOCZAK
Drexel University, Department of Mechanical Engineering & Mechanics*
Department of Materials Engineering, Philadelphia, PA 19104

ABSTRACT

The elevated temperature modulus and strength of aluminum, titanium, and hybrid aluminum/titanium metal matrix composites were investigated. Aluminum (6061-F) and titanium (Ti-6Al-4V) metal matrix composites reinforced with AVCO silicon carbide or boron fibers were vacuum hot pressed and their tensile properties evaluated to temperatures in excess of 300°C. Microstructure, fracture modes and mechanical properties were characterized to assess the effect of fibers and matrix on composite strength and modulus as a function of temperature. Finally, a comparison of specific strength and modulus is provided as a function of temperature. In general, the metal matrix composites exhibited low density (< 2.8 g/cm^3), high modulus (200 GPa), and strengths equivalent to 1250 MPa at 250-300°C. The effect of fiber orientation on axial stiffness was investigated using boron fiber reinforced aluminum (6061-F).

INTRODUCTION

The superiority of metal-matrix composites over resin matrix composites with regard to transverse strength, shear properties, impact resistance, and most importantly, their elevated temperature properties has been realized for many years. Several experimental studies in the past have been directed toward determining strength and modulus of the composites under various loading environments [1-10]. Limited work has been reported on the degradation of the mechanical properties of unidirectional metal matrix composites at elevated temperatures [11], and it has been shown that even though there was a dramatic reduction in the tensile strength after prolonged exposure of the composite to a temperature of 600°C, the axial modulus reduction was still insignificant.

In the present study, a comprehensive data base for the axial stiffness and strength of silicon carbide fiber and boron fiber reinforced laminates at room and elevated temperatures is presented.

MATERIALS & SPECIMEN GEOMETRY

The materials tested in this study were AVCO silicon carbide/aluminum-6061F and AVCO boron/aluminum-6061F manufactured by DWA Composite Specialties, and a hybrid AVCO silicon carbide/aluminum - 6061F - titanium Ti-6Al-4V manufactured by AVCO Specialty Materials. For the silicon carbide reinforced composites the orientations were unidirectional. For the boron/aluminum, six different lay-ups, namely, $[0]_8$, $[90]_8$, $[0/\pm45]_{2s}$, $[0/\pm45/90]_s$, and $[0/\pm45/0]_s$ were tested. Unnotched tensile specimens having dimensions of 200 mm x 12.7 mm (8.0 x 0.5 inch) were cut from plates having initial dimensions of 200 x 200 x 1.3 mm (8.0 x 8.0 x 0.052 inch). Aluminum end-tabs 25 x 1.3 mm (1.0 x 12.0 x 0.05 inch) were adhesive bonded to the specimens in the grip sections.

MECHANICAL PROPERTY EVALUATIONS

All specimens were tested on a closed loop servo-hydraulic Instron testing machine (Model 1331). Quasi-static tensile tests were performed under stroke control mode at a rate of approximately 0.05 mm/min (0.002 in/min). In order to ensure axial loading, specimens were strain-gaged in such a manner that two axial strain gages were located across the width and at the same distance from the grips. Identical outputs of these two gages with an increasing applied load confirmed a uniform axial loading.

Testing procedures for determination of laminate elastic constants and strength as a function of temperature are described as follows. In order to ensure that elastic limits were not exceeded while determining the elastic properties, the specimens were loaded to not more than 10% of their average tensile strength and then unloaded, yielding load-strain plots. The load-displacement plots were also obtained simultaneously by means of a compliance gage with a 25.4 mm (1.0 in) gage length. The loading-unloading cycle was repeated at least three times to ensure the reproducibility of the data, and to check that the elastic limits were not exceeded. Afterwards, the specimens were loaded to failure and ultimate unnotched strength was obtained. In the case of elevated temperature testing, the specimens were first loaded at room temperature to less than 10% of their expected ultimate strength, unloaded, the temperature was raised to a predetermined value, and the loading-unloading cycle was repeated at this temperature, assuring that the maximum loads did not exceed the elastic limits of the subject laminate. This process was repeated and the temperature was successively increased incrementally in 71°C increments to 371°C. The specimens were then loaded to failure at the last testing temperature. By this process the change in stiffness with temperature was obtained, and the scatter arising due to using different specimens of the same lay-up was reduced. Loading and unloading did not cause any measurable residual strains upon unloading, thus ensuring once again that elastic limits were not exceeded.

MECHANICAL PROPERTY RESULTS

The effect of temperature on the axial stiffness of the composites evaluated is shown in Figure 1. In Figure 1a, the axial stiffness of the SiC fiber reinforced aluminum and the boron fiber reinforced aluminum do not show significant changes as a function of temperature. The hybrid SiC reinforced aluminum 6061F-titanium (Ti-6Al-4V) composite showed a trend toward decreasing stiffness above 200°C. However, the data points are within the scatter of measurement.

In Figure 1b, the change in axial stiffness with temperature as a function of fiber orientation is shown for boron fiber reinforced aluminum-6061F laminates. The results indicate that at all temperatures, the axial stiffnesses of the $[0]_8$ and $[90]_8$ laminates form the upper and lower bounds, respectively. The axial stiffnesses of all other lay-ups fall between these bounds. While the scatter in the data make conclusions difficult, it is generally seen that there was little decrease in axial stiffness up to 200°C for any of the laminates. Those laminates containing off-axis reinforcement exhibited a trend in decreasing stiffness above about 200°C. This is attributed to the behavior of the 6061 aluminum matrix which has significant degradation in stiffness reported at that temperature [12]. Significant reduction in the stiffness of boron fibers is anticipated only above 400°C [13,14].

The elevated temperature strength response of aluminum, titanium and hybrid aluminum/titanium metal matrix composites serving as candidate materials for high temperature performance are depicted in Figure 2. In terms of absolute

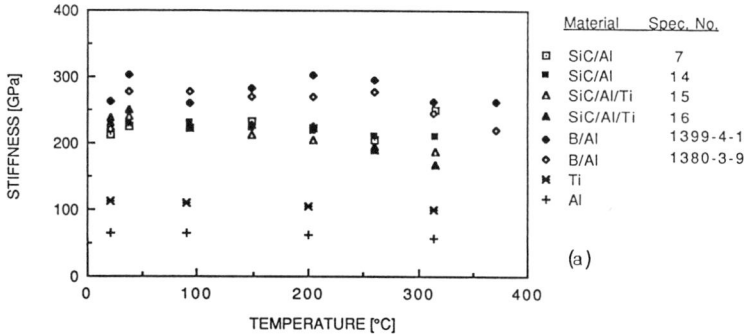

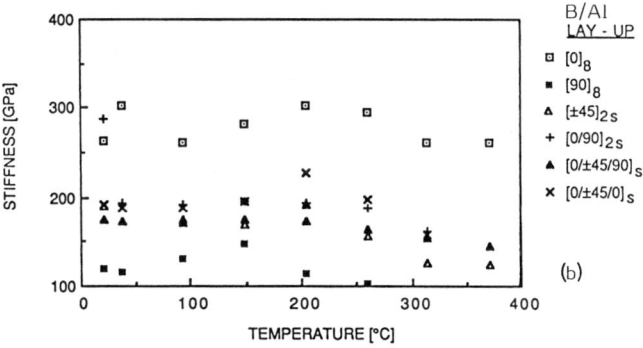

Figure 1. STIFFNESS VERSUS TEMPERATURE FOR SCS-6 AVCO SiC(V_f= 0.36)/ 6061 ALUMINUM; SCS-6 AVCO SiC (V_f= 0.50) / 6061 ALUMINUM (Y. Minoda et al., 16); AVCO BORON (V_f= 0.50); & HYBRID SCS-6 AVCO SiC (V_f=0.29) /6061- ALUMINUM (V_f= 0.43) / Ti-6Al-4V (V_f= 0.28)

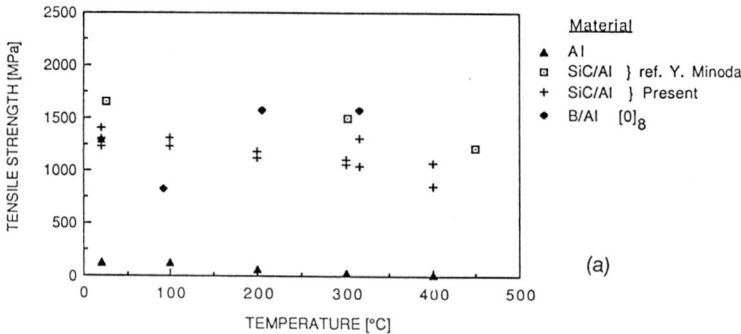

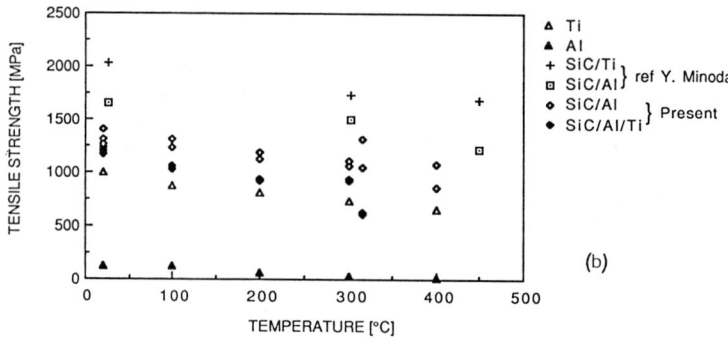

Figure 2. STRENGTH VERSUS TEMPERATURE FOR SCS-6 AVCO SiC(V_f= 0.36)/ 6061 ALUMINUM; SCS-6 AVCO SiC (V_f= 0.50) / 6061 ALUMINUM (Y. Minoda et al., 16); AVCO BORON (V_f= 0.50)/6061 ALUMINUM; HYBRID SCS-6 AVCO SiC (V_f=0.29) / 6061- ALUMINUM (V_f= 0.43) / Ti-6Al-4V (V_f= 0.28); & SCS-6 AVCO SiC (V_f= 0.37) / Ti-6Al-4V (Y. Minoda et al., 16)

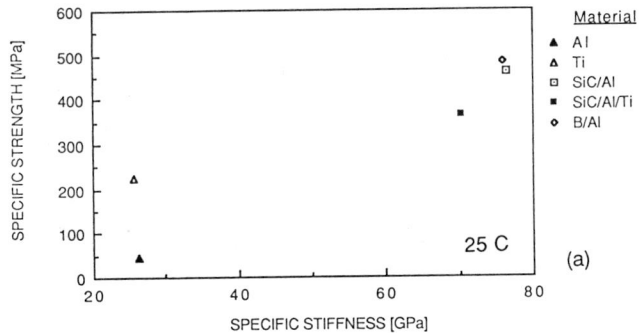

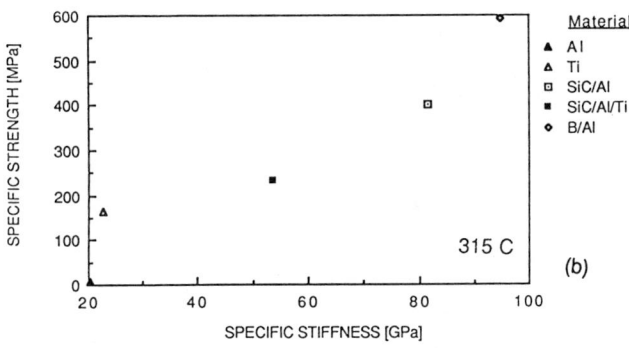

Figure 3. SPECIFIC STRENGTH VERSUS SPECIFIC STIFFNESS AT 25 C & 315 C FOR SCS-6 AVCO SiC(V_f= 0.36)/ 6061 ALUMINUM; AVCO BORON (V_f= 0.50)/6061 ALUMINUM; & HYBRID SCS-6 AVCO SiC (V_f=0.29) /6061- ALUMINUM (V_f= 0.43) / Ti-6Al-4V (V_f= 0.28)

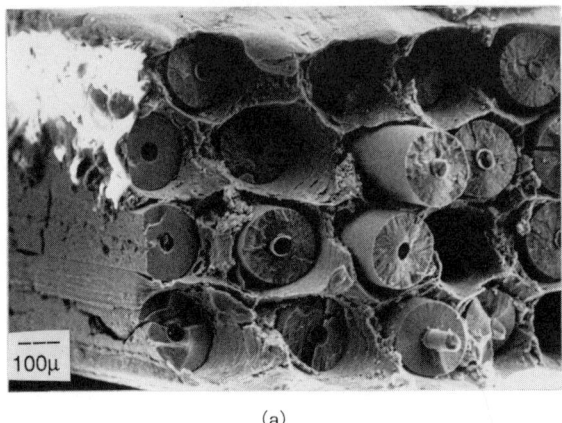

(a)

(b)

FIGURE 4. FRACTURE MORPHOLOGY AT 300°C FOR (a) SCS-6 AVCO SiC (V_f = 0.36)/6061 ALUMINUM AND (b) HYBRID SCS-6 AVCO SiC (V_f = 0.29)/ 6061 ALUMINUM (V_f = 0.43)/Ti-6Al-4V (V_f = 0.28)

strength the titanium reinforced composites with SiC reinforcement exhibited the highest strength values at temperatures to 300°C. Boron and silicon carbide reinforced aluminum while somewhat lower in absolute strength are comparable in terms of specific strength at equivalent volume fractions of reinforcements. The general trend of decreasing composite strength with increased temperature parallels the loss in tensile strength of the matrix alloys. The specific strength versus specific stiffness at 25 and 315°C are depicted in Figures 3a and 3b, respectively. At ambient temperatures, boron and silicon carbide reinforced systems are comparable at similar volume fractions of reinforcements, with the hybrid reinforced composite providing good specific properties as well as improved transverse properties. At 315°C, the boron and silicon carbide reinforced aluminum exhibited significantly higher specific properties than the hybrid reinforced composite.

The fracture morphology of silicon carbide/aluminum-6061F and a hybrid silicon carbide/aluminum-6061F-titanium Ti-6Al-4V were also determined, respectively. In general the fracture morphology at ambient and elevated temperatures revealed extensive fiber pull-out indicative of a weak interfacial bond and good fiber strengths. Typical fracture surfaces at a test temperature of 300°C are shown in Figure 4.

ACKNOWLEDGEMENT

The authors gratefully acknowledge program support from the Office of Naval Research as well as the Strategic Defense Initiative Office.

REFERENCES

1. J. W. Mar and K.Y Lin, , J. of Composite Materials, 11, 405-421 (1977).
2. R.B. Pipes, R.C. Wetherhold, and J.W. Gillespie, Jr., J. of Composite Materials, 13, 148-160, (1979).
3. J.H. Steele and H.W. Herring, Failure Modes in Composites, I.J. Toth, Ed., The Metallurgical Society of AIME, New York, New York, 1, 343-356, (1972).
4. W.S. Johnson, C.A. Bigelow, and Y.A. Bakei-El-Din, NASA TP-2187, (1983).
5. K.G. Kreider, L. Dardi and K. Prewo, AFML-TR-71-204, (1971).
6. C.C. Poe and J.A. Sova, NASA TP-1707, (1980).
7. J. Awerbuch and H.T. Hahn, J. of Composite Materials, 13, 82-107, (1979).
8. D.D. Daily, Prediction of Fracture Toughness for Specially Orthotropic Composite Laminates, M.Sc Dissertation, School of Engineering, Air Force Institute of Technology (AFIT), (1974).
9. M.A. Wright, D. Welch, and J. Jollay, Proceedings of First USA-USSR Sympsium on Fracture of Composite Materials, G.C. Sih and V.P. Tamuzs Eds., Riga, USSR, 4-7, (1978), Sijthoff & Noordhoft, 221-238, (1979).
10. J. Awerbuch, Madhu S. Madhukar, J. of Reinforced Plastics and Composites", 4, 3-159, (1985).
11. M.A. Wright and B.D. Intwala, J. of Materials Science, 8, 953-963, (1973).
12. L.F. Mondolfo, Aluminum Alloys: Structures and Properties, Butterworths Publishers Inc., Boston, Mass. (1979).
13. V.S. Erasov, E.N. Pirogov, V.P. Konoplenko, V. A. Akimkin, A.P. Marukhin, A.M. Tsirlin, E.A. Shchetilina, and N.M. Balagurova, Mechanics of Composite Materials, 18, No.2, 127-130, (1982).
14. J.A. DiCarlo, Composite Materials: Testing and Design (Fourth Conference), ASTM STP 617, American Society of Testing and Materials, 443-465, (1977).

THERMAL, VISCOPLASTIC ANALYSIS OF COMPOSITE LAMINATES

E. KREMPL and K.D. LEE
Rensselaer Polytechnic Institute, Mechanics of Materials Laboratory, Troy,
NY 12180-3590

ABSTRACT

For the modeling of ply deformation behavior the orthotropic, thermal viscoplasticity theory based on overstress is used. It can represent creep, relaxation and rate sensitivity as well as monotonic and cyclic loadings. The theory is "unified" since creep and plasticity are not separately modeled. No yield surfaces and loading/unloading conditions are employed. The laminate theory for in-plane loading maintains the geometric assumptions of classical laminate theory. The elasticity law, however, is replaced by the thermal, orthotropic viscoplasticity law. Numerical experiments illustrate the predictions of the theory for an angle-ply and a cross-ply laminate subjected to a temperature increase, temperature hold and subsequent return to the original temperature. The ply and laminate stresses are calculated as a function of time for unconstrained and constrained conditions using postulated properties close to a real metal matrix composite. Redistribution of ply stresses and relaxation are found. In some cases, nearly permanent residual ply stresses are present after completion of the temperature cycle.

INTRODUCTION

Metal matrix composites are increasingly used in primary structures which are subjected to severe conditions of loading and environment. Included are variable temperature services such as occur in a satellite in orbit or during flight of the space plane. Other examples are components in propulsion systems, jet engines and rockets. To ensure safe operation, stress and life-time analyses must be performed long before the part is built.

The high degree of anisotropy present in composite structures requires the development of new analysis techniques which account for the variation of the material properties with direction. When metal matrix composites are used, inelastic deformation cannot be ruled out even if the service is at low homologous temperature and if the overall loads are within the nominal elastic limit [1]. For high homologous temperature service, inelasticity is found in the form of time-dependent deformation, even in monolythic materials.

Traditionally high temperature stress analysis is performed by combining elasticity with time-independent plasticity and creep theories. For each element, a separate constitutive equation is postulated. Except for initial conditions, creep and plasticity are treated as separate phenomena with no interaction between them.

With this approach, the exact identification in experiments of creep and plastic strains is problematic, see [2]. Further, material science shows that dislocations and other changes in the defect structure are responsible for inelastic deformation which is considered to be time dependent. As a consequence, several new constitutive equations have been proposed during the last two decades which do not separate creep and plastic deformation. They are called "unified" constitutive equations; a recent review of some of these is given in [3].

The viscoplasticity theory based on overstress (VBO) is one of the unified theories. It was developed in response to the observed time-dependence of engineering alloys at ambient temperature [4-7]. Subsequently, an orthotropic version of the theory was formulated [8,9] and applied to the modeling of the in-plane deformation of metal matrix composite laminates [10].

The purpose of this paper is to introduce and to apply a thermal version of the orthotropic VBO to metal matrix composite laminates subjected to thermal and mechanical loadings. Numerical examples are given for constrained and unconstrained, angle-ply and cross-ply laminates subjected to a temperature increase followed by a temperature hold and subsequent return to the original temperature. Laminate and ply stress components are calculated as a function of time in a simple theory which is patterned after the classical laminate theory (CLT), see [11]. The geometric assumptions of CLT are maintained such as constant strain through the laminate and satisfaction of the stress boundary condition for the whole laminate only. The linear orthotropic elasticity law of CLT is, however, replaced by the thermal, orthotropic viscoplasticity theory based on overstress. It is now possible to model hysteresis, creep, relaxation and rate sensitivity as well as time dependent stress redistributions between plies due to thermal and/or mechanical loadings. The present theory assumes the ply to be an orthotropic continuum with its properties represented by the thermal VBO. No interactions between fiber and matrix are modeled. However, this can be done without any difficulty and will be pursued in the future.

PLY CONSTITUTIVE EQUATIONS

An orthotropic version of the viscoplasticity theory based on overstress (VBO) is used. In this theory, the total small strain rate is the sum of the elastic, inelastic and thermal strain rates. For the elastic strain rates, the rate form of Hooke's law in orthotropic form is employed , i.e. the time derivative of the product of the orthotropic compliance with the stress. The inelastic strain rate is only a function of overstress, which is the difference between the current stress and the equilibrium stress, the state variable of the theory. The equilibrium stress is the stress which can be indefinitely sustained after deformation when all rates have returned to zero. Initially, the equilibrium stress is zero but evolves with deformation according to a separately postulated orthotropic growth law. It is responsible for modeling almost linear elastic regions and hysteresis. In the present theory, no recovery terms are included. An extension of the theory to recovery is under development. The thermal strain rate is the time derivative of the product of the orthotropic coefficient of thermal expansion with the temperature difference reckoned from a reference temperature. All material properties of the theory can be functions of temperature and must be determined from suitable experiments on a ply. This includes tests in the fiber and transverse directions. Rate change tests are essential in determining the viscous (time-dependent) properties of the theory.

The VBO assumes that inelastic deformation is basically rate dependent. Rate dependence is always present and can change with temperature. Normally, it increases with rising temperature. This property can be modeled by making certain constants in the repository for rate dependence a function of temperature. The normally encountered decrease in the flow stress with increasing temperature is also modeled easily. In fact, since no trend of the temperature dependence is presumed by the theory, even anomalous trends such as an increase in strength with increasing temperature can be represented by this VBO.

The theory does not use a yield surface and loading/unloading conditions. Inelastic strain rates are always present but are extremely small in the elastic regions. On a stress-strain graph, the linear elastic region predicted by the theory can be a perfect straight line. This is accomplished by the growth law for the equilibrium stress.

The equations of the theory need detailed explanations which cannot be included in this paper because of space limitations. The theory is presented in [12].

IN-PLANE LAMINATE BEHAVIOR

The orthotropic VBO described above is now specialized for the case of plane stress and used as a constitutive equation for a particular ply in a simple theory of in-plane laminate behavior. The theory retains all the geometric assumptions of CLT, see [11]. In this paper, the orthotropic, linear elasticity law used in CLT is replaced by the thermal, orthotropic VBO. As a consequence, rate dependence, creep, relaxation and hysteresis can be modeled, in addition to the effects of changing temperature. The theory includes the modeling of stress redistributions between plies during deformation.

This theory has been developed for angle-ply laminates and a computer program has been written for the numerical integration of the resulting simultaneous nonlinear, ordinary differential equations, see [12]. Once the material constants and functions of the theory are known, the program can be used for any thermal and/or mechanical history imposed on the laminate. Included are uniform temperature changes for a constrained or an unconstrained laminate, as well as simultaneous thermal and mechanical loadings.

For the sake of brevity, these equations are not given here. They can be found in [13].

NUMERICAL SIMULATION OF LAMINATE BEHAVIOR UNDER A TEMPERATURE CHANGE

To illustrate some aspects of the capability of the theory without listing the governing equations, the following procedure is adopted. Hypothetical but realistic material properties are assumed in the thermal, orthotropic VBO. These properties result in a certain stress-strain behavior in the fiber and in the transverse directions. These diagrams are taken to be an indication of the material properties of each ply. Then a $[+45/-45]_s$ angle-ply and a $[0/90]_s$ cross-ply laminate are "built theoretically" and their responses to a temperature history with and without mechanical constraints are computed. The uniform temperature excursion (every part of the laminate sees the same temperature) is a $200^{\circ}C$ increase followed by a temperature hold and a subsequent decrease to the reference temperature as depicted in Fig.1a. In the first case, the laminates are free to expand and only thermal stresses between the plies develop due to the differences in the orientation and the coefficients of thermal expansion. The laminate boundaries are stress free. In the second case, the laminates are constrained in the one-direction but are free to expand in the two-direction, see Fig.1b. Thermal stresses are now due to constraint and due to the differences in orientation of the plies. This exercise is to demonstrate the capabilities of the theory under small temperature changes. Of course, larger temperature changes can be simulated, as long as the material data are known. Examples are the computation of the residual stresses that may develop during manufacturing when the laminate cools down from the working temperature.

The stress-strain diagrams of a ply in the fiber and the transverse directions at the reference temperature and at the maximum temperature of the cycle are shown in Figs. 2a and 2b, respectively. The difference in the strengths of the matrix and the fibers is obvious as is the increased rate sensitivity of the matrix as compared to the fibers. Overall, the rate sensitivity modeled by this hypothetical material is not very pronounced. Other stress-strain relations and rate sensitivities can be modeled easily by adjusting the material constants, see [10].

For the simulation of the temperature cycles imposed on the laminate, it is assumed that the positive coefficient of thermal expansion in the fiber direction is about one sixtieth of that of the transverse direction. Such relations are found in some metal matrix plies.

Case 1: Unconstrained Laminates

In this case, no external stresses act on the laminate. Owing to the assumptions of CLT, stresses can act at the boundary of individual lamina as long as their sum is zero.

Fig. 3 shows the computed results for both the angle-ply and the cross-ply laminates. Owing to the small temperature change, the stresses are modest and are within the elastic behavior of the fibers, see Fig. 2a. Due to symmetry, only ply shear stresses exist in the $[+45/-45]_s$ laminate. The stresses in the two plies have opposite signs. In the cross-ply laminate, no shear stresses are found and the ply stresses in the one- and the two-directions are equal. Equilibrium requires that the stresses in the zero and ninety degree ply add up to zero. For these reasons, only one curve is shown in Fig. 3. It is seen that relaxation occurs during the temperature hold and that residual stresses exist when the temperature returns to its original value. They decrease slightly in magnitude with time.

Case 2: Constrained Laminates

Due to the constraint, see Fig. 1b, laminate stresses exist in the one-direction. In the two-direction, the laminate stresses are zero.

The results for the angle-ply laminate are depicted in Fig. 4. Compared to Fig. 3, compressive stresses in the one-direction develop upon heating which are almost zero when the temperature excursion is finished. Due to symmetry, the ply stresses in the one-direction are equal to each other and equal to the laminate stress. Shear ply stresses develop also in the constrained case and their magnitude is smaller than in Fig. 3 due to the presence of the normal stresses and the nonlinearity of the theory. At first glance, it is surprising that the shear stress magnitude is higher than the stress in the one-direction due to constraint. This outcome is largely due to the coefficients of thermal expansion chosen in this case. As mentioned above, the coefficient of thermal expansion is sixty times higher in the two-direction than in the one-direction.

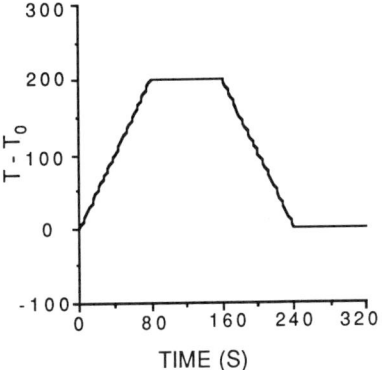

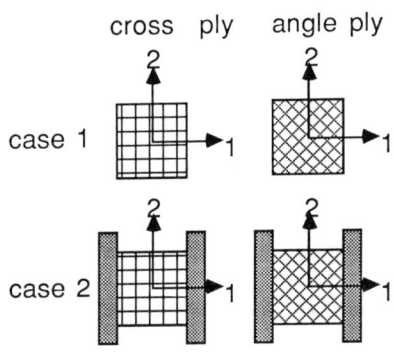

Fig. 1a. Temperature history imposed uniformly on the laminates.

Fig. 1b. Schematic showing the loading conditions. In Case 1, the laminates are free to expand but are fixed in the one-direction in Case 2. A $[0/90]_s$ cross-ply and a $[+45/-45]_s$ angly-ply are considered.

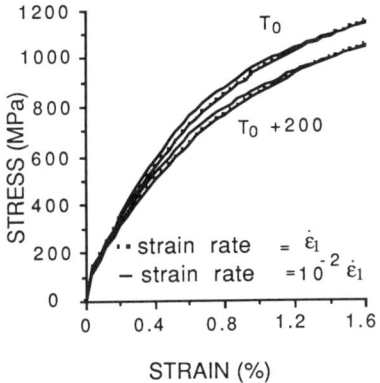

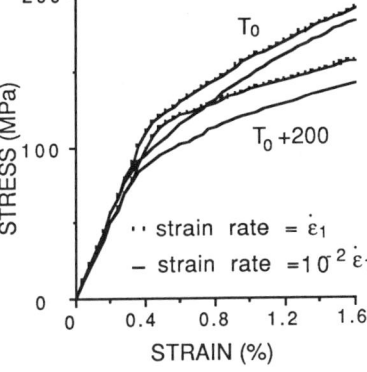

Fig. 2a. Stress-strain diagrams in the fiber direction at the reference temperature T_o and at $T_o + 200°C$ for two strain rates differing by two orders of magnitude. $\dot{\varepsilon}_1 = 4 \times 10^4$ 1/s.

Fig. 2b. Same as Fig. 2a except that the direction of straining is perpendicular to the fibers. In essence, these stress-strain diagrams represent matrix behavior.

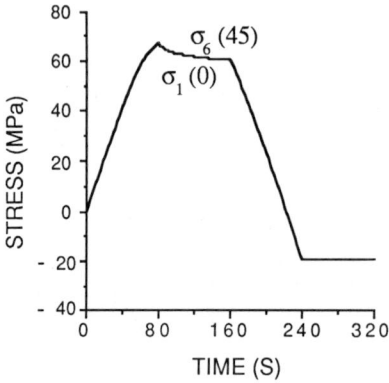
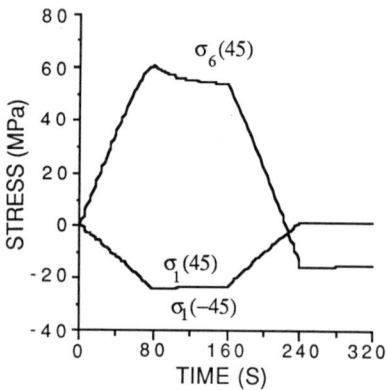

Fig. 3. Stresses in the laminates which are free to expand, see Case 1 in Fig. 1b. In the $[+45/-45]_s$ laminate only the shear stresses σ_6 exist and $\sigma_6[45] = -\sigma_6[-45]$. All other stresses are zero. For the cross-ply laminate, $\sigma_1[0] = \sigma_2[90] = -\sigma_1[90] = -\sigma_2[0]$. The variation of $\sigma_1[0]$ in the cross-ply is equal to the variation of the shear stress $\sigma_6[45]$ in the angle-ply due to symmetry.

Fig. 4. Stresses in the constrained angle-ply laminate, Case 2 in Fig. 1b. In addition to the self equilibrating shear stress $\{\sigma_6[45] = -\sigma_6[-45]\}$, stresses in the one-direction are induced. The ply stresses are equal to the laminate stress in the one-direction. All other stress components are zero.

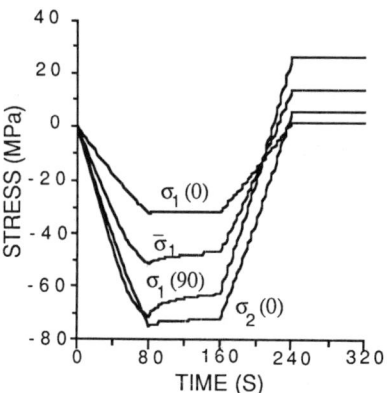

Fig. 5. Stresses in the constrained cross-ply laminate, Case 2 in Fig. 1b. The stresses in the one-direction are shown ($\bar{\sigma}_1$ is the laminate stress). Self equilibrating stresses in the two-direction are induced, $\sigma_2[0] = -\sigma_2[90]$. All other stress components are zero.

The variation of the stress components with time is plotted in Fig. 5 for the cross-ply laminate. The shapes of the curves are very similar to each other and to those of Figs. 3 and 4. It is clearly seen that the laminate stress in the one-direction is the sum of the corresponding ply stresses. Due to the differences in the coefficient of thermal expansion, the stress in the one-direction of the 90-degree ply is higher than the stress in the one-direction of the zero-degree ply. For the same reason, the stress in the one-direction of the 90-degree ply is a little higher than the stress in the two-direction of the zero-degree ply in the elastic region. When inelasticity sets in, this trend is reversed. After the temperature excursion is over, tensile stresses are present in the laminate which, on the graph, do not appear to relax with time. A check of the numerical data, however, reveals a slight decrease in time.

DISCUSSION

The above examples have shown that the theory can model simple cases of thermal stresses in laminates. This includes relaxation and the development of residual stresses. The theory predicts smooth variation of the stresses with temperature history. (The ragged appearance of some of the curves is due to the PC graphics package employed in making the figures.)

This paper uses fictitious material properties to illustrate, in principle, the capability of the theory on some simple examples. For a practical application, the material functions and constants of the theory must be determined by suitable experiments as a function of temperature, see [4,5]. In addition, off-angle tests are necessary, see [10], where some of the pertinent literature is cited. The theory has many flexibilities, such as almost linear elastic behavior in the fiber direction but viscoplastic behavior transverse to it. However, the major question is, what are the minimum number of constants and functions necessary to model a given behavior? This aspect has been considered in [9,10]. There, a "minimal" theory for isothermal deformation is shown which can reproduce the behavior of Borsic/Al metal matrix composites. Similar studies must be performed with the present thermal VBO and additional experience must be gained through further theoretical and experimental research.

The geometric limitations of CLT carry over to the present theory. From Figs. 3-5, it is seen that the ply stresses at free edges are not always zero, only the laminate stresses must vanish there. These ply stresses can be rather large. In Case 2, they are higher than the stresses caused by the constraint. When different sets of thermal expansion coefficients are used, this trend can be altered and the constraint stress magnitude can become the largest. This has been verified by separate computations.

The small temperature excursion and the small coefficient of thermal expansion in the fiber direction are responsible for the small thermal stresses found in the laminate. They are within the elastic region of the stress-strain diagram in the fiber direction. The redistribution of the stresses with time is thought to be due to the chosen "soft" matrix properties. By selecting different "viscous" properties in the material model, the redistributions can be enhanced or retarded. These properties will have to be explored by future numerical experiments.

ACKNOWLEDGEMENT

This research was supported by DARPA/ONR Contract N 00014-86-K0700 to Rensselaer Polytechnic Institute.

REFERENCES

1. G. J. Dvorak, in <u>Mechanics of Composite Materials: Recent Advances</u> (Proc. IUTAM Symposium on Mechanics of Composite Materials, Pergamon Press, Inc., New York, NY, 1982).

2. E. Krempl, Welding Research Council Bulletin No. 195 (Welding Research Council, New York, NY, 1974).

3. A. K. Miller, editor, <u>Unified Constitutive Equations for Creep and Plasticity</u> (Elsevier Applied Science, London and New York, 1987).

4. D. Kujawski and E. Krempl, J. Appl. Mech. $\underline{48}$, 55 (1981).

5. D. Kujawski, V. Kallianpur and E. Krempl, J. Mech. Phys. Solids $\underline{28}$, 129 (1980).

6. D. Yao and E. Krempl, Int. J. of Plasticity $\underline{1}$, 259 (1985).

7. E. Krempl, J. J. McMahon and D. Yao, Mech. of Materials $\underline{5}$, 35 (1986).

8. M. Sutcu, PhD Thesis (Rensselaer Polytechnic Institute, Troy, NY, December 1985).

9. M. Sutcu and E. Krempl, Rensselaer Polytechnic Institute Report MML87-8, September 1987, submitted for publication.

10. E. Krempl and B.-Z. Hong, Rensselaer Polytechnic Institute Report MML87-9, September 1987, submitted for publication.

11. S. W. Tsai and H. T. Hahn, <u>Introduction to Composite Materials</u> (Technomic Publishing Co., Westport, CT, 1980).

12. K. D. Lee and E. Krempl, Rensselaer Polytechnic Institute Report MML 88-1, April 1988.

13. K. D. Lee and E. Krempl, Rensselaer Polytechnic Institute Report MML 88-2, May 1988.

ENHANCED PLASTICITY OF MECHANICALLY ALLOYED ALUMINUM IN90211

T.R. Bieler*, T.G. Nieh**, J. Wadsworth**, and A.K. Mukherjee*
*Department of Mechanical Engineering, Division of Materials Science
University of California, Davis CA 95616
**Lockheed Missiles and Space Co. Inc., Research and Development
Palo Alto, CA 94304

ABSTRACT

The tensile behavior of IN90211 was characterized at strain rates between 0.0001/sec and 340/sec at temperatures between 425 and 475 °C. At strain rates below 0.1/sec, the strain rate sensitivity m is about 0.027, with corresponding low elongation (<100%). At strain rates above 0.1/sec, the strain rate sensitivity increases to 0.26. A maximum elongation of 500% was obtained at 475 °C at a strain rate of 2.5/sec. Grain boundary sliding and rotation was observed on the highly elongated specimens and fracture surfaces exhibited intergranular fracture. Experimental data in the high strain rate regime (superplastic) revealed the existence of a temperature dependent threshold stress that seemed unrelated to the low stress deformation regime. This result is consistent with stress relaxation experiments. These threshold stresses are generally lower than those typically observed in other oxide dispersion strengthened (ODS) alloys. This observation is not expected from conventional superplastic creep theory.

INTRODUCTION

One of the drawbacks of superplastic forming of commercial alloys is the relatively slow strain rates at which optimum superplastic elongations are found. In general, superplastic strain rate varies inversely with grain size, the principal microstructural feature governing superplastic flow. Recently, superplastic-like deformation of aluminum alloy-ceramic composites (Al 2124/SiC$_w$, IN9021 and IN90211) has been demonstrated at unusually high strain rates resulting from their small grain size [1-3]. This type of high strain rate phenomenon has also been observed by Gregory, et al., in a study of the mechanically alloyed, nickel-based superalloy IN 6000 [4], where 300% elongation occurred at a strain rate of 0.3/sec. The transition in strain behavior (associated with the dramatic change in the strain rate sensitivity) of IN90211 will be described in the context of threshold stress theories commonly used to describe creep in oxide dispersion strengthened (ODS) alloys.

EXPERIMENTAL PROCEDURES

Specimens of IN90211 were produced by Novamet Aluminum. Their composition is shown in Table I. (IN9021 or AL9021 as currently produced by Incomap.) The alloy was manufactured by mechanical alloying and contains about 7 vol% oxide and carbide particles approximately 30 nm in diameter with an inter-particle spacing of about 60 nm. As the alloy is similar to 2124, there are also Al$_2$Cu (θ phase) precipitates in the matrix, depending upon the exact thermomechanical processing history. The material had been subjected to a complex thermomechanical processing procedure to reduce extruded plate of 127 mm thickness to a sheet 2.5 mm thick. The plate was annealed and water quenched after the rolling. Specimens with a 2.5 mm square cross section and a gage length of 6.7 mm were machined from sheets with the tensile axis parallel to the rolling direction. Texture was evident in X-ray diffraction of undeformed specimens. Details of tensile tests are described in reference 5.

TABLE I: COMPOSITION OF IN90211

IN90211:		Mg	Cu	C	O	Al_2Cu	Al_4C_3	Al_2O_3	Al
	wt%	2.0	4.4	1.1	0.8				91.7
*	wt%	2.0	0.9			6.5	4.4	1.7	84.4
	mol%	2.2	1.9	2.5	1.4				92.1
*	mol%	2.2	0.4			4.5	5.8	2.3	84.8
Density						4.36	2.95	3.99	2.70
Vol % *						10.2	4.7	2.5	82.6

* Assuming 80% of Cu in Al_2Cu, O and C in Al_4C_3, Al_2O_3
17 vol% hard particles (including Al_2Cu) at room temperature
7 vol% oxides and carbides at elevated temperature

RESULTS AND DISCUSSION

The elongation-to-failure for IN90211 alloy, as a function of strain rate, is shown in Figure 1a for temperatures between 425 and 475°C. At strain rates between 0.0001 to 0.01/s (where most aerospace aluminum alloys exhibit superplasticity), IN90211 shows modest elongation of 30-40%. A maximum elongation-to-failure of 505% was recorded at a test temperature of 475°C and a strain rate of 2.5/s. The peak true stress is plotted as a function of true strain rate in Figure 1b. Each of these regimes has a particular value of stress dependence upon strain rate. For superplastic studies it is convenient to use the simple equation $\sigma = k\dot{\varepsilon}^m$ where σ is the true flow stress at a true strain rate, $\dot{\varepsilon}$, k is a constant, and m is the strain rate sensitivity exponent. At strain rates between 0.0001 and 0.1/sec, m ≈ 0.027, but at strain rates above 0.1/s m ≈ 0.26. The exponent changes to m ≈ 0.15 at strain rates above 50/s.

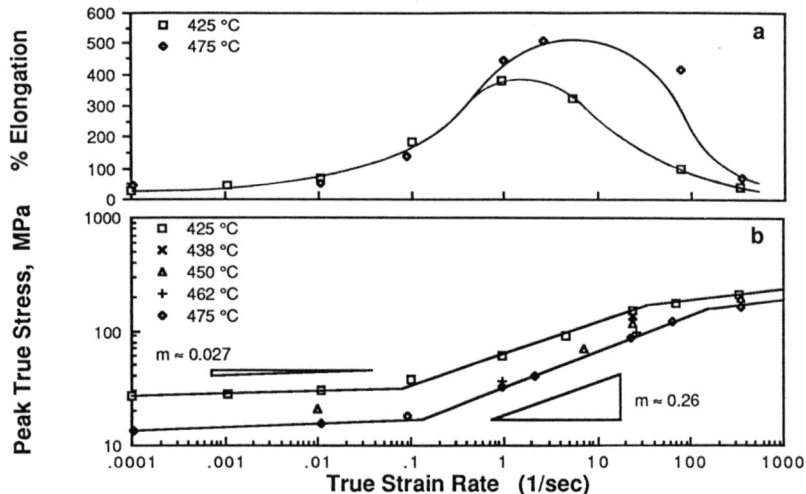

Figure 1 Elongations-to-failure (a) and peak true stress (b) vs. strain rate

Threshold stresses are commonly invoked in the analysis of stress-strain behavior of ODS materials, especially at low stress and strain rates [5-10]. Theoretically the strain rate can be expressed in terms of an applied stress σ, and a threshold stress σ :

$$\frac{\dot{\varepsilon}kT}{D_l Eb} = A\left(\frac{\sigma-\sigma_0}{E}\right)^n, \tag{1}$$

where $\dot{\varepsilon}$ is strain rate, kT are Boltzman's constant times the absolute temperature, D_l is the temperature dependant lattice bulk diffusivity (145 kJ/mol), E is temperature dependant Young's modulus, b is the Burger's vector, A is a mechanism dependant constant that includes the grain size, and n is the stress exponent (where n=1/m). A special effort was made to determine whether a threshold stress is operative by conducting a stress relaxation experiment at 450°C. After a few seconds, an apparent steady state stress of 9 ± 1 MPa was reached and it remained constant for several hours.

The apparent threshold stress can also be estimated from the data of Figure 1b using the approach described by Mohamed [5], where a threshold stress is extrapolated from a plot of the cube root (i.e. n=3.0) of strain rate versus stress, as shown in Figure 2b, using only the data from the high strain rate regime of Figure 1b (the low strain rate data fell off the line). These data were linear, permitting threshold stress values of 5 and 16 MPa for 475 and 425°C data, respectively. Using the measured stress exponents of 3.94 at 425 and 3.71 at 475 °C, the data were not linear. These extrapolated values are consistent with the 9 MPa value obtained from the stress relaxation measurement at 450°C. The introduction of this threshold stress (to replace the flow stress with an effective stress) in the normalized plot of Figure 3 does not appear to affect the abrupt change in the stress exponent, contrary to the results of others [5,8]. A threshold stress that does affect the low stress regime is illustrated in [11], where the stress exponent of the applied stress varies smoothly between the high and low stress regimes. Thus, the threshold stress obtained above in IN90211 is not associated with the change in slope between the low and high stress regimes. This is corroborated by the lack of a threshold stress using the same method for the low strain rate data in Figure 2a. Apparently no threshold stress exists for the low stress data, but a threshold stress exists for the high stress data.

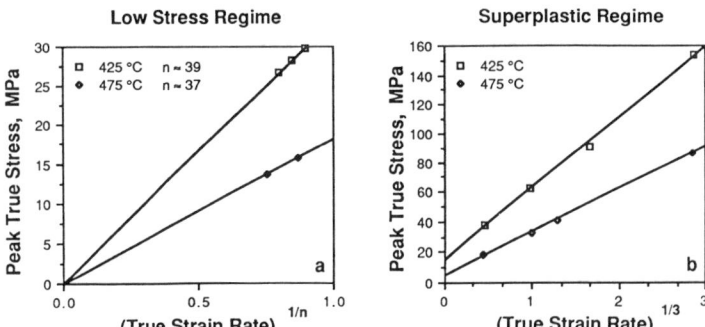

Figure 2 (Peak true stress) versus (true strain rate)$^{1/n}$ plots (note n=1/m).
a) There is no apparent threshold stress in the low stress regime.
b) 5 and 16 MPa threshold stresses for 475 and 425 °C, in the superplastic regime.

Threshold stresses in other ODS systems are typically about half of the Orowan stress value [7,9]. These threshold stresses are thought to be related to localized climb [6] around particles and/or departure side pinning of dislocations [8,9]. Using 30 nm particles 60 nm apart, an Orowan stress of about 200 MPa can be estimated, which is 10-40 times the measured and calculated threshold stresses [12]. The smaller measured threshold stress is thus related to other mechanisms. The stress sensitivity change at high stresses occurs at a stress near 200 MPa. The change in deformation mechanism indicated by the highest rate data suggests a stress exponent around 7, possibly resulting from dislocation generation and motion by Orowan looping [13]. The change in behavior may also be due to power-law breakdown. More data are needed in this third regime to clearly understand the deformation mechanism.

The apparent activation energy of the two regimes was determined using constant strain rate data (including temperature dependant shear modulus corrections) and the appropriate n value. In the superplastic regime, the activation energy obtained using the peak stress and the effective stress (peak stress minus threshold stress) data are 125 ± 25 and 160 ± 40 kJ/mol, respectively. The activation energy for bulk diffusion in pure aluminum (145 kJ/mol) is within the range of the experimental superplastic data. The data in the high strain rate regime superposed using the bulk diffusivity but not with the grain boundary diffusivity value of 84 kJ/mol. Neither the threshold stress, modulus temperature correction nor the bulk diffusivity compensated for the temperature dependance in the low strain rate data, as these data did not superpose. Since there is no apparent threshold stress in the low stress regime, the experimentally measured (peak) stress was used to determine the apparent activation energy. An extremely high value of 2000 ± 50 kJ/mol was obtained. Similar work on SiC whisker reinforced 2124 aluminum indicates an apparent activation energy of about 900 kJ/mol in the high stress exponent deformation regime [14]. It is noted that the grain size and the composition of the $2124/SiC_w$ alloy and IN90211 are similar.

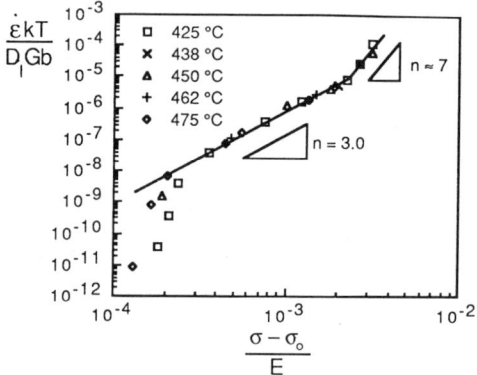

Figure 3 Normalized stress as a function of normalized strain rate.

Microscopy was reviewed in detail in [3], and is summarized as follows: From transmission electron microscopy the grain size was determined to be 0.5 ± 0.3 μm. The grain size and microstructure of samples after deformation appeared to be similar to the undeformed microstructure, indicating high stability at elevated temperatures. Second phases were visible, including 30 nm oxides and carbides, and precipitates of θ and S phases (Al_2Cu and Al_2CuMg), notably on grain boundaries. Grain boundary sliding and rotation was observed in the deformed specimens by scanning electron microscopy. Fracture surfaces from superplastically deformed IN90211 specimens exhibited an intergranular fracture mode. Granular features on the fracture surface were similar to the grain size. observed by TEM. This fracture surface is quite different from the observation in IN9021 by Shaw [15], where a fibrous fracture surface was observed. This differences is difficult to explain since the testing temperature used by Shaw was similar to the temperatures used in [3].

X-ray analysis was conducted on deformed and undeformed specimens to determine which phases were present. A Siemens Diffraktometer using Cu-Kα radiation was used. Specimens were oriented so that the X-ray beam was perpendicular to the tensile axis, so that diffraction was from the originally rolled surfaces. In addition to the strong peaks for aluminum, a number of small peaks were noted. Al_2Cu (θ phase), was clearly identified, and correspondence was made for Al_4C_3, Al_2O_3, and Al_2CuMg (S phase, but accounting for no more than 20% of the 2 wt% Mg). Except for the small amount associated with the $CuMgAl_2$, Mg was not apparent in compounds identified in X-ray or TEM diffraction. Phase diagrams indicate that the 2 wt% Mg is soluble in Al at ambient temperatures, but that Cu will form precipitations. At 400-500°C, the Cu and Mg are soluble in Al, and thus at elevated temperatures the precipitates would dissolve and only the oxides and carbides would remain. Although energetically favored over Al_2O_3, MgO was not observed, as has been reported for other aluminum-magnesium alloys with oxide phases. In cases where molten aluminum is present, Mg in solution reacts with Al_2O_3 to form MgO at interfaces [16-17], but since IN90211 processing was solid state, this reaction was precluded. The TEM observation of S phase on grain boundaries [3] is consistent with an Auger spectroscopy study of Al-Mg alloys indicates that Mg increasingly segregates to surfaces with decreasing temperature [18]. With increasing temperature, the segregation ratio of Mg in the boundaries decreases, putting more of the Mg in the lattice. Thus at elevated temperature, the phases present are only oxides and carbides of Al, and the Mg is predominantly in solution in the lattice.

High temperature deformation usually occurs by means of slip and/or diffusion. Mg solute atoms cause locking and unlocking behavior of dislocation slip, resulting in serrated flow curves. Serrations occur in elevated temperature superplastic deformation, and also solution treated samples deformed at room temperature [19]. (Serrations are not apparent in aged IN9021 [19], suggesting that the solute concentration is low.) The effects of slip are also evident in the texture indicated by the relative X-ray intensities of the IN90211 sample, compared to a powder sample. The relative intensity of {200} planes was 1.3 times higher in the undeformed sample compared to the Al powder standard. This preferred orientation of {001} poles parallel to the rolling direction increases the number of {111} planes oriented for slip. The superplastically deformed {200} peak intensity was twice the Al {200} powder standard. The serrated flow, n=3, and the increase in texture that facilitates slip during superplastic deformation, indicate that slip is important in the superplastic deformation of IN90211.

The deformation mechanisms that operate in ODS materials are varied and uncertain. Shaw [15] has suggested that the phenomenon may be a result of dynamic recrystalization from the high strain rate of testing, rather than due to the fine grain size. His theory however, does not explain the observation of grain boundary sliding and the lack of grain growth after superplastic deformation. Gregory, et al., [4] have argued that a combination of slip with Coble creep (in which the Coble creep exhibits a threshold stress) can explain the phenomenon in a mechanically alloyed nickel based superalloy. However, the IN90211 has a stress dependance of 3 in the superplastic regime, while Gregory measured a slope near 1, which is consistent with Coble creep theory. A similar phenomenon may exist in IN90211, where the stress is determined by the slip associated with solute drag, and the threshold stress may represent the stress necessary for lattice dislocations breakaway from solute pinning sites. Extrinsic grain boundary dislocations (from lattice slip) have been noted to contribute effectively to regions II and III in superplastic materials [20]. The majority of theories for superplastic deformation often have grain boundary sliding as a major feature, but predict a stress dependance of 2 [21]. In the case of IN90211, the combination of slip and grain boundary sliding is necessary, but slip is the rate controlling event in the deformation process, since n=3.

CONCLUSIONS

Tensile elongation in excess of 500% were obtained with IN90211 mechanically alloyed aluminum at strain rates between 1 and 10/s at 475°C. There was evidence of a threshold stress associated with superplastic behavior, but this threshold stress does not account for the observed low stress-low strain rate behavior. The superplastic-like behavior at high strain rate is associated with grain boundary sliding of the fine grained (0.5 μm) microstructure. Instead of the stress exponent usually associated with grain boundary sliding, a stress exponent of 3 was obtained in the superplastic-like regime. This stress exponent is consistent with the serrated flow observed and solute drag of lattice dislocations. Current superplastic theories do not fully account for the combination of slip with solute drag and grain boundary sliding observed in this alloy.

ACKNOWLEDGEMENTS

The authors acknowledge the support of this research by the Air Force Office of Scientific Research under grant No. AFOSR-860091 monitored by Dr. Alan Rosenstein. TGN and JW acknowledge support of the Lockheed Independent Research Program. The authors gratefully acknowledge Dr. R. Shelton and T. Folkerts of the UCD Physics department for their help with X-Ray diffraction.

REFERENCES

1. T.G. Nieh, C.A. Henshall, J. Wadsworth, Scripta Met., 18, 1405, (1984).
2. T.G. Nieh, P.S. Gilman, J. Wadsworth, Scripta Met., 19, 1375, (1985).
3. T.R. Bieler, T.G. Nieh, J. Wadsworth, A.K. Mukherjee, Scripta Met., 22, 81, (1988).
4. J.K. Gregory, J.C. Gibeling, W.D. Nix, Met. Trans., 16A, 777, (1985).
5. F.A. Mohamed, J. Mat. Sci., 18, 582, (1983).
6. R.S.W. Shewfelt, L.M. Brown, Phil. Mag., 35, 945, (1977).
7. M. McLean, Acta Met., 33, 545, (1985).
8. V.C. Nardone, D.E. Matejczyk, J.K. Tien, Acta Met., 32, 1509, (1984).
9. E. Arzt, J. Roesler, J.H. Schroeder, in Creep and Fracture of Engineering Materials and Structures, edited by B. Wilshire, R.W. Evans, (Institute of Metals, London, 1987), pp. 217-230.
10. A.H. Clauer, N. Hansen, Acta Met., 32, 269, (1984).
11. R.W. Lund, W.D. Nix, Acta Met., 24, 467, (1976).
12. E. Orowan, Internal Stresses in Metals and Alloys, (Institute of Metals, 1948), 451.
13. J.E. Bird, A.K. Mukherjee, J.F. Dorn, in Quantitative Relation Between Properties and Microstructure, (Israel University Press, Haifa, 1969), p. 255.
14. A.H. Chokshi, T.R. Bieler, T.G. Nieh, J. Wadsworth, A.K. Mukherjee, in Superplasticity in Aerospace Aluminum, Phoenix, AZ, AIME-TMS, 1988, in press.
15. W.J.D. Shaw, Mat. Letters, 4, 1, (1985).
16. J.E. Hack, R.A. Page, R. Sherman, Met. Trans. 16A, 2069, (1985).
17. S.R. Nutt, R.W. Carpenter, Mat. Sci. & Eng., 75, 169, (1985).
18. C. Lea, C. Molinari, J. Mat. Sci., 19, 2336, (1984).
19. S.J. Bane, J. Bradfield, M.R. Edwards, Mat. Sci. & Tech., 2, 1025, (1986).
20. P.R. Howell and G.L. Dunlop, in Creep and Fracture of Engineering Materials and Structures, edited by B. Wilshire and D.R.J. Owen, (Pineridge Press, Swansea, U.K., 1981), pp. 127-140.
21. M. Suery, A.K. Mukherjee, in Creep Behavior of Crystalline Solids, edited by B. Wilshire and R.W. Evans, (Pineridge Press, Swansea, U.K., 1984), p. 137.

PART III

Ceramic Composite Microstructural Development

GLASS AND CERAMIC MATRIX COMPOSITES
PRESENT AND FUTURE

KARL M. PREWO
United Technologies Research Center, Silver Lane, East Hartford,
CT 06108

INTRODUCTION

During the past 25 years materials scientists have been able to make a major change in the way materials are considered for application. In the past designers have worked with data representing the properties of homogeneous, isotropic materials and designed their components to fit written accepted ranges of "design allowables". More recently, however, the concept of composite materials has permitted almost limitless tailoring of composites to create entirely new designs never previously possible. By choice of types of material constituents, their relative percentages and their orientation the designer can now work closely with the materials scientist to optimize system performance. This philosophy has firmly taken hold in the family of fiber reinforced polymer matrix composites and more recently has made metal matrix composites an industrial reality.

The widespread acceptance of this philosophy, combined with bold new challenges to create engineering advances in aerospace and commercial areas have created tremendous opportunities for new composites possessing greater environmental stability and higher temperature capabilities. This is an excellent environment to support the development of ceramic matrix composites (CMC).

It is the purpose of this paper to address the opportunities for CMC and demonstrate the following several key points.

- CMC materials are here now and available for potential demonstration applications.

- While CMC materials have superb potential to meet high temperature performance needs, they can also compete successfully with other materials at low temperatures.

- A knowledge base has already been established to give direction to a next generation of even higher performance CMC than now exist.

Why Ceramic Matrix Composites

There are many potential reasons for CMC. The following three, however, are foremost among them.

(1) <u>Toughness:</u> The concern for the toughness of ceramic materials has severely limited their application. By incorporating second phase constituents it is possible to increase their toughness substantially or, even more promising, is the ability to completely change their failure mode. Figure 1, for carbon fiber reinforced cement composites, compares simple bend test load-deflection curves [1]. It is important to note that this overriding concern for toughness by ceramists has been shared by those who first introduced fiber reinforced polymer matrix composites (PMC). In particular the acceptance of carbon fiber reinforced epoxy composites possessing stress-strain curves with failure strains of 1% or less has required a complete rethinking of the concept of fracture toughness. The lack of self-similar crack growth and the occurrence of multiple competing failure modes assured both concern on the part of the user and ultimately provided reliability even beyond that of many metals.

(2) Environmental Stability and Density

Polymers decompose and burn while metals corrode and melt under conditions that have little or no effect on many ceramics. The simple statement, combined with the relative low density of ceramics assures CMC's a hearty welcome in many potential applications. The potential use temperatures and densities of many notable CMC matrix candidates are compared in Fig. 2 with the performance of notable metals and polymers.

(3) Cost

One of the major reasons why ceramic materials are the largest single class of materials used by man, despite their brittleness, is their low cost. Cement and concrete, bricks and ceramic tiles, porcelain and plaster, glass and glass-ceramics are all around us in our daily lives. Preserving this low cost and increasing their usability through toughening provides the opportunity for materials usage which dwarfes both polymer matrix (PMC) and metal matrix composites (MMC).

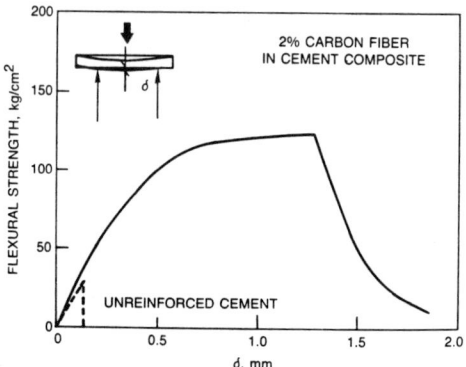

Fig. 1. Comparison of bend tests for unreinforced cement and cement matrix composites containing 2% of chopped carbon fiber.

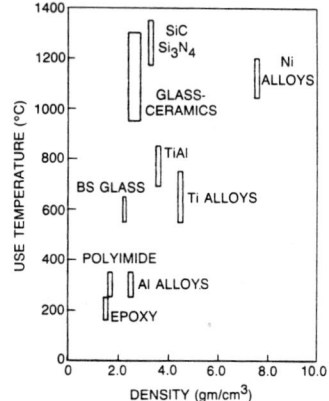

Fig. 2 Densities and use temperatures of potential composite matrices.

CMC Candidate Systems

While the potential range of combinations of matrices and reinforcements is almost limitless, the number of systems that can be created with desirable properties is far fewer. Most CMC's will require fabrication conditions that can potentially degrade either or both constituents to eliminate any advantage to their use. However, it is anticipated that the continued development of new reinforcements and processes for matrix densification will significantly exceed the range currently known.

It must also be stated that CMC's do not simply consist of a combination of reinforcement and matrix after fabrication. More typically, either through intent or fiber precoating, or through reactions during processing, the resultant composites will contain a region separating the reinforcement and matrix that, as will be shown, is most important in controlling resulting composite properties. This region can be very narrow, or very broad, as determined by the chemical composition and residual stress state of the final composite.

As a simplified way of beginning, the properties of some currently available fibers and matrices are enumerated in Figures 3-4. The fibers listed are whiskers, multifilament yarns and large diameter monfilaments. The compositions of all of these may be more complex than this limited discussion can cover. In particular the yarns and monofilaments can be very complex in chemistry and in the case of carbon yarns it will be noted that properties can also be highly anisotropic.

In the case of the matrices, simple descriptions may also be misleading since fabrication procedures are many and can be varied depending on the ultimate application. The final matrix may, however, differ significantly from its unreinforced counterpart, due to the presence of porosity, microcracks, degree of crystallinity and chemistry. The processes most common at present are listed in Table I.

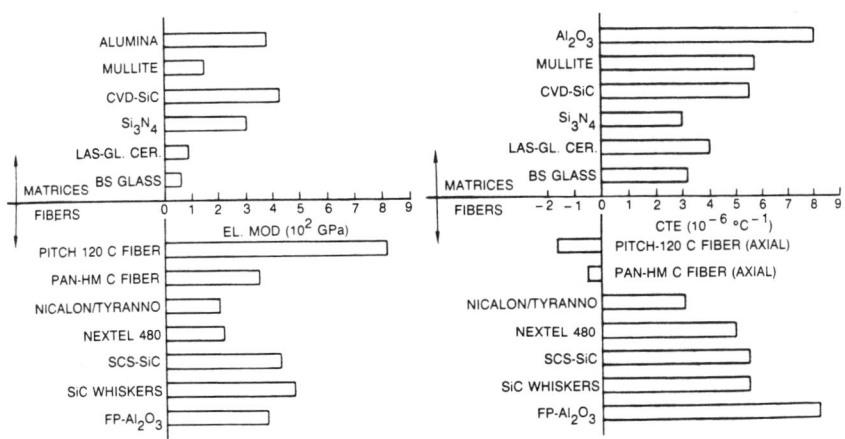

Fig. 3 The elastic moduli of candidate ceramic matrices and reinforcing fibers.

Fig. 4 The coefficient of thermal expansion of candidate ceramic matrices and fibers.

The combination of starting materials properties and fabrication procedure ultimately determine the properties of the resultant composite. This simple and obvious statement encompasses too many facets to be considered in detail here. Two, however, can be singled out for comment. First, the fibers should not be greatly degraded during processing either by handling or chemical reaction and second the resultant fiber-matrix interface must have the characteristics necessary to prevent excessive fiber-matrix bonding. The Nicalon fiber reinforced LAS glass-ceramic system has met these requirements through the development of a

Table I

Manufacturing Processes
for Ceramic Matrix Composites

Viscous Glass Consolidation	Examples (Fiber/Matrix)
- Ply lay up and hot press - Matrix transfer mold - Injection mold	Nicalon/LAS Carbon/Glass
Chemical Vapor Deposition	
- Infiltrate prewoven structures	Nicalon/SiC Carbon/SiC
Polymer Conversion	
- Infiltrate & pryolysis	Carbon/Carbon Nicalon/SiC
Sol-Gel	
- Infiltrate & sinter	Carbon/Glass
Powder & Hot Press	
- Traditional ceramic processing	SiC_w/Al_2O_3 $FP\ Al_2O_3/Al_2O_3$
Gas-Liquid Metal Reaction	
- "Lanxide"	$Nicalon/Al_2O_3$

carbon rich fiber-matrix interface, [2], and a retention of fiber strength [3]. Another example, however, has clearly also shown the importance of the relative values of matrix and fiber coefficients of thermal expansion (CTE), [4]. When a silicon carbide fiber was embedded in a cordierite matrix the effect of CTE could be examined through composite heat treatment. Heat treatments below 1000°C resulted in a marix whose CTE was greater than that of the fiber. This caused brittle failure and low strength while heat treatment at higher temperatures reversed the relative CTE's and resulted in a low interfacial strength, high composite strength and a tough, fibrous failure mode.

Toughened Ceramics

One of the most important classes of CMC can be readily classified as toughened ceramics. In this case the fibrous addition serves to increase the toughness through crack deflection. Important examples of this type of composite have been developed utilizing primarily SiC whiskers and significantly differing matrices. The matrices range from glass ceramics with elastic moduli of only 1/5 that of the whiskers, to mullite, to alumina, to Si_3N_4 which has an elastic modulus approximately equaling the whiskers. The use of the high modulus whiskers in a lower modulus glass-ceramic serves to both reinforce the matrix (increasing elastic modulus and strength) and toughness [5]. In contrast the higher elastic modulus matrices are primarily toughened [6]. The levels of performance can be readily understood using two concepts.

First, in a well bonded composite system the elastic stiffness of the composite will be increased over that of the matrix to a degree dependent on the volume fraction, distribution and aspect ratio of the whiskers. If the composite ultimate fracture is controlled by the failure strain of the matrix then, even for these strongly bonded systems the composite strength will be increased, i.e. for the same failure strain the higher stiffness material will exhibit a higher failure stress. This strengthening may be enhanced even further by the presence of fibers effectively increasing the matrix failure strain, however, it can also be decreased by the fibers acting as stress concentrations within the matrix.

The toughness increase in these composites has been predicted Fig. 5 [7] and it can be seen that significant increases are possible even though the overall crack growth path is similar to that of the parent matrices. This toughness increase can also be readily translated into an effective increase in usable strength. In engineering applications the combination of material strength and toughness enhancement will result in greater structural performance.

Composites of these types have already found application as cutting tools and will undoubtedly increase significantly the potential scope of traditional ceramics usage.

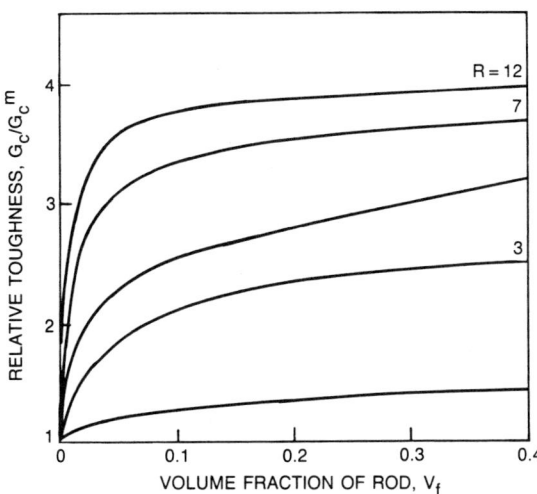

Fig. 5 Relative toughness predictions from crack deflection model for rod-shaped particles of three aspect ratios and spherical particles.

Toughened and "Structurally Strengthened" Composites

While those designing with metals have come to "fear" the triaxial stress state at the tip of a sharp crack, those creating CMC materials have learned to "appreciate" and rely on these stresses. In a metal the occurrence of triaxial stresses at the crack tip inhibits plasticity and can lead to "embrittlement" of the metal (particularly in plane strain) since the yield criteria in metals are based on the magnitudes of the differences in principal stress. In contrast, for traditional ceramics the occurrence of fracture is simply based on the magnitudes of the maximum tensile stresses which are enhanced by the presence of stress concentrations. Thus, if interfaces are present in the path of an advancing crack these interfaces can be caused to fail and blunt the crack. This effect, traditionally recognized by those working with ceramics for centuries was formalized by Cook and Gordon [8] Fig. 6. Without the added stress component caused by the presence of the crack there would not be the possibility of significant blunting. So, if we use the addition of fibers to provide interfaces in the path of the crack we can blunt the crack tip and significantly alter cracking patterns. Cook and Gordon attempted to estimate the level of interfacial strength necessary to provide crack blunting. They found that, for an elliptical crack, an interfacial strength of 1/5 or less of the principal strength would cause interfacial failure in advance of the crack.

A formidable example of this is the use of small (1-3% by volume) percentages of discontinuous carbon fibers to toughen cement. The loosely bonded carbon fibers alter the fracture morphology to permit multiple microcracking to take place and change the deformation behavior. By this change in fracture characteristic the effective "toughness" of this material becomes very high and also difficult to quantify because cracks are diverted from their original path i.e. no longer self similar. At very low volume percentages of fiber, even through the tensile strength of the composite is not significantly enhanced, the ability to alter the fracture mode can cause very significant increases in reliability and "structural strength". The comparison of load deflection traces for this material in Fig. 1 demonstrate this point. Stable redistribution of load occurs on the tensile side of the composite flexural beam. This has been so successful that it has led to the development of a highly useful light weight concrete, Fig. 7, [1].

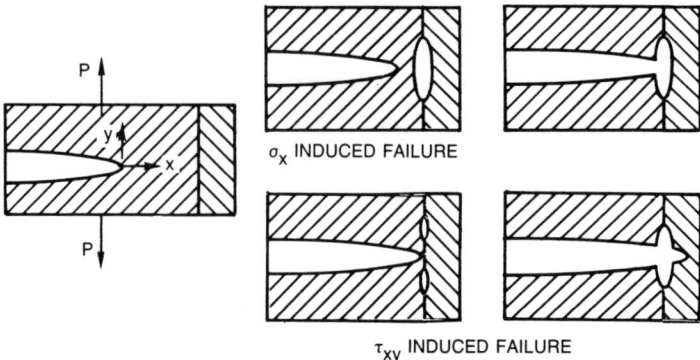

Fig. 6   Interaction between an advancing crack and a plane of weakness. Under the load P, both tensile and shear stresses ahead of the crack can cause crack blunting.

- 33% LIGHTER THAN CONCRETE
- COST SAVINGS (LESS STEEL FRAMING)
- 3 MONTH TIME SAVING

Fig. 7 37 story building in Tokyo with carbon fiber reinforced concrete exterior.

High Performance CMC

It is in this range of highly reinforced composites where the major emphasis in R&D has been placed in the recent past. By chosing high strength, relatively high modulus, continuous (or relatively long) fibers and incorporating them into matrices without damage it has been possible to create high strength composites if one key additional feature is achieved. As stated previously, that key is to create a fiber-matrix interface region which is weak enough to divert cracks. The ultimate strength of such a composite is then controlled by the "in situ" fiber strength and a composite tensile strength well beyond that associated with matrix cracking can be achieved. Such a composite is shown in Fig. 8 [3] where the tensile stress-strain curves for Nicalon SiC fiber uniaxially reinforced glass-ceramic are shown. While other prominent examples of high performance CMC have been developed, this system will be used to illustrate various key aspects of CMC behavior. In Figure 9 both the as pressed and "ceramed" composites exhibit nearly identical stress-strain behavior and both fracture with tough and fibrous fracture surfaces. The point of matrix microcracking is clearly evident and has been predicted by Aveston, Cooper and Kelly, [9] and varified by the observations of Marshall and Evans, [10]. This point can be referred to as the "proportional limit" in that it represents a first deviation from linearity in the stress-strain curve. The "ACK" derived expression for the composite tensile strain at which this limit is reached is given by the following expression.

$$\varepsilon_{PL} = \left[\frac{24\tau\gamma_m E_f V_f^2}{E_c E_m^2 d_f V_m} \right]^{1/3}$$

Where d_f, E_f, V_f are fiber diameter, elastic modulus and volume fraction, τ is the fiber-matrix interfacial shear strength, and γ_m is the unreinforced matrix fracture energy.

Once matrix microcracking has started to take place, the contribution of the matrix to the composite elastic modulus is significantly reduced. As shown in Figure 9, this reduction becomes more significant with increasing strain and eventually the composite stiffness appears to be solely dependent on the fibers. It should be noted that while this cracking indicates significant reduction in composite integrity, it does not cause a precipitous reduction in composite strength. Fatigue testing of these composites at 22°C, [11], has determined that full tensile strength is maintained despite extensive cyclic loading.

By examining the ACK model it can be seen that the matrix microcracking strain is not limited by the original matrix failure strain; instead it can be significantly enhanced by the presence of stiff reinforcing fibers. Jamet, [12], performed a detailed series of calculations to demonstrate this affect for Nicalon/LAS, Fig. 10. His curves can be used, along with the measured value of Nicalon fiber-LAS matrix interfacial shear strength of approximately 2.0-2.5 MPa obtained by Marshall and Evans, [10], to predict a composite proportional limit of approximately 250 MPa. This somewhat underestimates the experimentally observed value of 400 MPa, however, Jamet's values of starting matrix performance may also be somewhat underestimated.

The reason for this low fiber-matrix interfacial strength is due to the presence of a carbon rich layer formed during composite fabrication, [2]. Luh and Evans [13] examined the importance of this interfacial strength, τ, relative to the strength of the reinforcing fibers, S, in determining composite fracture toughness, K and fracture morphology, Fig. 11. As anticipated, the maintenance of a low interfacial strength is critical to the achievement of composite toughness. However, while providing the desired toughness, this also causes low off-axis composite strength and the need for multiaxial fiber reinforcement for just about all conceivable engineering applications. The examination of the performance of 0/90 cross ply composite properties is therefore critical.

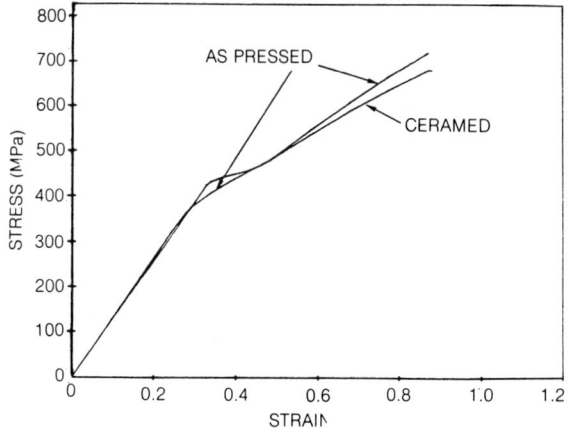

Fig. 8 Tensile stress-strain curve for 0°-Nicalon fiber reinforced LAS-II in the as pressed and ceramed conditions at 22°C.

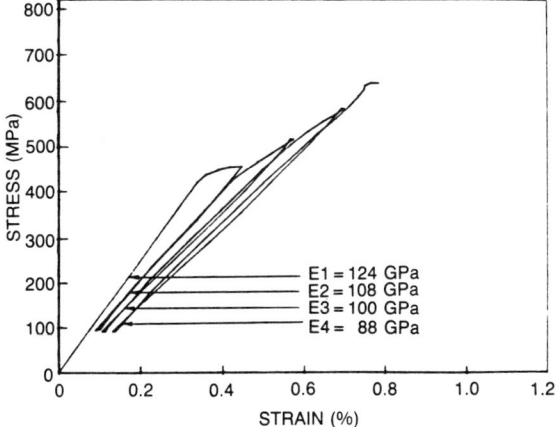

Fig. 9 Cycled tensile stress-strain curves for 0°-SiC reinforced LAS-II (ceramed) at 22°C.

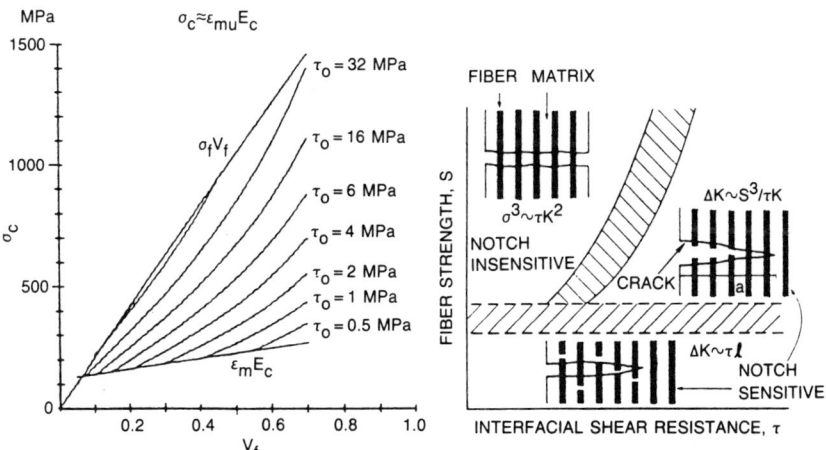

Fig. 10 Critical stress diagram of the unidirectional Nicalon/LAS system.

Fig. 11 Fracture mechanism map for a uniaxial fiber reinforced ceramic matrix composite.

A typical tensile stress-strain curve for a 0/90 cross ply reinforced composite is shown in Fig. 12, [14]. This curve is considerably more complex than that for the unidirectionally reinforced composite discussed above because the effects of both 90° ply failure and 0° ply matrix cracking are both evident. It is also important to note that a fibrous, tough composite fracture was achieved even though the 90° plys severely constrain cracking of the 0° plys. Thus, multiaxial reinforcement does provide strength and toughness, however, it also adds a new factor indicating that composite failure processes are occurring at even lower stresses than the 0° ply proportional limit.

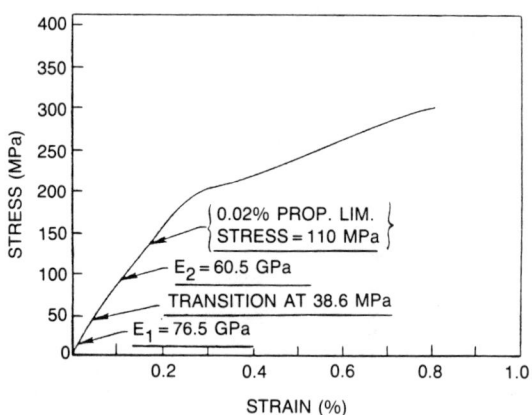

Fig. 12 Tensile stress-strain curve for 0/90-Nicalon fiber reinforced LAS-III at 22°C.

The necessity to cross ply and this observation of early internal damage have led to concern as to which stress (strain level) determines composite environmental stability. For this purpose measurements of composite stress-strain behavior and ultimate tensile strength have been made at temperatures of up to 1300°C in argon and air [14]. The results, Fig. 13 clearly show that strength is severely reduced by the presence of air. The composite fracture mode is also less fibrous indicating a higher level of fiber-matrix interfacial strength in agreement with measurements made by Luh and Evans [13] which indicated a 20 fold increase in interfacial strength at 1000°C in air to a level of 40 MPa. Examination of tensile stress-strain curves at 1000°C indicate that composite strength is limited by the 0° ply proportional limit point and not the 90° ply failure point, Fig. 14. This was confirmed by 100 hour stress rupture tests and fatigue tests performed within the same program.

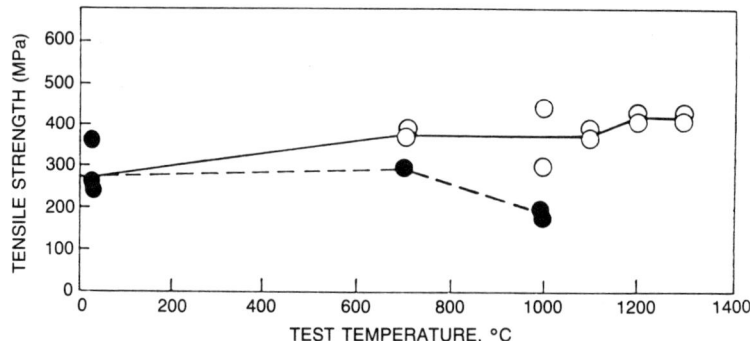

Fig. 13 Tensile strength of 0/90-Nicalon fiber reinforced LAS-III tested in argon (○) and air (●).

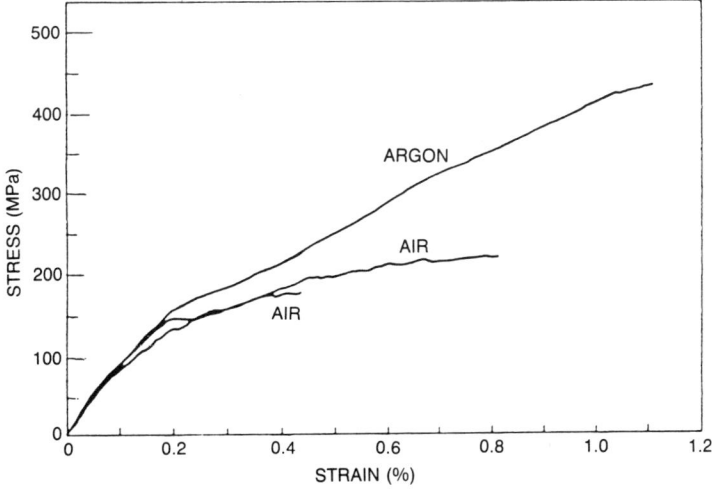

Fig. 14 Tensile stress-strain curves for 0/90-Nicalon fiber reinforced LAS-III at 1000°C.

Challenges for the Future Development of High Performance CMC

While our knowledge is by no means complete, the extensive testing performed to date on existent CMC systems has served to focus our attention on issues important to the creation of newer improved materials and structures.

First among these is to recognize that in almost all applications we can expect some matrix cracking to take place. This being the case we must have fiber-marix interfaces which are environmentally stable. While not an easy task, the approach of coating the reinforcing fibers prior to compositing into a matrix has clearly been chosen by many to achieve this goal. Developing coatings that can be applied to a variety of fibers is probably an easier method to achieve desired surface chemistry than to expect fibers to be synthesized to meet each of many varied needs. Large diameter fibers, such as the Avco CVD-SiC monofilament, have already been used to demonstrate the validity of this approach for metal matrix composites.

Second, the structural limitation of the proportional limit stress (strain) is very significant. The matrix cracks which occur at this point not only increase the environmental stability problem described above, but also provide the designer with the difficulty of a material elastic modulus which changes with increasing strain. Such materials have not been used in the past and hence most structural design stresses will be limited to below the proportional limit. Approaches to increasing this value will probably be best in the area of improving matrix toughness and strength. If fiber-matrix interface strength is increased to accomplish the same result it unfortunately may also cause significant losses in composite toughness, as noted in the above discussion on Nicalon/LAS.

Finally, composite structural integrity will require, for many applications, reinforcement in all three principal directions. The development of weaving and braiding concepts addresses this need, however, they can also significantly increase cost and fabrication difficulties.

REFERENCES

1. Kajima Corporation and Kurhea Chemical Industries, Tokyo, Japan.

2. J. J. Brennan, Proc. Conf. on Tailoring of Multiphase and Composite Ceramics, Penn State Univ., July 1985, Plenum Press, 1985.

3. K. M. Prewo, Jl. Mat. Sci., 21, p 3590, 1986.

4. J. Aveston, "Strength and Toughness in Fiber Reinforced Ceramics", Proc. Conf. on Properties of Fibre Composites", IPC Science and Technology Press, 1971.

5. K. Gadkaree and K. Chyung, Am. Cer. Soc. Bull., vol 65, 1986 p 370-376.

6. B. F. Becker and T. N. Tiegs, Jl. Am. Cer. Soc., 70, 9, 651-654, 1987.

7. K. T. Faber and A. G. Evans, Acta Met., vol 31, #4, pp 565-576, 1983.

8. J. Cook and J. E. Gordon, Proc. Royal Soc., London, A-282, 1964.

9. J. Aveston, G. Cooper and A. Kelly, Natl. Physical Lab. Conf. Proceedings, "The Properties of Fiber Composites", London, Nov. 1971.

10. D. B. Marshall and A. G. Evans, Jl. Am. Cer. Soc., 68, vol 5, 225, 1985.

11. E. Minford and K. M. Prewo, Proc. Conf. on Tailoring of Multiphase and Composite Ceramics, Penn State Univ., July 1985, plenum Press 1985.

12. J. Jamet, Proceedings of the Fifth Intl. Conference on Composite Materials, Aug. 1985, AIME.

13. E. Y. Luh and A. G. Evans, J. Am. Cer. Soc., 70, 7, 466, 1987.

14. K. M. Prewo, to be published, Jl. Mat. Sci., 1988.

PROCESSING OF POLYMERIC PRECURSOR, CERAMIC MATRIX COMPOSITES

R. J. DIEFENDORF AND R. P. BOISVERT
Materials Engineering Department, Rensselaer Polytechnic Institute, Troy, NY 12180-3590

ABSTRACT

Ceramic matrix composites have been produced by utilizing polymer pyrolysis as the processing technique. The precursor, polyvinylsilane, is a viscous, thermosetting polymer which yields a predominantly SiC ceramic material when pyrolyzed. This organometallic polymer in combination with SiC fibers and SiC whiskers was used to fabricate ceramic matrix composites. One of the major problems with a brittle/brittle composite system in which strong coupling exist between the fibers and matrix is the characteristic catastrophic failure that occurs once the strain to failure of one of the constituents is exceeded. This brittle behavior can be altered by the application of a suitable barrier layer between the fiber and matrix. Due to the success of a barrier layer between fiber and matrix in producing higher performance composites, multiple barrier layers were used to further improve the performance of the ceramic composite.

INTRODUCTION

Fiber reinforced ceramic matrix composites are a promising class of structural materials for applications which require high strength, high stiffness, low thermal expansion, low density and high temperature stability. One of the major limitations with ceramic materials lies in their brittle, non-ductile behavior which renders these materials notch-sensitive and prone to catastrophic failure. Hence, there has been an explosion of activity in the search for tougher, damage resistant ceramic materials. Attempts to increase the strength and work of fracture of ceramic materials has revolved around the addition of discrete particles or fibers into ceramic matrices to act as barriers to crack propagation thereby improving the fracture behavior. This was first demonstrated for carbon fiber reinforced vitreous silica by Crivelli-Visconti and Cooper [1] in 1969. More recently, the availability of continuous SiC fibers has lead to the development of tough glass and glass-ceramic composites which are more resistant to high temperature oxidation than the carbon fiber composites.[2]

The fracture behavior of ceramic matrix composites is different from polymer matrix, and metal matrix composites, due to the characteristics of the individual constituents. A polymeric matrix typically has a lower modulus but higher failure strain compared to the reinforcing fibers. As a result, the role of the matrix is simply to transfer load to the fibers and hold the composite together. Failure originates in the brittle, higher modulus fibers. In the SiC ceramic/ceramic system [3,4] reported in this paper, the modulus of the fiber is higher than the matrix but the strain to failure of the matrix is lower. This composite does benefit from load transfer from the matrix to the fibers, but instead of the fibers initiating failure as in the polymeric composite, the ceramic matrix will fail before the fibers. If perfect coupling exists between fiber and matrix the composite is found to catastrophically fail at the failure strain of the matrix thereby limiting the strength and work of fracture of the composite. Substantial increases in strength and toughness are possible by controlling the fiber/matrix interface. An innovative approach to controlling brittle composite failures was investigated which consisted of layering the matrix of the composite. These multiple layers introduce several crack barrier interfaces which dramatically affect the performance of these ceramic matrix composites.

EXPERIMENTAL PROCEDURE

Ceramic/ceramic composite development via polymer pyrolysis is a low pressure, low temperature processing route. A vinylic polysilane, developed by Schilling, et al. [5], was investigated as a ceramic matrix precursor. The vinyl and SiH groups of the vinylic polysilane provides an efficient thermal crosslinking mechanism. A vacuum bag process was used to form the laminate which was cured in a press at 200°C for 6 hours under 0.3 MPa (50 PSI) pressure. After cure, laminates were cut into an appropriate sample size and pyrolyzed to 800°C at 1°C/min under a nitrogen atmosphere. Reimpregnation of pyrolyzed bars was necessary to increase the density and mechanical properties. The reader is referred to references [3] and [4] for a more detailed description of composite fabrication. Nicalon in the form of eight harness satin weave was used as the reinforcement.

Carbon coatings were placed on Nicalon SiC fibers prior to lamination in an attempt to improve the mechanical performance of the ceramic composites. Carbon was chosen as the interfacial layer due to its weak layered structure, inertness to SiC, and excellent high temperature capability. The coatings were deposited via carbon polymeric precursors called Ashland 240 pitch and the Toluene Insolubles of a heat treated Ashland 240 pitch. Uniformity of the carbon coating relies on the type of solvent used along with the concentration of the solution. The first step in producing a carbon coating involved removing the sizing on the reinforcement by subjecting the fibers to a temperature of 700°C under argon for 2 hours. Appropriate solutions of precursor and solvent were prepared. Ceramic fabric was dipped in the solution, removed and allowed to air dry between 16 to 24 hours. The cloth was then placed in a tube furnace and heated using a controlled conversion cycle to obtain the desired carbon coating.

Ceramic composite specimens were impregnated with a carbon polymeric precursor between polysilane densifications as a means of fracture control. Various cuts and concentrations of Ashland 240 pitch or Toluene Insolubles in 1,2,4-trichlorobenzene were carefully prepared and heated to 100°C to ensure complete dissolution of the precursor. The ceramic composites were then added to the hot solution. The submerged specimens and solution were placed in an autoclave for 20 minutes under 1 MPa (150 PSI) of argon gas. Samples were removed and allowed to air dry for 4 hours. Pyrolysis followed to 800°C, 1°C/min under flowing nitrogen gas in a tube furnace.

RESULTS AND DISCUSSION

Fiber/Matrix Interface Control

Preliminary ceramic matrix composites were prepared utilizing uncoated Nicalon SiC fibers with a polyvinylsilane/15 wt % SiC whisker resin mixture. The 15 wt % of whiskers was found to be optimum for controlling shrinkage and microcracking during the initial pyrolysis cycle. Subsequent impregnations of composite specimens were performed with pure polyvinylsilane. Volume fraction of reinforcing fibers was determined to be ~40%. Ceramic composites were mechanically tested utilizing a three-point bend fixture. Table I outlines the results obtained for these ceramic matrix composites.

Samples tested after the initial pyrolysis cycle failed in a shear mode by delamination between plies. This is attributed to the lack of matrix resulting in poor load transfer and low interlaminar strengths. Failure mode changed to a tensile failure after two impregnations. Brittle failure occurred for specimens with two or more densifications. Composite test specimens were found to fail catastrophically at the failure strain of the matrix. No fiber pull-out was identified on composite fracture surfaces. Poor mechanical performance was attributed to good bonding between fiber and matrix. The decomposing polymer

TABLE I: Uncoated Nicalon Reinforced SiC Whisker/Polyvinylsilane Composites

Impregnation #	Preform	0*	1*	2	3	4	5
Density (gm/cm^3)	1.59	1.51	1.72	1.86	1.95	2.00	2.04
Modulus (GPa)	**		20.2	31.1	38.7	53.0	58.7	57.4
Flexural Strength (MPa)	**		42.6	52.3	56.7	61.2	63.2	59.7

*Indicates shear failure.
**No data available.

and highly reactive Nicalon surface form a good chemical bond during processing which ultimately leads to catastrophic failures in the composite system.

To determine the effect of a weak fiber/matrix interface, composites were manufactured with Nicalon SiC fibers which were carbon coated utilizing the toluene insolubles of A240 pitch in trichlorobenzene (Table II), as well as other coating solutions. Improvements in flexural strength of 60% were noted for these composites. Non-brittle failures occurred for specimens which underwent up to 4 densification cycles. The failure strain for these composites increased over non-coated specimens resulting in a higher load carrying capability. Fiber pull-out was evident in composite fracture surfaces up through the fifth densification. Brittle failure occurred after 5 impregnations resulting in strength decreases.

TABLE II: Carbon Coated Nicalon (Toluene Insolubles of A240 in Trichlorobenzene, .0075 gm/cm^3) Reinforced SiC Whisker/Polyvinylsilane Composites

Impregnation #	Preform	0*	1*	2	3	4	5
Density (gm/cm^3)	1.54	1.40	1.60	1.73	1.81	1.89	1.92
Modulus (GPa)	**	6.0	21.2	33.9	36.1	39.0	45.7
Flexural Strength (MPa)	**	19.4	76.5	128	128	157	126

*Indicates shear failure.
**No data available.

Multilayered Matrix

Attempts to improve ceramic composite performance have been based on manipulating the interface between the fiber and matrix. Cracks which originate in the brittle matrix propagate relatively unimpeded until they encounter a fiber/matrix interface where they are either blunted or propagate through the fiber. The amount of energy stored in the crack tip depends on the distance the crack travels before it encounters the fiber. A technique has been used to limit the size and number of these cracks reaching the reinforcing fibers by placing additional crack deflecting layers within the matrix of the composite. This layered structure was developed by alternating polyvinylsilane and a carbon polymeric precursor impregnations. Tables III and IV list results gathered for composites which utilized different carbon polymeric precursors.

Non-catastrophic failures occurred for all composite samples tested. Increases in failure strain over previous specimens resulted in composites with higher load carrying capability. Flexural strengths continuously increased over the whole range of densification cycles (Figure 1). Large amounts of fiber pull-out were clearly evident on all fracture surfaces. The importance of the number of these carbon interfacial layers on influencing the mechanical performance of ceramic matrix composites was studied. Carbon impregnations were alternated with polysilane densifications in the beginning stages of composite fabrication. Initial carbon layers were found to have the most impact on composite performance. Very little benefit is gained when alternating carbon layers are added after the second polyvinylsilane impregnation.

TABLE III: Nicalon Reinforced Polyvinylsilane Composites. Fiber Coating Using Toluene Insolubles of A240 in TCB (.0075 gm/cm^3). Carbon Impregnations Using Toluene Insolubles of A240 in TCB (.01 gm/ml).

# Carbon and PVS Impregnations	Preform	0*	1*	2	3	4
Density (gm/cm^3)	1.60	1.50	1.74	1.87	1.88	1.93
Modulus (GPa)	**	7.0	25.9	34.2	37.6	41.2
Flexural Strength (MPa)	**	20.3	98.6	186	213	234

*Indicates shear failure.
**No data available.

TABLE IV: Nicalon Reinforced Polyvinylsilane Composites Fiber Coating Using Coral Products Toluene Insolubles in TCB (.01 gm/ml). Carbon Impregnations Using the Toluene Insolubles of Heat Treated A240 Pitch in TCB (.05 gm/ml).

# Carbon and PVS Impregnations	Preform	0*	1*	2	3	4	5
Density (gm/cm^3)	1.40	1.22	1.47	1.66	1.80	1.87	1.93
Modulus (GPa)	**	**	**	33.6	25.3	29.0	37.8
Flexural Strength (MPa)	**	**	**	85.5	161	235	291

*Indicates shear failure.
**No data available.

The carbon coatings and multilayered matrix not only improved the flexural strengths, but also had a dramatic effect on the toughness of these composites as evident by the area under their stress/strain curves (Figure 2). The uncoated specimens displayed catastrophic failures at the failure strain of the matrix resulting in a brittle composite with very little energy absorption during fracture. Improvements in overall load carrying ability and work of fracture resulted with specimens composed of carbon coated Nicalon. A change in slope of the initial linear section of the curve indicates the onset of matrix microcracking and the forgiving nature of the composite. The matrix microcrack strain also is observed to increase dramatically using a multilayered matrix. A major portion of the increased toughness can be attributed to the phenomenon of fiber pull-out. The largest increase in performance is noted when fiber coatings and a multilayered matrix are employed within the ceramic matrix composite. A detailed analysis of the effect of processing and structure on the stress/strain behavior will be presented in a future publication.

CONCLUSIONS

Interfacial layers on fibers and dispersed through the matrix have been shown to be effective in improving the mechanical performance of ceramic matrix composites. Carbon has been shown to be a suitable interface material, but is inadequate when not protected in an oxidizing atmosphere. Carbon polymeric precursors can be used to apply interfaces on fibers or interlayers within the matrix both cheaply and quickly. The actual gains in composite be-

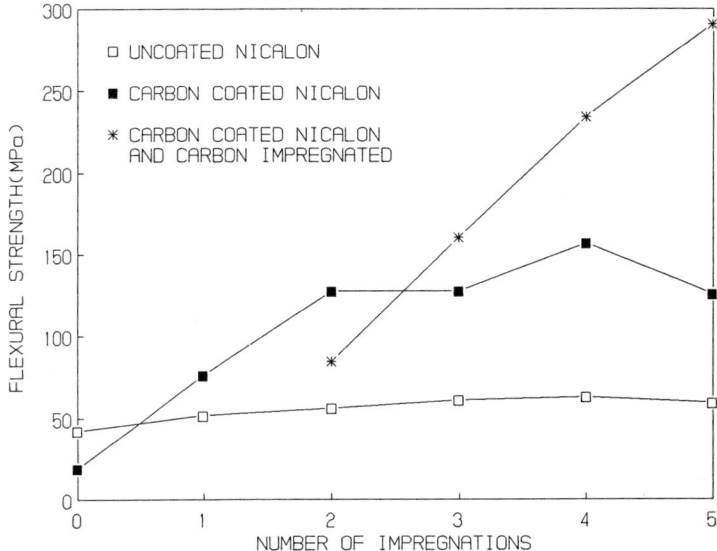

FIGURE 1: Mechanical Strength of Nicalon/Polyvinylsilane Composites.

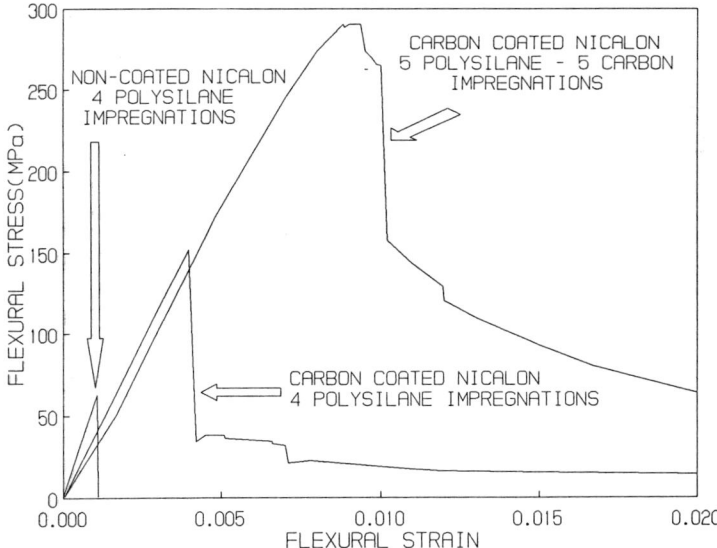

FIGURE 2: Comparison of Stress/Strain Curves for a Variety of Composites Employing Various Fracture Control Techniques.

havior depends critically on the type of carbon precursor utilized.

ACKNOWLEDGMENT

This research was supported by Kaiser Aerotech, the Office of Naval Research-Defense Advanced Research Projects Agency (ONR/DARPA) Contract #N00014-86-K0700.

REFERENCES

1. I. Crivelli-Visconti and G. A. Cooper, Nature, 221, 754 (1969).

2. K. M. Prewo and J. J. Brennan, J. Mater. Sci., 15, 463 (1980); 17, 1201 (1982); 17, 2371 (1982).

3. R. P. Boisvert, MS thesis, Rensselaer Polytechnic Institute, 1988.

4. R. P. Boisvert and R. J. Diefendorf, presented at the ACS 12th Annual Conference on Composites Materials and Structures, Cocoa Beach, FL, 1988.

5. C. L. Shilling and T. C. Williams, Polym. Prepr. (Am. Chem. Soc., Div. Polym. Chem.), 25(1), 1 (1984).

FRACTURE BEHAVIOR OF
3-D BRAIDED NICALON/SILICON CARBIDE COMPOSITE

J.-M. YANG[*], J.-C. CHOU[*] and C. V. Burkland[**]
*Department of Materials Science and Engineering, University of
California, Los Angeles, CA 90024
**Amercom, Inc., Chatsworth, CA 91311

ABSTRACT

The fracture behavior of a 3-D braided Nicalon fiber-reinforced SiC matrix composite processed by chemical vapor infiltration (CVI) has been investigated. The fracture toughness and thermal shock resistance under various thermo-mechanical loadings have been characterized. The results obtained indicate that a tough and durable structural ceramic composite can be achieved through the combination of 3-D fiber architecture and the low temperature CVI processing.

INTRODUCTION

The demand for strong, tough, defect-free and near net shape structural ceramic composites for various high temperature applications is growing rapidly. Considerable amount of innovative approaches have been pursued to develop new reinforcements [1], new processing methods to infiltrate the matrix [2] and novel reinforcement architectures [3]. Chemical vapor infiltration (CVI) processing is one of the promising composite fabrication processes being considered for producing ceramic composites with a variety of reinforcement geometries. Various unidirectional and 2-D woven fabric reinforced ceramic matrix composites densified by CVI technique with improved properties have been reported [4-9]. However, these conventional laminated composites suffer from a low interlaminar strength, and they are difficult to form into complex structural shapes. In order to develop a new toughened structural ceramic composite for turbine engine applications, various 3-D reinforcement geometries have been developed [10]. 3-D braided structure not only offers a multidirectional reinforcement, but also has the capability for the direct formation of complex structural shape in an integrated manner. With the combination of 3-D braided reinforcement and the low temperature CVI processing, we have been able to produce a ceramic composite with significant improvements in performance [11].

The objective of this paper is to characterize the fracture resistance and damage tolerance of a 3-D braided Nicalon fiber reinforced SiC matrix composite by CVI technique under various environmental exposures. The failure modes and toughening mechanisms will also be identified.

EXPERIMENTS

The 3-D braided Nicalon preform has a braided pattern of 1 x 1, and the angle between the fiber bundle and the braiding axis was ±20. A thin layer of carbon was applied to the surface of the Nicalon fibers to control the interface bonding. The composite was densified with a silicon carbide matrix by CVI technique. A detailed description of the CVI process can be found in reference 11. The as processed composite contains 35 % fiber volume fraction, with a open porosity of 16.5 %.

The fracture toughness was measured by using single edge notched beam (SENB) technique in three-point bending. The specimen has a dimension of 0.5 by 0.5 by 25.4 mm. The notch was all perpendicular to the braiding axis and introduced by a 12 mil thick diamond blade. The toughness was calculated according to the following equation:

$$K_I = \frac{3PS}{2bw^2} Y(a)^{1/2} \quad \text{where } Y = 1.96 - 2.75\left(\frac{a}{w}\right) + 13.66\left(\frac{a}{w}\right)^2 - 23.98\left(\frac{a}{w}\right)^3 + 25.22\left(\frac{a}{w}\right)^4$$

where, a is the notch length, W the specimen thickness, P the load, b the specimen width and S the moment arm. In order to study the thermal stability of the composite, some of the specimens were heat treated at a given temperature, in air without stress for 100 hours. Testing was conducted in an Instron machine at room temperature, with a crosshead speed of 0.5 mm/min.

For the thermal shock test, the specimens were heated slowly to a given temperature and held for 24 hours, and then dropped into cold water rapidly. The retained flexural strength was determined by a four-point bend test. The fracture toughness after thermal shock was tested under the same conditions as described above. Optical and scanning electron microscopic examinations were conducted on fracture surface to study the failure characteristics.

RESULTS AND DISCUSSIONS

Fracture Toughness

The fracture toughness of the composite measured by the single edge notched beam bend test as a function of the thermal exposure temperatures is shown in Figure 1. An average fracture toughness of 15 MPa m$^{1/2}$ has been obtained for the as processed specimen tested at room temperature. A very significant increase in fracture toughness has been obtained as compared with unreinforced SiC which has typical value of 3 to 4.6 MPa m$^{1/2}$. The fracture toughness increases slightly after thermal exposure at 500°C. After high temperature exposure (between 1200 to 1500°C), the toughness decreases, but the composite still retained a fracture toughness of 11 MPa m$^{1/2}$. It is clear that the

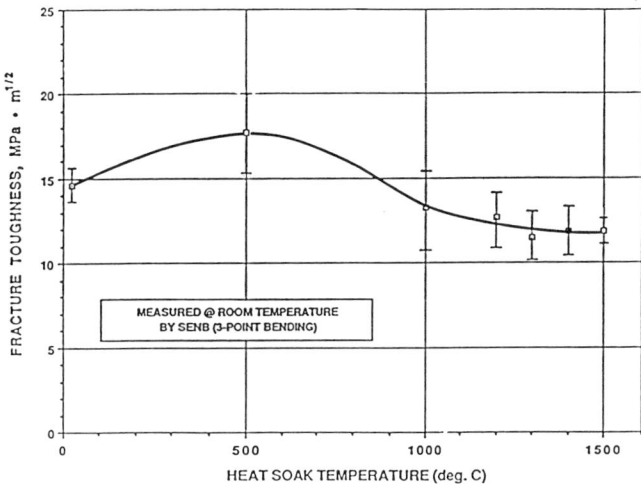

Fig. 1 The fracture toughness of 3-D braided Nicalon/SiC composite after thermal exposure

composite can retain high fracture toughness at oxdizing environment over a long period of time. The stress-deflection curve from the SENB bend test for both as processed and heat treated specimens consistently exhibited non-catastrophic failure.

SEM micrographs of a fracture surface of the SENB fracture toughness specimen are shown in Figure 2. A significant amount of fiber pullout was observed on the fracture surface, which contributes to the high toughness of the composite. This also indicate that a weak fiber/matrix interfacial bonding was developed through the carbon coating. Matrix cracking and delamination were also observed on the fracture surface, which will also contribute to the high toughness of the composite. The delamination was due to matrix density gradient from the external surfaces to the core which is inherent to the CVI processing. However, the delamination was stopped by the interwined fiber bundle, hence, the 3-D integrated fiber structure can provide structural toughening of the composite.

After thermal exposure in air above 1000°C, the composite became more brittle with less fiber pullout. This might due to the differences in fiber/matrix bonding conditions and fiber strength degradation. The formation of a silica layer on the surface of an exposed Nicalon fiber when heated at temperature above 1200°C will promote the formation of strong fiber/matrix bonding. As a result, a crack can easily propagate through the fiber/matrix interface. Fiber strength degradation is another factor which causes the reduction of fracture

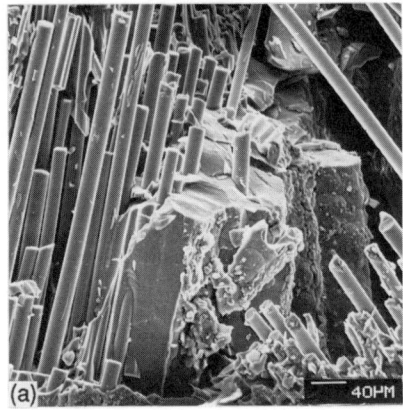 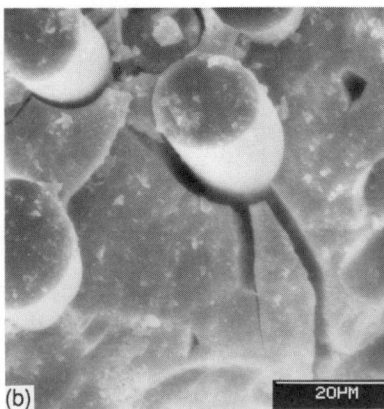

Fig. 2 Fracture surface of 3-D braided Nicalon/SiC composite, (a) fiber pullout, (b) matrix cracking

toughness after high temperature thermal exposure. Some evidence of fiber degradation can be found in Figure 3. It has been found that Nicalon fiber exhibit tensile strength reduction following thermal aging in air, due to fiber coalescence at the silica surface, a poorly adhered and cracked crystalline silica crust and trapped bubbles at fiber/silica interface [12-13].

Thermal Shock

The retained flexural strength after thermal shock has been reported earlier[11]. Excellent thermal shock resistance has been demonstrated. For example, no loss in flexural strength occurred after water quench from temperatures up to 1000°C. This performance is believed to be attributable to the high fracture toughness of the braided composite. The retained fracture toughness after thermal shock is shown in Figure 4. A reduction of fracture toughness has been observed after thermal shock as compared with the as processed specimen. This might be attributed to several damage mechanisms including the presence of thermal stress gradient, matrix microcracking, fiber strength degradation and changes in fiber/matrix interface bonding. Furthermore, the retained fracture toughness decreases as the thermal shock temperature increases. Microstructure observation has indicated that the amount of fiber pullout on the fracture surface decreases as the thermal shock

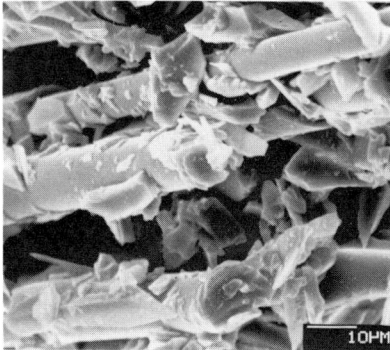

Fig. 3 Fracture surface of 3-D braided Nicalon/SiC composite after heat treatment at 1300°C

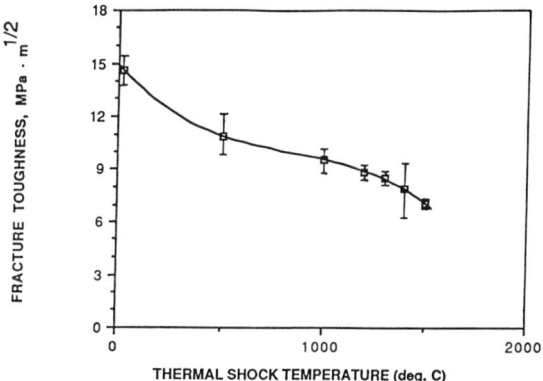

Fig. 4 The retained fracture toughness of 3-D braided Nicalon/SiC composite after thermal ahock

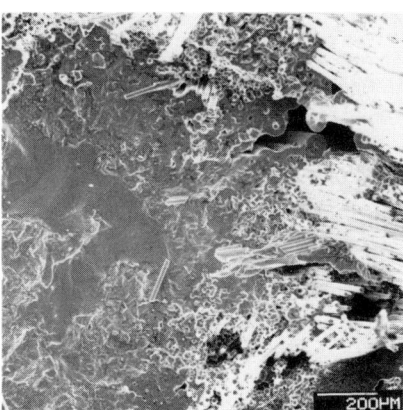

Fig.5 Fracture surface of 3-D braided composite after thermal shock

temperature increases. A brittle zone near the outer surfaces of the specimen without any fiber pullout was observed when the composite was subjected to thermal shock test at 1500°C (figure 5). Further examination of the fracture surface of the brittle zone shows that a uniform thickness of silica layer adhered to the fiber. The presence of the oxide interfacial layer would also lead to the brittle fracture of the composite.

CONCLUSION

(1) A tough and durable structural ceramic matrix composite can be produced with the combination of structural toughening from 3-D braided reinforcement and the CVI processing technique.
(2) The fracture toughness of the composite has been measured after processing and after thermal exposure in air at various temperature. The composite can retain high toughness in an oxdizing environment over an extended period of time.
(3) The retained flexural strength and fracture toughness after thermal shock at different temperature has been characterized. Excellent thermal shock resistance has been demonstrated.

REFERENCES

1. T.I. Mah, M.G. Mendiratta, A.P. Katz, and K.S. Mazdiyasni, Am. Cera. Soc. Bull. 66, 304 (1986).
2. J.A. Cornie, Y.-M. Chiang, D.R. Uhlmann, A. Mortensen, and J.M. Collins, Am. Cera. Soc. Bull. 65, 293 (1986).
3. F. Ko, Cera. Eng. Sci. Proc. 8, 822 (1987).
4. R.E. Fisher C.V. Burkland, and W.E. Bustamante, Cera. Eng. & Sci. Proc. 16, 806 (1985).
5. P.J. Lamicq, G.A. Bernhart, M.M. Dauchier, and J. G. Mace, Am. Cera. Soc. Bull. 65, 336 (1986).
6. A.J. Caputo, D.P. Stinton, R.A. Lowden, and T.M. Beamann, Am. Cera. Soc. Bull. 66, 368 (1987).
7. R. Colmet, I. Lhermitte-Sebire, and R. Naslain, Adv. Cera. Mat. 1, 185 (1986).
8. E. Fitzer, and R. Gadow, Am. Cerm. Soc. Bull. 65, 326 (1986).
9. J.Y. Rossignol, J.M. Quenisset, H. Hannache, C. Mallet, R. Naslain and F. Christin, J. Mat. Sci. 22, 3240 (1987).
10. J.-M. Yang, "Modeling and Characterization of 2-D and 3-D Textile Structural Composites", Ph. D. Thesis, University of Delaware, 1986.
11. C.V. Burkland and J.-M. Yang, Proc. of the 12th Meeting on Advanced Ceramics and Composites, 1988.
12. L.C. Sawyer, R.T. Chen, F. Haimbach, P.J. Harget, E.R. Prack and M. Jaffe, Cera. Eng. Sci. Proc. 7, 914 (1986).
13. T.J. Clark, R.M. Arons and J.B. Stamatoff, Cera. Eng. Sci. Proc. 7, 576 (1985).

MICROSTRUCTURAL CHARACTERIZATION OF A SIC WHISKER-REINFORCED HIPped REACTION-BONDED SI_3N_4

S.C. FARMER, P. PIROUZ, AND A.H. HEUER
Department of Material Science and Eng.
Case Western Reserve University
Cleveland, Ohio 44106

ABSTRACT

A SiC whisker reinforced HIPped RBSN material fabricated with a Y_2O_3 sintering aid was characterized using TEM. The matrix is > 90% β-Si_3N_4 with a Y-Si-O-N glassy phase at the Si_3N_4 grain boundaries and about the SiC whiskers. The SiC whiskers are heavily faulted and have a well defined core. Si_3N_4 precipitates are observed in the core region after composite fabrication. A preliminary mechanism for the growth of the SiC whisker, based on the VLS mechanism is proposed.

INTRODUCTION

SiC whisker-reinforced Si_3N_4 represents one of the promising high temperature ceramic matrix structural composites. Microstructural characterization of one such material,[*] with a room temperature toughness of ~5 $MPa\ m^{1/2}$ and containing 20 v/o of randomly distributed TATEHO (rice hull-derived) whiskers has been made in this study. The reaction-bonded Si_3N_4 (RBSN) matrix contains 4 w/o Y_2O_3 and was nitrided at $T < 1400°C$. The composite was then HIPped to full density at higher temperatures using a proprietary glass encapsulation procedure.[1]

EXPERIMENTAL PROCEDURE

Samples for transmission electron microscopy were ground and polished to a thickness of < 75 μm and argon ion milled to perforation. TEM analysis was performed on a Philips 400T analytical microscope operated at 120 kv and a JEOL 200CX microscope operated at 200 kV.

RESULTS

The overall microstructure is shown in Fig. 1 and reveals that the Si_3N_4 matrix is generally fine-grained, <0.5 μm, while the whisker diameter ranges from 0.2 to 1 μm, with aspect ratios of 5-10. Other details of the matrix microstructure are shown in Fig. 2. The Si_3N_4 is >90% β-Si_3N_4 and occurs both as fine grains and as large prismatic crystals. These growing β grains occasionally engulf small SiC particles as shown in the example of Fig. 1. The Si_3N_4 grains are set in a Y-Si-O-N glassy phase;

*provided by Norton Co., Worcester, Mass.

no evidence for crystallization of this intergranular phase has been seen in TEM. The prismatic Si_3N_4 grains are crystallographically quite perfect, as shown by HREM (such images are not included here); the hexagonal morphology is maintained to quite large sizes, and even when the growing crystals engulf SiC and other impurity phases (WC milling media and Y-Si-O-N glass) during nitridation and coarsening of the β phase. In addition, some of the Si_3N_4 grains assume the morphology of the transient liquid that was present prior to nitridation. Large ~100 μm β-SiC particles, presumably included in the whisker batch, are also present.

The whiskers are heavily faulted so that identification of a single α-SiC polytype is not possible. The whiskers grow with a [0001] growth axis and are a mixture of α and β-SiC polytypes. The faults along the whisker axis arise during VLS growth, probably due to fluctuations in vapor phase transport. The whiskers also possess a core structure which is shown in the end-on micrograph of Fig. 3.

Whiskers within a single batch vary in diameter (0.2 to ~1.0 μm), external symmetry (triangular, hexagonal, or irregular cross-sections), and relative core size (from <1/10 to ~1/3 the whisker diameter). The triangular whiskers are predominantly β-SiC with a <111> growth axis and are facetted on {211}[3]. The hexagonal whiskers are predominantly α-SiC with a [0001] growth axis and are facetted on {10$\bar{1}$0}[3,4]. Whiskers of all external symmetries exhibit the pronounced radial faulting shown in Figures 3 and 4. The radial faults are not linear defects but rather planar defects bounding regions of slightly different orientation or of a different polytype. For example, the whisker in Fig. 4 contains a high density of defects, some of which are in strong contrast. One particular planar defect is arrowed in Fig. 4a. Upon tilting, the change in orientation across the boundary becomes apparent from the sharp change in diffraction contrast, Fig. 4b.

A study of these defects using HREM is in progress. In a preliminary analysis, some of these planar faults show features which are consistent with double positioning boundaries[5], defects common in epitaxial growth on {111} or (0001) substrates. In the dark field image of a longitudinal whisker section, the nature of the radial faults is more clearly seen (Fig. 5). The inclined planar faults give rise to fringe contrast; most faults in this foil, being normal to the electron beam, are out of contrast in this figure. The pairs of partial dislocations which bound the planar faults are visible.

The whiskers have a definite core region over at least a portion of their length. The core has discrete boundaries as can be seen in Fig. 6, and contains many inclusions. No direct diffraction evidence has been obtained to show this region is crystalline. It should be noted that the cores of some as-received whiskers are hollow.

During composite fabrication, the core region is replaced by the transient Y-Si-O-N liquid, Fig. 6a and b. In the whisker of Fig. 6, a portion of the core contains Y-rich glass and a Si_3N_4 single crystal. In the central section of the whisker (arrowed) the enclosing SiC "shell" has been broken through during ion beam thinning and the Si_3N_4 shows clearly. Above and below the arrowed region a portion of the original core structure is visible. The dark field image of Fig. 6b (taken using a Si_3N_4 reflection) shows the internal Si_3N_4 crystal, a Si_3N_4 "cap" at the whisker extremity and Si_3N_4 inclusions in the original core material.

Precipitation of a Si_3N_4 crystal internal to the SiC whiskers is commonly observed. The Si_3N_4 crystal is facetted on {10$\bar{1}$0} and grows with a well-defined orientation relationship to

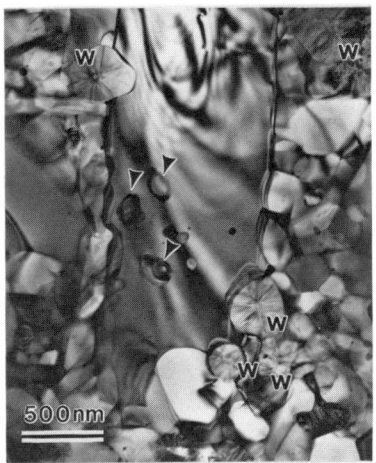

Figure 1 Typical microstructure of the SiC Whisker-reinforced composite. The faulted regions labelled W are the whisker cross-sections. The grains are Si_3N_4; the large one at center contains SiC inclusions (arrowed).

Figure 2 A typical whisker observed longitudinally. Note the high density of planar faults perpendicular to the whisker axis.

Figure 3 A SiC whisker viewed end on in which both prominent radial faulting and a well defined core region are evident.

a) b)

Figure 4 An end-on view of a SiC whisker showing numerous radial planar faults, one of which is arrowed. In b) the whisker is oriented such that the planar nature of the boundary can be distinguished by an abrupt change in diffraction contrast across the fault.

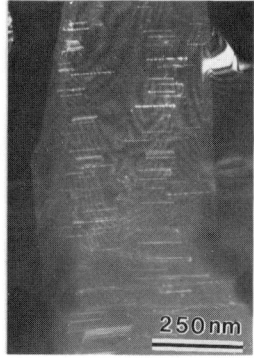 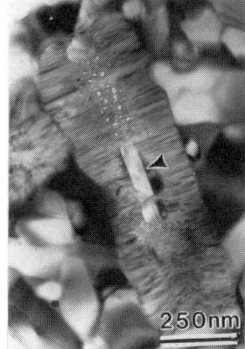 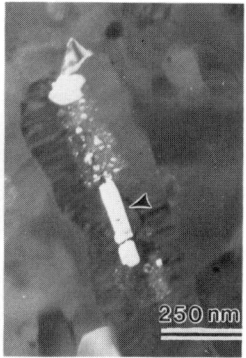

a) b)

Figure 5 A dark field image of a longitudinal whisker section showing the partial dislocations which bound the planar faults.

Figure 6 A longitudinal whisker section showing portions of the original core and core regions replaced by the Y-Si-O-N glass (dark contrast), from which a Si_3N_4 crystal precipitates. The dark field micrograph b) taken using a Si_3N_4 reflection images both the Si_3N_4 crystal and smaller Si_3N_4 inclusions in the original core material.

the SiC whisker in which it forms:

$$(30\bar{3}3)_{Si_3N_4} \parallel (10\bar{1}3)_{SiC}$$

$$[1\bar{2}10]_{Si_3N_4} \parallel [1\bar{2}10]_{SiC}$$

(Indices for the 2H polytype of SiC were arbitrarily chosen.)
The mismatch in interplanar spacings is small ~0.7 per cent, the Si_3N_4 reflections occurring at 1/3 and 2/3 of the SiC spacing.

DISCUSSION

The Si_3N_4 matrix microstructure of the composite material does not differ substantially from what has already been presented in the literature for polycrystalline Si_3N_4 materials. The complex microstructure of the whiskers proved of greater interest. The present results suggest a possible mechanism for the whisker formation process, which will be discussed here, although research is continuing.

It is generally proposed that the rice hull-derived whiskers grow by the VLS mechanism, Fig. 7. VLS grown whiskers are frequently facetted on {10$\bar{1}$0} α-SiC or {211} β-SiC planes[3,4,6]. Because of the short diffusion lengths, and the thermal gradient, it is likely that solidification starts at the periphery of the solid/liquid interface[3]. Once nuclei form, they grow radially inward, following the thermal gradient, and "pushing" the impurities in the liquid towards the center of the whisker. The highly impure front, containing the various impurities in the pyrolized rice hull, finally solidify at the central region, thus forming the core of the whisker.

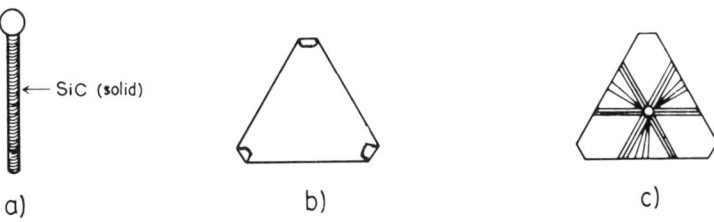

Figure 7 A schematic of VLS whisker growth. Nucleation occurs at the periphery of the liquid-solid interface. Independently nucleated regions b) grow inward, impinge, and create radial planar faults the majority of which lie parallel to the external faces.

Each nucleus could be i) in a parallel epitaxial and identical stacking sequence with the underlying layer or ii) in a parallel epitaxial relation but having a different stacking. In addition, independently nucleated regions may be rotated by 60° with respect to a neighboring nuclei, resulting in the formation of double positioning boundaries when they meet. The nuclei themselves tend to grow as facetted plates of different widths. In this manner, many different layers form during whisker growth each having a potentially different stacking sequence (3C, 6H, etc.) with each layer having planar boundaries due to the different solid fronts which were independently nucleated.

CONCLUSIONS

A SiC whisker reinforced HIPped RBSN fabricated with a Y_2O_3 sintering aid has been examined using TEM. The matrix microstructure consists of > 90% β-Si_3N_4 occuring both as small grains and large prismatic crystals. There is a Y-Si-O-N glassy phase at all Si_3N_4 grain boundaries and between the SiC whiskers and the matrix grains. The SiC whiskers are heavily faulted and contain α and β polytypes and a well defined core. In addition to faulting in the stacking sequence of the close-packed SiC planes, planar faults which extend from the whisker exterior to the core region are observed. During processing, the core is replaced by Y-Si-O-N glass, from which Si_3N_4 crystals may precipitate. A preliminary mechanism, based on the VLS process, has been proposed for the formation of whiskers, which accounts for the axial polytypic faulting, the radial occurrence of planar faults, and the formation of a core region.

ACKNOWLEDGEMENTS We would like to thank N. Corbin for useful discussion and thank N. Corbin and M. Ruhle for the samples received. This work is supported by DARPA #N00014-86-K-0773.

REFERENCES

1. C.A. Wilkins and N.D. Corbin, CCM Meeting Proceedings October 1987, to be published.
2. N.D. Corbin (private communication).
3. G.A. Bootsma, W.F. Knippenberg and G. Verspui, J. Crystal Growth 11, 297 (1971).
4. S.R. Nutt, J. Am. Ceram. Soc. 67[6], 428 (1984); 71[3], 149 (1988).
5. E.W. Dickson and D.W. Pashley, Phil. Mag. 7, 1315 (1962).
6. J.V. Milewski, F.D. Gac, J.J. Petrovic and S.R. Skaggs, J. Mat. Sci. 20, 1160 (1985).

SUSPENSION PROCESSING OF SiC WHISKER-REINFORCED COMPOSITES

MICHAEL D. SACKS, HAE-WEON LEE, AND OSWALDO E. ROJAS
Department of Materials Science and Engineering, University of Florida, Gainesville, FL 32611

ABSTRACT

Suspension processing was used to prepare Al_2O_3/SiC whisker and Al_2O_3/ZrO_2/SiC whisker compacts with high green density, fine pore sizes, and homogeneous microstructure. Procedures for optimizing particle/whisker co-dispersion and subsequent consolidation by slip casting are reviewed. Compacts prepared by slip casting showed significantly higher sintered densities compared to dry-pressed samples.

INTRODUCTION

Ceramic-matrix composites reinforced with SiC whiskers are of considerable interest for applications requiring improved mechanical properties, such as fracture toughness, creep resistance, and thermal shock resistance [1-5]. However, the addition of whiskers to powder compacts tends to inhibit densification and, in most studies, hot pressing has been used to produce samples with high relative density. In order to fabricate complex shapes more readily (as well as to reduce processing costs), it is important to develop methods for pressureless sintering of whisker-containing compacts. A variety of processing strategies have been investigated, such as (1) using additives which promote liquid-phase sintering [5], (2) using whiskers with lower aspect ratio in order to reduce the development of shrinkage-inhibiting whisker-network structures [5-7], and (3) preparing compacts with improved green microstructures [6-8]. The latter approach has been emphasized by the present authors. This paper reviews the use of suspension processing to produce whisker-containing powder compacts with homogeneous microstructure, fine pore sizes, and high relative density. It will be demonstrated that compacts with these characteristics show enhanced densification compared to dry-processed samples.

PROCESSING CONCEPTS

Powder compact densification can be enhanced by using samples with improved green microstructural characteristics, i.e., with more homogeneous particle packing and higher overall relative density. These characteristics are more readily achieved using powder/liquid forming techniques, i.e., compared to dry processing methods. In the dry state, fine particles cluster together and form agglomerates. These agglomerates tend to retain their identity during dry powder consolidation, resulting in green compacts with variations in packing density (poor homogeneity), relatively large interagglomerate pores, and lower overall relative density. In contrast, it is often possible to achieve good dispersion of powders (in which the "primary" particles are separated) by suspension-processing methods. Consolidation of well-dispersed particle/liquid systems generally results in compacts with greatly improved green microstructures.

There are several requirements for optimum dispersion of a powder in a liquid medium: (1) There must be good wetting of the powder by the liquid. (2) Agglomerates must either be broken down or removed from the system. (3) It is necessary to overcome the tendency of the particles to flocculate (i.e., to re-agglomerate in the liquid medium). These requirements are discussed in more detail below.

Wetting Behavior

The wetting characteristics of a liquid on a solid surface (e.g., ceramic particles) depends on three specific interfacial energies: solid-vapor (γ_{sv}), liquid-vapor (γ_{lv}), and solid-liquid (γ_{sl}). In the absence of chemical reactions between the solid and liquid, the contact angle, θ, for the liquid on the solid surface is given by the Young-Dupre equation:

$$\cos \theta = \frac{\gamma_{sv} - \gamma_{sl}}{\gamma_{lv}}$$

Thus, good wetting and spreading behavior (i.e., $\theta \rightarrow 0°$) is promoted by high γ_{sv}, low γ_{lv}, and low γ_{sl}. The first two conditions are readily met in most ceramic particle/liquid suspensions. The third condition (low γ_{sl}) is also satisfied in many aqueous suspensions of oxide powders because the particle surfaces are often well-covered with hydroxyl groups.* However, in non-oxide powders (e.g., SiC whiskers), surface hydroxylation is often more limited and difficulties in wetting may arise. The present authors have observed that aging (in water) of as-received SiC whiskers results in substantial improvements in the wetting and dispersion behavior [7].

Agglomerate Breakdown/Removal

The method used to break down agglomerates depends on the strength of the interparticle bonds of the agglomerates. "Soft" agglomerates, in which particles are held together by relatively weak surface interactions (e.g., van der Waals forces), generally can be broken down by moderate shearing action. For example, Fig. 1 shows the effect of increasing ultrasonication time on the rheological (i.e., viscosity vs. shear rate) behavior of suspensions prepared with 50 vol% Al_2O_3/50 vol% water. The viscosity decreases and the flow behavior becomes less shear thinning, indicating that agglomerate breakdown occurs with continued sonication. (In suspensions containing agglomerates, some of the liquid is immobilized in the inter-particulate pore spaces. This results in a higher "effective" solids concentration, and a higher viscosity, relative to a suspension in which the "primary" particles are well-separated.)

"Hard" agglomerates (also called "aggregates"), in which the primary particles are bonded together by solid bridges, must be subjected to more energy-intensive grinding processes (e.g., ball milling, attrition milling, etc.) to be broken down. Figure 2 shows the effect of ball milling time on the suspension viscosity and powder sediment density obtained for 30 vol% Al_2O_3/70 vol% methanol suspensions. Initially, the viscosity was high (Fig. 2(A)) and bulk density upon sedimentation was low (Fig. 2(B)) because the powder contained hard agglomerates. The large decrease in viscosity and increase in sediment density during the first few hours of milling reflect the breakdown of these agglomerates. Most of the agglomerates were broken down after milling for approximately 4 h, as indicated by the leveling off in the viscosity and sediment density values.

Stabilization Against Flocculation

Although powder agglomerates may be initially broken down by an input of mechanical energy, there is still a tendency for particles to flocculate

---

* Wetting and dispersion of oxide powders in water may become difficult when the powders are calcined at temperatures high enough to cause substantial surface dehydroxylation.

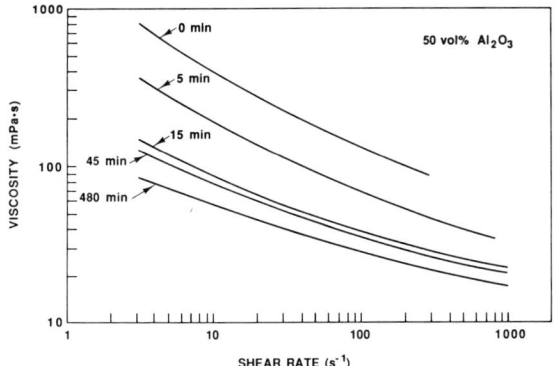

Fig. 1. Plots of viscosity vs. shear rate for 50 vol% Al_2O_3 suspensions prepared with the indicated sonication times.

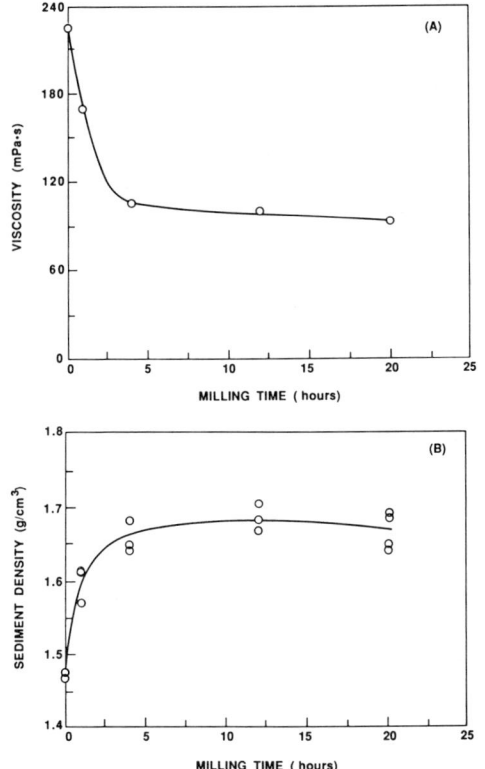

Fig. 2. Plots of (A) viscosity (measured at a shear rate of 37.5 s^{-1}) and (B) sedimentation bulk density vs. milling time for Al_2O_3 suspensions.

due to interparticle attractive forces, e.g., van der Waals forces. To maintain the dispersed state, it is necessary to develop interparticle repulsive forces of sufficient magnitude to overcome the attractive forces. One of the common mechanisms for "stabilizing" suspensions against flocculation is to impart a surface charge (either positive or negative) on the particles. Surface charge development occurs when ionic species either adsorb from the solution phase onto the particle surfaces or dissolve from the particle surfaces into the solution phase. In practice, this is usually accomplished by adjusting the pH (i.e., acid or base additions) and/or by adding "dispersants" (e.g., simple electrolytes or polyelectrolytes).

In order to electrostatically stabilize suspensions containing several particulate materials, it is necessary to develop a sufficiently high surface charge of the same sign (either positive or negative) on each component. In the present study, this was accomplished by pH adjustment. Figure 3 shows plots of zeta potential (i.e., the near-surface electrical potential) vs. pH for Al_2O_3, ZrO_2, and SiC whisker suspensions. For each material, it was possible to achieve a high zeta potential (absolute value) at both low and high pH values.

To achieve the maximum stability against flocculation by the electrostatic mechanism, it is also important to avoid excessive ionic strength of the solution phase. According to DLVO theory [9], the repulsive force between particles decreases as the concentration and valence of ions in solution increase. Therefore, it is important to avoid extreme values of the suspension pH in processing because the acid or base that is used to adjust the pH (i.e., to produce the surface charge on particles) also contributes ions to the solution phase. Furthermore, as the ionic strength increases, the electrical double layer is "compressed" and the zeta potential tends to decrease. This is illustrated in Figs. 3(A) and 3(B) for the Al_2O_3 and ZrO_2 suspensions, respectively, by the decreases in the absolute values of the zeta potential at the extreme pH values. Figure 4(A) shows that very high acid/base additions were required to adjust the pH to extreme values in concentrated Al_2O_3 suspensions and, thus, the ionic strengths became very high. Similar results were obtained for ZrO_2 suspensions [8].

In contrast to the behavior observed with Al_2O_3 and ZrO_2, the zeta potentials for the SiC whiskers (Fig. 3(C)) did not exhibit local maxima (in the absolute values) at the extreme pH values. This reflects the fact that the suspensions prepared for these measurements were very dilute (<0.1 vol%) and, therefore, the amount of acid or base needed to adjust the pH was minimal. Hence, the ionic strengths remained low for these suspensions over the range of pH values used.

The results described above illustrate that it becomes increasingly important to exercise control over the ionic strength as the suspension solids concentration increases. As discussed in the section below, it is generally desirable to use as high a suspension solids loading as possible during the consolidation (i.e., shape forming) step. Thus, in most electrostatically-stabilized suspensions used in ceramic processing, it is important to strike a balance between high zeta potential and moderate ionic strength in order to maximize stability against flocculation. In the present study, the optimum pH for processing was chosen as pH≈4 since concentrated suspensions for each material showed excellent stability at this value. This was indicated by measurements of suspension viscosity and slip cast green density. For example, results for Al_2O_3 suspensions show that the viscosity was minimized (Fig. 4(B)) and the green relative density was maximized (Fig. 4(C)) at pH≈4.

It was somewhat fortuitous that, in the present study, each component showed excellent stability in the same pH range. In cases where this does not occur, alternative methods of imparting surface charge (e.g., polyelectrolyte additions) or alternative methods of achieving stability against flocculation (e.g., "steric" stabilization) may be used.

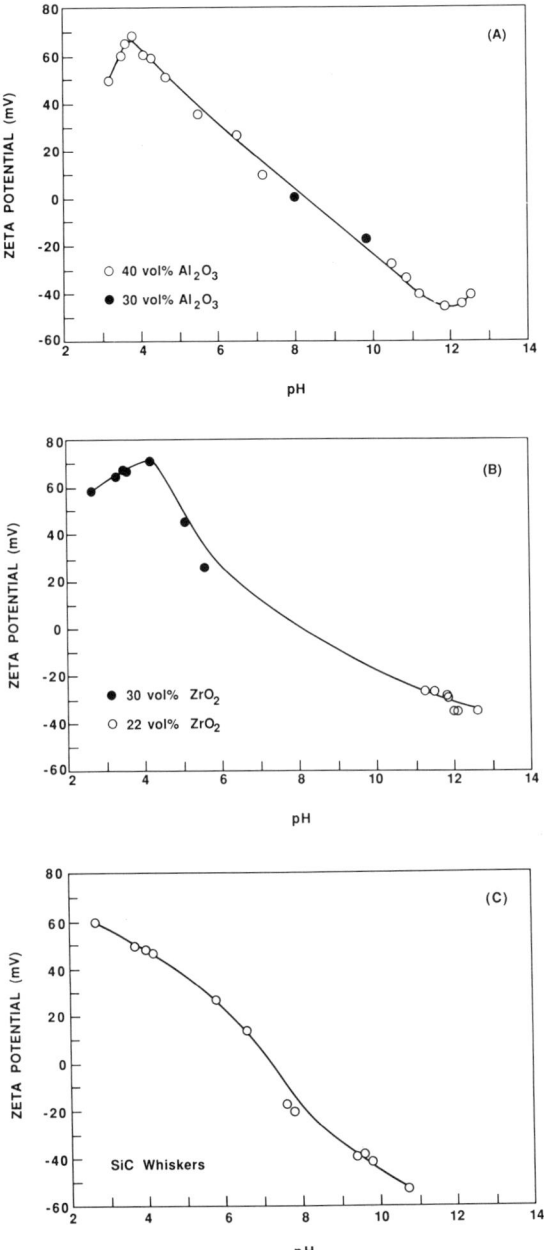

Fig. 3. Plots of zeta potential vs. pH for (A) Al_2O_3, (B) ZrO_2, and (C) SiC whiskers.

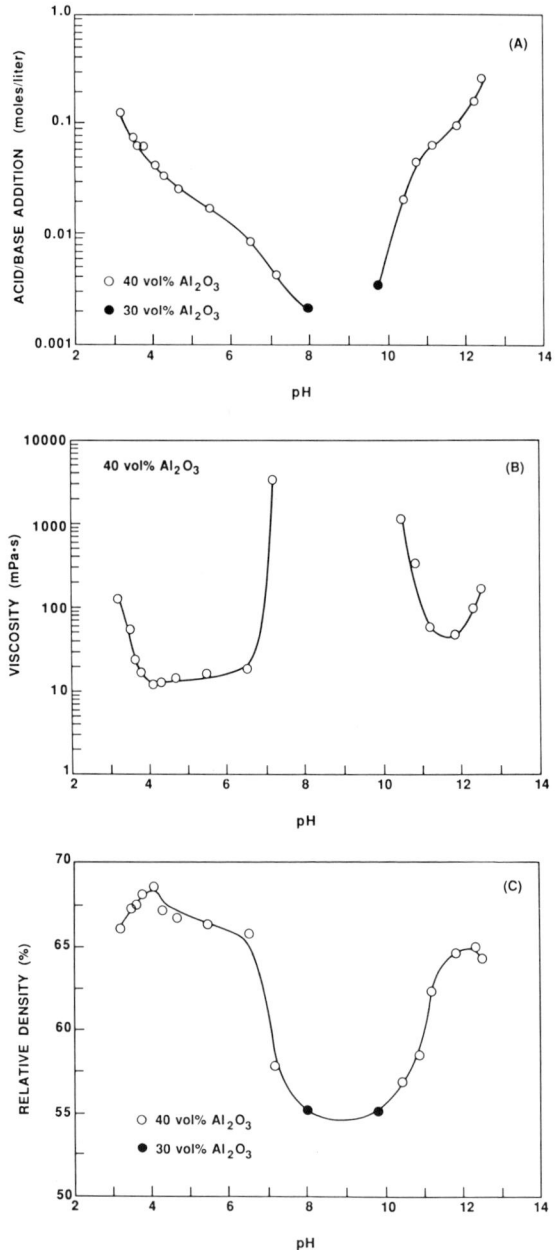

Fig. 4. Plots of (A) acid or base concentration vs. pH, (B) suspension viscosity vs. pH, and (C) relative density of slip cast samples vs. pH of Al_2O_3 suspensions.

Green Body Formation

In addition to using well-dispersed suspensions, the development of green compacts with optimum microstructure requires that the suspensions remain homogeneous during the consolidation process. There is a tendency in both single component and multicomponent suspensions for particles to segregate due to differences in particle size or specific gravity. Thus, particles with larger size and/or higher density, which settle at a faster rate, tend to accumulate toward the bottom of a cast sample, while particles with smaller size and/or lower density tend to remain near the top of the sample. There are a variety of approaches for overcoming this problem, such as (1) using suspensions with higher solids concentration in order to promote hindered settling, (2) consolidating suspensions more rapidly to minimize the amount of time available for segregation to occur, (3) using suspensions with higher solution viscosity to minimize the settling of all the solid phases during consolidation, and (4) using dispersion conditions which cause components to heterocoagulate. (The last approach, however, has the disadvantage that cast samples tend to have lower green density compared to samples prepared from co-dispersed suspensions.) In the present study, the first and second strategies were used. Suspensions were prepared with solids loadings ≥ 45 vol% and consolidation was carried out by slip casting. Figure 5 shows the pore size distributions and relative green density values obtained by mercury porosimetry for Al_2O_3/SiC whisker and $Al_2O_3/ZrO_2/SiC$ whisker compacts. All of the samples had high relative densities (~66-69%) and small median pore channel radii (≤ 50 nm).

Sintering

Densification of whisker-containing composites was considerably enhanced by using homogeneous, high green density samples, such as those described in Fig. 5. Figure 6 shows the green and sintered relative densities for both dry-pressed and slip-cast Al_2O_3/SiC whisker compacts having several whisker concentrations. The slip-cast samples have much higher green densities and this results in much higher sintered densities. However, it is also evident from Fig. 6 that the SiC whiskers retard densification considerably. At 1450°C, the sintered densities of all the slip-cast Al_2O_3/SiC whisker compacts remain <85%, while the slip-cast Al_2O_3 sample has a sintered density ≈99%. However, the densities for the samples with 5 and 15 vol% whiskers were increased considerably (see Fig. 7) by using a higher sintering temperature (i.e., 1800°C for 30 min).

The data from this study was compared to results reported by Tiegs and Becher [5] on Al_2O_3/SiC samples which were liquid-phase sintered with MgO/Y_2O_3 additives. (To enhance densification, the SiC whiskers used in the study by Tiegs and Becher were also ball milled for 4 h to reduce the aspect ratio.) Figure 7 shows that similar sintered density values were obtained in the two studies for the lower whisker concentrations ($\leq 15\%$), while the trend at higher whisker loadings indicates that suspension processing produces higher densities. However, the sintered density for the sample with 30 vol% whiskers is still relatively low.

The sintering behavior of slip-cast $Al_2O_3/ZrO_2/SiC$ whisker compacts was also investigated. Table I gives the results for several compositions which were sintered at 1700°C for 30 min. As in the case of the Al_2O_3/SiC whisker compacts, it was possible to achieve high relative densities (~94-99%) for samples with ≤ 15 vol% whiskers.

SUMMARY

Suspension processing was used to prepare homogeneous Al_2O_3/SiC whisker and $Al_2O_3/ZrO_2/SiC$ whisker green bodies with high relative density (~66-69%) and small median pore channel radii (≤ 50 nm). Although densification was

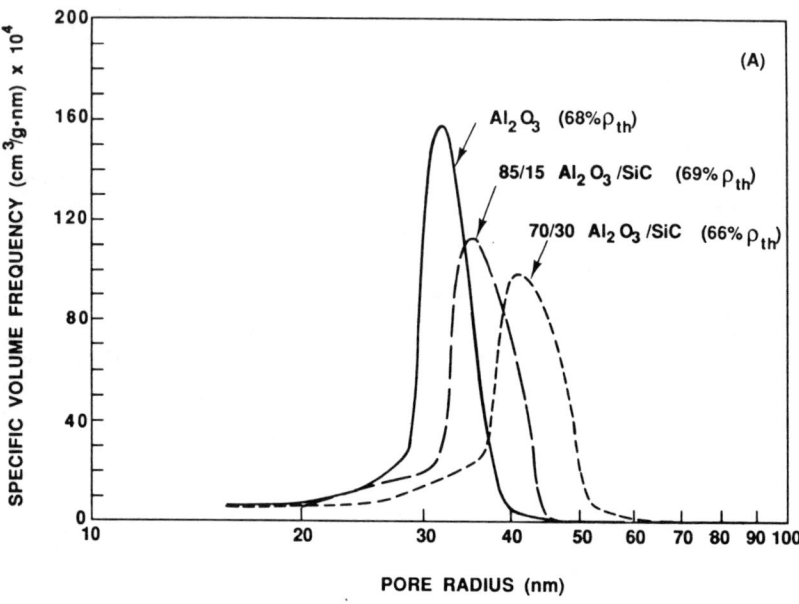
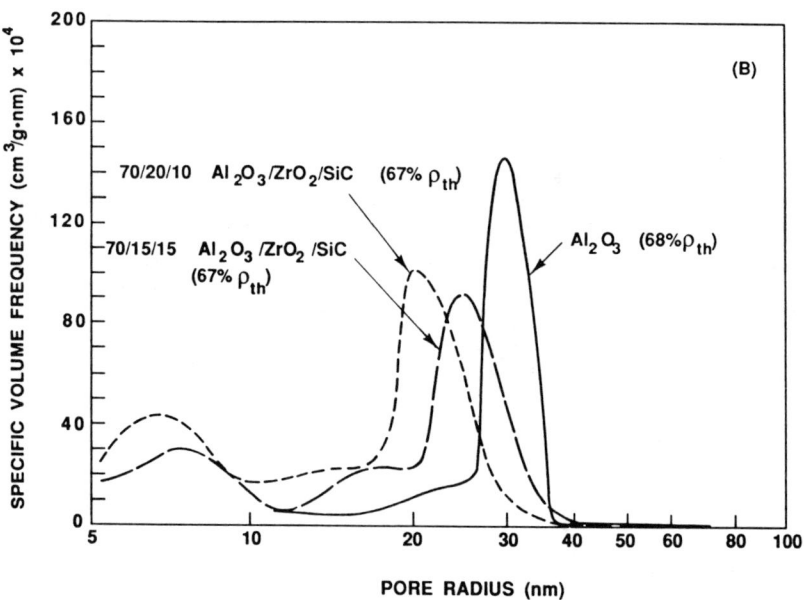

Fig. 5. Plots of specific pore volume frequency vs. pore radius obtained by mercury porosimetry for slip-cast compacts containing several ratios of (A) Al_2O_3/SiC whiskers and (B) Al_2O_3/ZrO_2/SiC whiskers.

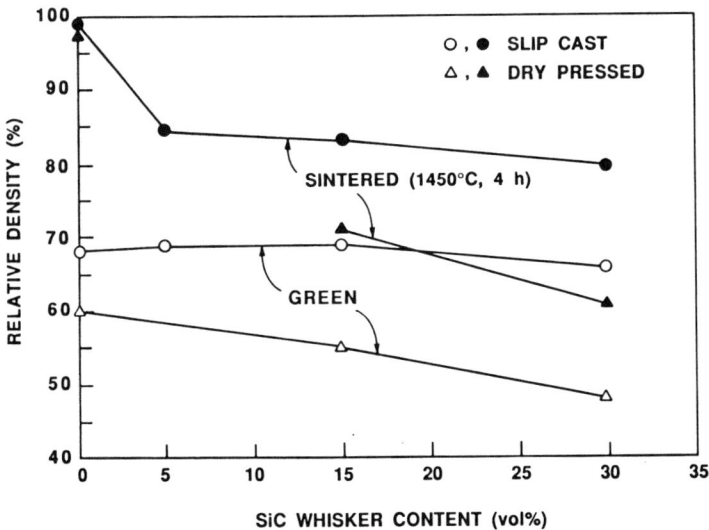

Fig. 6. Plots of green and sintered relative density vs. vol% SiC whiskers for Al_2O_3/SiC whisker samples prepared by slip casting and dry pressing.

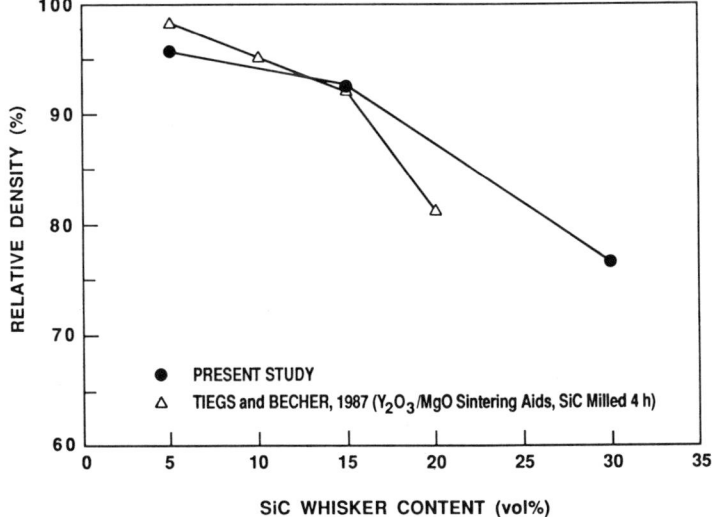

Fig. 7. Plots of sintered relative density vs. vol% SiC whiskers for Al_2O_3/SiC whisker samples. The results obtained in the present study are compared with results for liquid-phase sintered samples reported by Tiegs and Becher [5].

Table I. Green and Sintered Densities for $Al_2O_3/ZrO_2/SiC$ Whisker Samples

$Al_2O_3/ZrO_2/SiC$	Green Density (%)	Sintered Density (%)
85/10/5	68	99
70/25/5	66	97
70/20/10	67	96
70/15/15	67	94

severely inhibited by the SiC whiskers, slip-cast samples sintered to considerably higher densities compared to dry-pressed samples. Al_2O_3/SiC whisker and Al_2O_3/ZrO_2/SiC whisker compacts with $\leq$15 vol% whiskers could be sintered to ~93-99% relative densities at temperatures in the range 1700-1800°C for 30 min.

ACKNOWLEDGMENT

The authors gratefully acknowledge that support for this work was provided by the National Science Foundation, Division of Materials Research, Ceramics and Electronic Materials Program (DMR-8451916), the U.S. Army Research Office, Materials Science Division (DAAG29-85-K-0156), and the following NSF-PYI sponsors: E.I. du Pont de Nemours & Co., General Electric Co., General Motors, Corp., and Sohio Engineered Materials Co.

REFERENCES

1. G.C. Wei and P.F. Becher, Am. Ceram. Soc. Bull. 64 (2), 298-304 (1985).
2. A.H. Chokshi and J.R. Porter, Comm. Am. Ceram. Soc. 68 (6), C-144 - C-145 (1985).
3. P.D. Shalek, J.J. Petrovic, G.F. Hurley, and F.D.Gac, Am. Ceram. Soc. Bull. 65 (2), 351-356 (1986).
4. J.R. Porter, F.F. Lange, and A.H. Chokshi, Am. Ceram. Soc. Bull. 66 (2), 343-347 (1987).
5. T.N. Tiegs and P.F. Becher, Am. Ceram. Soc. Bull. 66 (2), 339-342 (1987).
6. F. Takao, W.R. Cannon, and S.C. Danforth, in Ceramic Powder Science, Advances in Ceramics, Vol. 21, edited by G.L. Messing, K.S. Mazdiyasni, J.W. McCauley, and R.A. Haber (American Ceramic Society, Westerville, OH 1987), pp. 699-708.
7. M.D. Sacks, H.W. Lee, and O.E. Rojas, to be published in J. Am. Ceram. Soc. 71 (1988).
8. M.D. Sacks, H.W. Lee, and O.E. Rojas, in Ceramic Powder Science II, Part A, Ceramic Transactions, Vol. 1, edited by G.L. Messing, E.R. Fuller, Jr., and H. Hausner (American Ceramic Society, Westerville, OH 1988), pp. 440-451.
9. J. Th. G. Overbeek, in Colloid Science, edited by H.R. Kruyt (Elsevier, Amsterdam, 1952), pp. 247-277.

THEORETICAL ANALYSIS OF CHEMICAL VAPOR INFILTRATION IN CERAMIC / CERAMIC COMPOSITES

NYAN-HWA TAI AND TSU-WEI CHOU
Center for Composite Materials and Department of Mechanical Engineering, University of Delaware, Newark, Delaware 19716

ABSTRACT

A model for the deposition of alumina and titanium carbide within a ceramic fiber bundle from the chemical reactions has been examined in this paper. The model considers vapor diffusion, chemical reaction on the inner surface of the capillary, deposition film growth, porosity, and effects of reactants composition at various reactor temperature and pressure. Binary, multicomponent diffusion and Knudsen diffusion have been taken into account for the different stages of the CVI process. Furthermore, both diffusion controlled and reaction controlled processes have been examined in order to determine the dominating process in chemical vapor infiltration.

INTRODUCTION

Chemical vapor infiltration (CVI) technique is a relatively new method for processing ceramic fiber-reinforced ceramic matrix composites; it minimizes mechanical, chemical, and thermal degradation of the fibers. The CVI process involves the infiltration of vapors into a fibrous preform and the deposition of a solid product from the chemical reaction of the vapor species to form the matrix of the composite. The deposition thickness is affected by the condition of the reactor and the composition of vapors. For example, Refs.[1-3], based on thermodynamic calculations, reported the effect of pressure ratio of inlet vapor to deposition components and suggested an optimum partial pressure of vapor species in chemical vapor deposition (CVD). Furthermore, the reactor temperature, total pressure, partial pressures of the vapor species, and the microstructure of fibrous preforms are all interrelated and will affect not only the deposition compositions but also the porosity of the composites[8]. Premature pore closures within the preform can be alleviated by proper control of the reactor conditions, and hence, achieve a better densified composite.

Analytical modeling of the CVI process has been reported in Refs.[4~9]. Ref.[4] predicts the optimum processing conditions, as well as the resulting composite structure and properties; a constant deposition thickness throughout a volume element of the preform is assumed. Ref.[5] studies an one dimensional model for a Nicalon cloth lay up preform with a SiC matrix processed by CVI. Ref.[6] examines the diffusion process in a two-dimensional preform, and a modified van den Brekel model is used to calculate the deposition profile of the porous media. In Ref.[7], both experimental and analytical results for TiC deposition on the inner surface of a capillary for infinitely long time are reported. In Refs.[8,9], a capillary model was used to simulate the CVI process for a unidirectional fiber reinforced ceramic composites. The growth history of alumina matrix within the capillary was analyzed. The proper reactor condition was recommend based upon the theoretical analysis.

This paper examines the CVI process for a unidirectional ceramic-ceramic composite by considering the CVD inside a cylindrical tube. The objectives of this paper are to predict (1) the deposition profile within the capillary, (2) proper reactor conditions for an isothemal CVI process, (3) the processing time influenced by the reactor conditions, and (4) the porosity - time - reactor condition relations of the CVI product. An approximate solution for this modelization has been obtained. Based on this solution as well as considerations of the chemical reaction rate and diffusivities in different reactor conditions, the deposition rate and thickness can be predicted. The deposition of both alumina and titanium carbide within the inner wall of a capillary has been considered. A good agreement has been achieved in the comparison of theoretical predictions and experimental data of Ref.[7] for the deposition profiles of TiC.

THEORETICAL ANALYSIS

Basic assumptions

The modeling analysis considers a unidirectional compact fiber bundle situated in a reactor. The following assumptions are made regarding the fiber preform and vapor deposition process:
1. The fibers of identical radius in the bundle are arranged in a hexagonal - close - packed array.
2. The cross-section of the void space between the fibers is circular in shape.
3. Vapors in the reactor are homogeneous.
4. The first order chemical reactions of the vapor species within the preform are heterogeneous.
5. No bulk flow in capillary during the CVI process.

The chemical vapor reactions for the deposition of Al_2O_3 and TiC within a preform are

$$H_{2(g)} + CO_{2(g)} = H_2O_{(g)} + CO_{(g)} \tag{1}$$

$$2AlCl_{3(g)} + 3H_2O_{(g)} = Al_2O_{3(s)} + 6HCl_{(g)}$$

and

$$TiCl_{4(g)} + CH_{4(g)} = TiC_{(s)} + 4HCl_{(g)} \tag{2}$$

respectively.

Governing equation and boundary conditions

Consider the cylindrical void space along which diffusion takes place (Fig.1). A control volume is shown in cylindrical coordinates (r, θ, z) with N denoting the molar flux (mole/m^2·sec). The mass conservation in the unsteady state leads to the following governing equation.

$$\frac{\partial C}{\partial t} + \frac{1}{r}\frac{\partial rN_r}{\partial r} + \frac{\partial N_z}{\partial z} = 0 \tag{3}$$

Here, C is the mole of the total vapor per unit volume (mole / m^3) and t denotes time (sec). N_r and N_z are, respectively, the molar flux in the r and z directions (mole / m^2·sec).

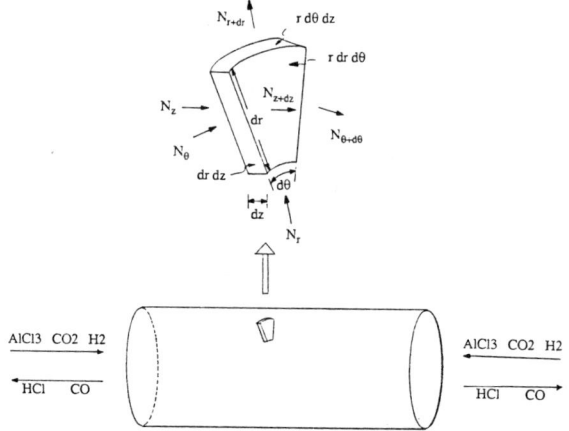

Fig.1 A control volume for the unidirectional fiber preform under CVI

is : The relationship between the molar flux N, and molar fraction x of a given vapor species

$$N_j = -CD\frac{\partial x}{\partial j} \qquad j = r, z \qquad (4)$$

D is diffusivity (m^2/sec). The governing equation for the i-th species of a multicomponent system becomes

$$\frac{\partial C_i}{\partial t} - \frac{CD}{r}\frac{\partial}{\partial r}\left(r\frac{\partial x_i}{\partial r}\right) - CD\frac{\partial^2 x_i}{\partial z^2} = 0 \qquad (5)$$

Here, x_i is the molar fraction of the i-th species. In the case of steady state and defining $x_i = C_i/C$, the governing equation is reduced to

$$\frac{\partial}{\partial r}\left(r\frac{\partial X_i}{\partial r}\right) + r\frac{\partial^2 X_i}{\partial z^2} = 0 \qquad (6)$$

and the boundary conditions of the problems are:

1. $X_i = 0$ at $z = 0$ and L \hfill (7a)

2. X_i is finite at $r = 0$ \hfill (7b)

3. $X_i|_{r=R} = \dfrac{N_{iR}(z)}{CK} - x_{i0}$ \hfill (7c)

Solution of the governing equation

$$X_i(r,z) = \sum_{m=0}^{\infty} E_m I_0\left(\frac{m\pi r}{L}\right) \sin\left(\frac{m\pi z}{L}\right) \qquad (8)$$

where

$$E_m = \frac{-4\dfrac{x_{i0}}{(2m+1)\pi}}{I_0\left(\dfrac{(2m+1)\pi R}{L}\right) + \dfrac{D}{K}\left(\dfrac{(2m+1)\pi}{L}\right)I_1\left(\dfrac{(2m+1)\pi R}{L}\right)} \qquad (9)$$

I_0 and I_1 denote the Bessel functions.

Diffusion coefficients

The diffusion coefficient D in the vapor deposition process varies with temperature, pressure, vapor composition, and the pore diameter. The dominating diffusion processes can be categorized by the ratio of the mean free path of molecule λ(m) to the average pore radius r(m), as following[10]

$$\frac{\lambda}{2r} \geq 10 \qquad : \text{Knudsen diffusion}$$

$$0.01 \leq \frac{\lambda}{2r} \leq 10 \qquad : \text{transition region diffusion}$$

$$0.01 \geq \frac{\lambda}{2r} \qquad : \text{molecular diffusion}$$

In Knudsen diffusion[10]:

$$D = D_K = 9700\, r \left(\frac{T}{M}\right)^{0.5}$$

In the transition region diffusion[6]:

$$\frac{1}{D} = \frac{1}{D_{mol}} + \frac{1}{D_K}$$

where D_{mol} denotes the molecular diffusion coefficient.
In molecular diffusion[11]:

$$D = D_{mol} = \frac{10^{-3}\, T^{1.75} \left(\frac{1}{M_1} + \frac{1}{M_2}\right)^{0.5}}{P\left[\left(\sum_i v_{i1}\right)^{\frac{1}{3}} + \left(\sum_i v_{i2}\right)^{\frac{1}{3}}\right]^2}$$

where M : molecular weight (kg/kg mole), and v_i : diffusion volumes (determined by experiments).

Reaction rate

The reaction rate adopted in the present calculation is based upon Ref. [12] and the Arrhenius equation:

$$K = A_0 \exp(-E_a/RT)$$

where A_0: constant (m^3/mole·sec), R: gas constant (8.3143 J/K mole), T: temperature (°K) and E_a: activation energy (J/mole). The data which concerning the reaction rate is deduced from the deposition rate of alumina and titanium carbide [12,7].

NUMERICAL RESULTS AND DISCUSSIONS

The ratios of concentrations of the CO_2 vapor inside and outside of the capillary tube at various temperature along the tube length are shown in Fig.2. At lower temperatures, the reaction among the vapors is less active, and hence, more vapor can infiltrate into the middle section of the fiber bundle. As a result, the concentration profile is less varied.

The deposition thicknesses profile of the final product at two temperatures are shown in Fig.3. A relatively thicker layer is obtained at lower deposition temperature, as expected.

The growth of the deposition film is shown in Fig.4. The growth rate is obviously higher near the ends of the fiber bundle than at the middle section due to the variation in vapor concentration. Also, at a given section of the fiber bundle, the growth rate decreases with time; this is due to the decrease in inlet area and consequently less vapor can reach this section.

Fig.5 shows the reaction rate of the vapor species is sensitive to temperature. A lower reaction temperature implies a lower reaction rate and hence lessens the premature pore closure, but longer time is necessary to complete the process. A comparison of curves C and D shows that a change of temperature from 1200 (k) to 1150 (k) induces 10.4% increase in porosity (45.2% vs. 34.8%) and 69.3% increase in deposition time (400 min vs. 677 min). Similar trend is obtained for TiC deposition (Fig.6).

When α (P_{H2} / P_{CO2}) is altered from 1 to 12, obvious changes in deposition profile occurs (Fig.7). This is due to the change of vapor compositions and reaction rate. As α increases further to 30, no obvious change in deposition profile is noticed, but longer time is needed to complete the process. Comparing the reaction rates of the cases of $\alpha = 12$ and $\alpha = 30$, nearly the same reaction rate is attained. This is due to the fact that the deposition rate difference was compensated by the difference in concentrations.

The relationship between Sherwood number ($Sh=KR/D$), which reflects the reactor condition, processing period, and porosity is shown in Fig.8. From this figure, the proper reactor conditions for specific product requirements can be determined.

The results of the present analysis which considers a steady state deposition and moving boundary condition agrees with the experimental data of van den Brekel et al [7]. This is shown in Fig.9.

ACKNOWLEDGMENTS

This work was partially supported by the Air Force Office of Scientific Research (contract no. AFOSR-87-0383, Dr. Alan Rosenstein is the program manager) and the Center for Composite Materials of University of Delaware.

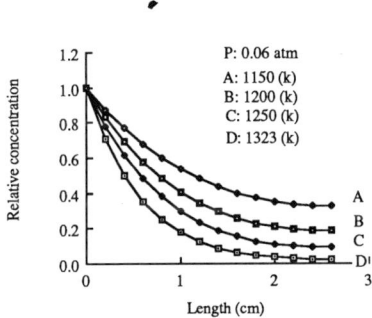

Fig.2 Effect of temperature on the concentration within the capillary

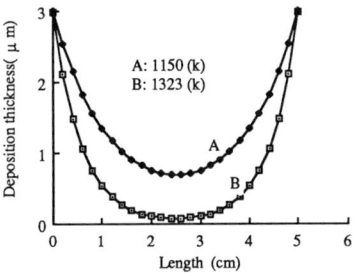

Fig.3 Alumina deposition profile of final product by CVI
for different reactor temperature at P=0.06 atm

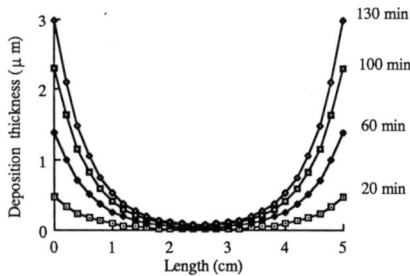

Fig.4 Alumina film growth history at T = 1323 (k) and P = 0.06 atm

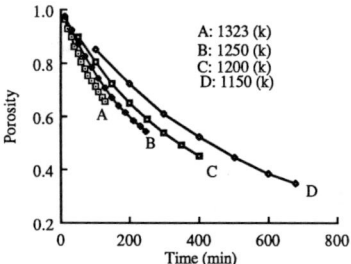

Fig.5 Effect of temperature on porosity and processing time for alumina deposition at P=0.06 atm and α (=P_{H2}/P_{CO2}) = 1

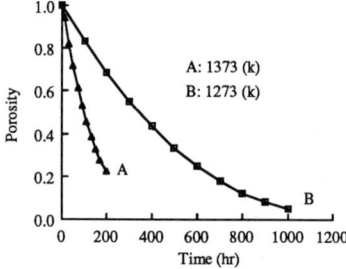

Fig.6 Effect of temperature on porosity and processing time of TiC deposition at p = 0.0888 atm

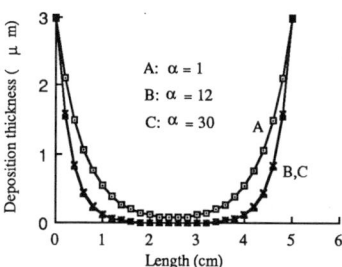

Fig.7 Effect of α ($=P_{H2}/P_{CO2}$) on alumina deposition profile by CVI at p = 0.06 atm and T = 1323 (k)

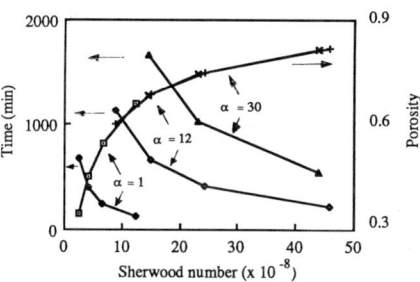

Fig.8 Porosity and processing time at different Sherwood numbers for alumina deposition by CVI

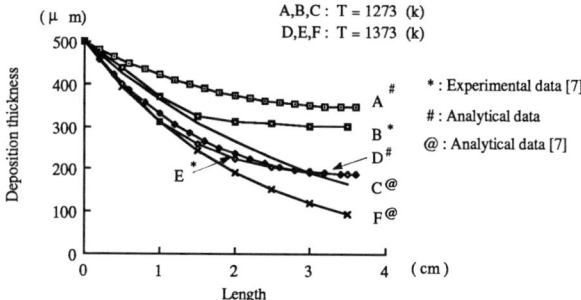

Fig.9 Comparing the experimental data with analytical results of TiC film thickness at P = 0.0888 atm

REFERENCES

1. R. Colmet; R. Naslain; P. Hagenmaller "Thermodynamic and experimental analysis of chemical vapor deposition of alumina from $AlCl_3$-H_2-CO_2 gas phase mixtures" J. Electrochemical Society V.129(6), P.1367, 1982

2. R.Colmet; R. Naslain "Chemical vapour deposition of alumina on cutting tool inserts from $AlCl_3$-H_2-CO_2 mixtures: influences of the chemical vapour deposition parameters and the nature of the inserts on the morphology and wear resistance of the coatings" Wear, V.80, P.221-231 (1982)

3. I. Lhermitte-Sebire; R. Colmet; R. Naslain "The chemical vapour deposition of alumina from $AlCl_3$-H_2-CO_2 on a stoichiometric TiC substrate: a thermodynamic approach" Journal of the Less - Common Metals, V.118, P.83, 1986

4. T. L. Starr "Model for CVI of short fiber preforms" Ceram. Eng. Sci. Proc., V. 8, P.951, 1987

5. T. L. Starr "Deposition Kinetics in Forced Flow/Thermal Gradient CVI " 12th Annual Conference on Composites and Advanced Ceramics,Cocoa Beach, Florida, 1988

6. J. Y. Rossignol, F. Langlais and R. Naslain " A tentative modelization of titanium carbide C. V. I. within the pore network of two-dimensional carbon-carbon composite preforms" Chemical Vapor Deposition, Proc. CVD-IX, The Electrochem.Soc.,Pennington, p. 596-614, (1984)

7. C.H.J. van den Brekel, R.M.M. Fonville, P.J.M. van der straten and G. Verspui," CVD of Ni, TiN, and TiC on complex shapes" Chemical vapor Deposition, Proc. CVD-VIII, The Electrochem.Soc. (1981)

8. Nyan-Hwa Tai and Tsu-wei Chou "Modeling of Chemical Vapor Infiltration(CVI) in Al_2O_3/SiC composites Processing" 12th Annual Conference on Composites and Advanced Ceramics,Cocoa Beach, Florida, 1988

9. Nyan-Hwa Tai and Tsu-wei Chou "Analytical Modeling of Chemical Vapor Infiltration(CVI) in Ceramic Matrix Composites Processing" submitted for publication.

10. Christie J. Geankoplis Mass Transport Phenomena, Holt, Rinehart and Winston, Inc. New York (1972)

11. Edward N. Fuller, Paul D. Schettler, J. Calvin Giddings "A new method for prediction of binary gas-phase diffusion coefficient" Industrial and Engineering Chemistry Vol.58, No.5, May 1966

12. R. Colmet, R. Naslain, P. Hagenmuller and C. Bernard " Thermodynamic and Experimental Analysis of Chemical Vapor Deposition of Alumina from $AlCl_3$ - H_2 - CO_2 Gas Phase Mixtures" Chemical Vapor Deposition, Proc. CVD - VIII, The Electrochem.Soc. (1981)

FIBERS AND GRIDS BY INTEGRATED CIRCUIT TECHNOLOGY

JAMES E. STEINWALL AND H. H. JOHNSON
Department of Materials Science and Engineering, Bard Hall, Cornell University, Ithaca, NY 14853

ABSTRACT

Ceramic fibers and grids of controlled geometry and composition were fabricated by electron beam evaporation of Al_2O_3 onto substrates patterned by optical lithography. The fibers were 4 and 5μm wide by 1μm thick. In addition, mixed metal oxide films from 0 to 10at% Ti, Zr, or Cr were produced by coevaporation with Al_2O_3. The compositions of the films were determined by Rutherford Backscattering Spectroscopy. TEM and electron diffraction examination showed all the films to be amorphous in structure.

The Al_2O_3 fibers had tensile strengths between 143 and 168 ksi and the Al_2O_3 film had a microindentation hardness of 8.4 GPa. Films with ≈1at% additions of Zr, Ti, and Cr had hardnesses of 11.0, 9.7, and 8.8 GPa respectively. The hardness then decreased with higher Zr, Ti, and Cr concentrations

INTRODUCTION

Microfabrication procedures offer a promising technology for the production of current and new materials as fibers, grids, and particulates. Fabrication of microcomposites with unprecedented control of reinforcement geometry may be possible. A range of deposition techniques may be used to produce reinforcements with close control of composition, structure, and properties

One method is to selectively deposit the desired material in a predefined pattern. Photolithography and physical vapor deposition are ideally suited for this route. Photolithography is a standard process and physical vapor deposition provides a unique method to control the fiber's composition, structure, and properties . Several deposition technologies can be utilized, including reactive evaporation, coevaporation, and sputtering.

EXPERIMENTAL

Liftoff Process

The microfabrication procedure for producing the fibers and grids follows a process developed by the microelectronics industry. The fiber design and layout is first created on a CAD system whose output is sent to a pattern generator. The pattern generator exposes a glass plate with a photographic emulsion to create a photo-mask. Next a three inch <100> silicon wafer is coated with a 3μm thick polyimide film followed by a 3.5μm thick positive photoresist film. The wafer and photo-mask are mounted and aligned in a contact aligner and the resist is exposed through the mask with 405nm UV radiation. The exposed wafer is soaked in chlorobenzene for the appropriate time to achieve a .5–1μm overhang thickness and is developed to remove the exposed resist. A liftoff patterned resist profile is shown in figure 1a. Material is then deposited onto the substrate with either an electron beam or thermal evaporator. Evaporation is necessary to ensure that no material coats the photoresist sidewalls.

After deposition, the wafer is soaked in acetone to dissolve the photoresist without dissolving the polyimide. This *lifts off* the unwanted material, leaving behind the desired structure (figure 1c). Finally, the polyimide is dissolved in methylene chloride creating a free floating structure (fibers or grids).

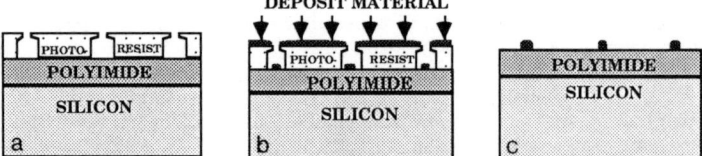

Figure 1: Schematic of the liftoff process. (a) illustrates the photo-resist patterned wafer and (b) shows the material being deposited on the pattern. (c) shows the resist after liftoff with the desired structure remaining.

Fiber Design and Test

Two basic fiber designs have been employed. The first was a straight fiber with a constant cross section the entire gauge length (figure 2b), and the second was a *reduced cross section* fiber designed to have its minimum cross section in the middle of the fiber (figure 2a). This design was specifically created for the brittle ceramic fibers that suffer from end loading effects and thus do not always break at their centers. These fibers have a minimum width of 4μm. Figure 2c is a SEM photograph of a two dimensional grid. Close examination of the figure 2a shows that the taper is not smooth but rather is stepped. This is an artifact of the mask fabrication. With the CAD software, a smooth arc can be created of almost any size; however the pattern generator must fracture the structure into individual exposures, the minimum size of which can be 2μm by 2μm. Thus to approximate a smooth arc, the CAD system overlays several small exposures which tend to overexpose the mask plate creating the steps.

The fibers were tested in a microtensile testing machine. One end of a fiber was glued to a hypodermic needle tip which was glued to a translation stage. The other end of the fiber was then glued to a 10g load cell. After the glue had set, the fiber was pulled at 0.6μm/s until failure. The tensile strength was determined from the load at failure and the area of the fiber.

The microindentation tests were perfomed using a microindenter developed at Cornell University by C. Y. Li[1].

Vapor Deposition

The Al_2O_3 films and fibers were deposited from an alumina source in a dual-gun electron beam evaporator which after being pumped down to a 1.0×10^{-7} torr vacuum, was backfilled with 1.0×10^{-5} torr of oxygen to maintain stoichiometry. The evaporation rates for pure Al_2O_3 were 10Å/s, and between 6–10Å/s when in combination with either Ti, Zr, or Cr. The final thicknesses were 1μm and films thin enough for TEM examination were made in the same run by evaporating 1000Å thick layers onto cleaved NaCl substrates.

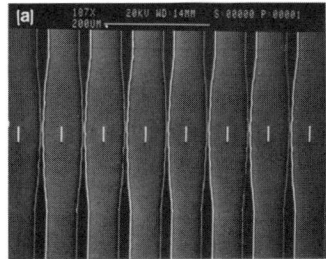
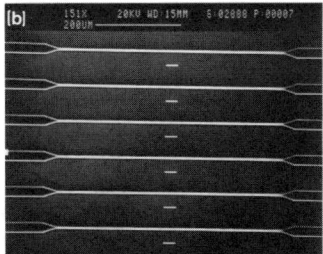
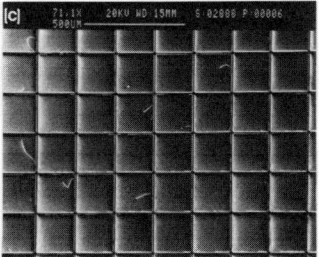

Figure 2 Al_2O_3 fibers (a) with reduced cross section and (b) with a straight cross section. The minimum width is 4μm. An Al_2O_3 grid is shown in (c) with lines 10μm wide and 200μm apart.

RESULTS AND DISCUSSION

Table I lists the compositions, density, structure, and hardness for the different films.

Composition

The compositions of the oxide and mixed oxide films were measured using RBS. All the films were found to be uniform in composition. Figure 3a and figure 3b show the experimental and simulated spectra for Al_2O_3 and Al_2O_3—7at% Cr films. RUMP[2,3] simulations showed the Al_2O_3 to be stoichiometric and determined the compositions of the mixed metal oxide films.. The density of the film can be calculated from the film's thickness measured by an Alpha-Step™ profilometer, the *RBS thickness* of the film in units of atoms/cm^2, and the film's composition. All of the films have a density considerably less than that of crystalline α-alumina's 3.97 g/cm^3 density.

Structure

Examination of the oxide films by electron diffraction showed all of the films to be amorphous in structure. Figure 4a is a bright field TEM photograph of an as-deposited Al_2O_3 film and figure 4b is the corresponding diffraction pattern. The bright field photograph reveals no crystalline structure at magnifications >50KX. This is confirmed by the appearance of broad, diffuse diffraction rings in the diffraction pattern. The spacing and number of rings is related to some short range ordering in the films. The bright field and diffraction patterns of the mixed oxide film are similar.

TABLE I Summary of oxide film compositions and properties.

FILM	COMPOSITION*			DENSITY g/cm^3	STRUCTURE	HARDNESS GPa**
	at%Al	at%O	at%Me			
Al2O3	40	60	—	3.1	amorphous	8.4
Al2O3 + low Cr	39.6	59.7	0.7	2.8	amorphous	11.0
Al2O3 + high Cr	35.9	56.7	7.4	3.0	amorphous	10.3
Al2O3 + low Ti	38.7	60.6	0.7	2.9	amorphous	9.7
Al2O3 + high Ti	36.2	54.6	9.2	3.2	amorphous	8.5
Al2O3 + low Zr	39.0	60.2	0.8	>3.2	amorphous	8.8
Al2O3 + high Zr	36.1	53.4	10.4	3.6	amorphous	8.1

*within 5% error **±.5GPa

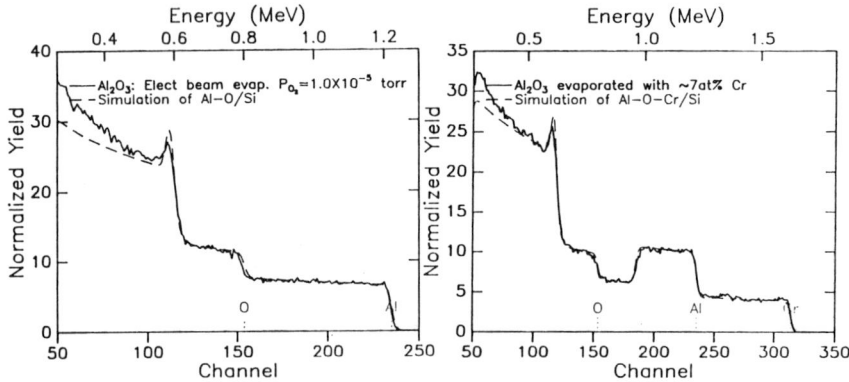

Figure 3a: RBS spectrum and simulation of stoichiometric Al$_2$O$_3$ with an incident beam of 2 Mev He^{++}.

Figure 3b: RBS spectrum and simulation of Al$_{35.9}$O$_{56.7}$Cr$_{7.4}$ with a 2 Mev He^{++} beam.

Mechanical Properties

The only oxide fibers tested have been the pure Al$_2$O$_3$ fibers. Before the reduced cross section fibers were designed, fibers with a constant cross section the entire length of the fiber were tested. The few fibers that did not break at the *grips* were recycled and exhibited increased tensile strength with decreasing length. These alumina fibers showed a maximum tensile strength of 168 ksi (~1.2 GPa). Several reduced cross section fibers did break at the region of minimum width despite the steps in the fiber. These fibers had a maximum tensile strength of 143 ksi (~1.0 GPa). These values are lower than for larger diameter polycrystalline Al$_2$O$_3$ fibers, with reported strengths between 200 and 300 ksi (1.4–2.0 GPa)[4].

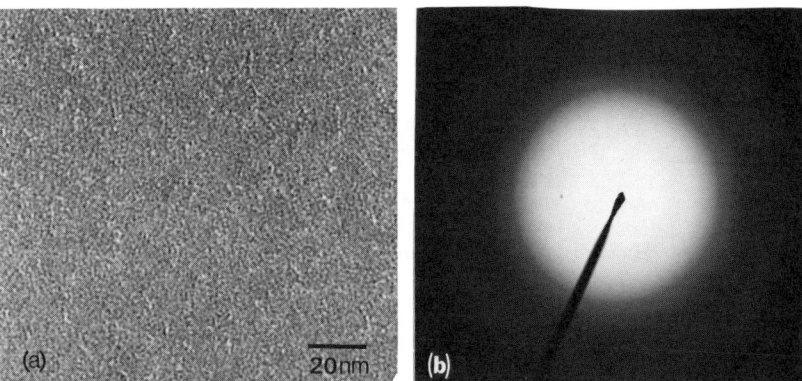

Figure 4 (a) shows a bright field TEM photograph of the amorphous as-deposited Al_2O_3 film and (b) is the corresponding selected area diffraction pattern from this film.

The results of the microindentation tests on the oxide films are reported in table I. The pure Al_2O_3 film had a hardness of 8.4 GPa. Other researchers have amorphized the surface of single crystal saphire with ion implantation and have determined the surface layer to have a hardness of 10 GPa[4]. The higher concentration mixed oxide films have hardnesses almost 1 GPa less than the corresponding lower concentration films. This is as yet unexplained and requires further examination. Figure 5a shows loading and unloading curves for the Al_2O_3 sample. The loading curves trace the same parabolic curve, indicating that this film is laterally homogeneous. The total depth of penetration and the unloading slope was used to calculate the plastic indentation depth. The hardness was then calculated from the maximum load, the indenter shape, and the plastic indentation depth. Results of hardness vs plastic depth are shown in fig. 5b. It can be seen that the hardness is independent of depth.

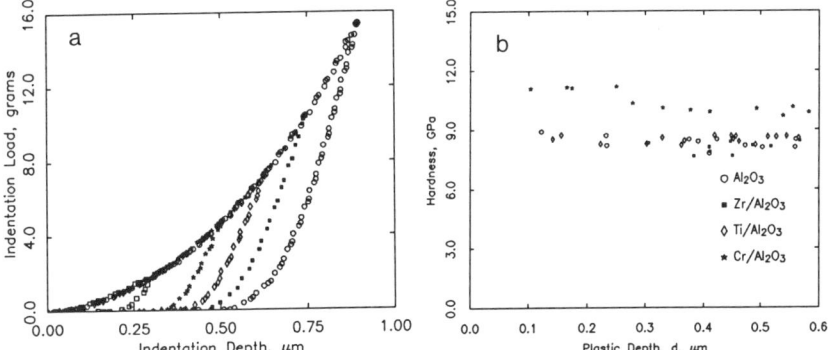

Figure 5 The loading and unloading curves for the pure Al_2O_3 film is shown in (a). (b) shows the hardness vs plastic depth curves for the high concentration films.

CONCLUSIONS

Alumina fibers were produced using microfabrication technology. Both straight and reduced cross section fibers were shown to have maximum tensile strengths of 168 and 143 ksi respectively. Mixed oxide films 1μm thick of Al_2O_3 and either Ti, Zr, or Cr were produced in the same manner as the fibers. RBS was used to measure the compositions and densities of these films. The Al_2O_3 film is stoichiometric and the mixed oxide films are homogeneous. All of the films had calculated densities less than the crystalline value of 3.97 g/cm^3 and were amorphous in structure as determined by TEM examination. The Al_2O_3 films had a hardness of about 8.4 GPa which is less than half the hardness of crystalline α-alumina. All of the mixed oxide films initially increased and then slightly decreased in hardness with increasing metal composition. The mixed films with chromium were the hardest.

ACKNOWLEDGEMENTS

Initial support for this research was provided by the National Science Foundation through the Materials Science Center at Cornell University. Project support was then provided by DARPA and DOE. The National Nanofabrication Facility and the Materials Science Center made critical experimental facilities available. The indentation tests were performed by Bill LaFontaine and the tensile testing was performed by Steve Ruoff; both are in the research group of Professor C. Y. Li. J. E. Steinwall is supported by a Unisys Doctoral Fellowship.

REFERENCES

[1] D. Stone, W. R. LaFontaine, P. Alexopoulous, T. W. Wu, and C. Y. Li, "An Investigation of Hardness and Adhesion of Sputter-Deposited Aluminum on Silicon by Utilizing a Continuous Indentation Test", J. Mater. Res. 3 (1), 141–147 (1988)

[2] L. R. Doolittle, "Algorithms for the Rapid Simulation of Rutherford Backscattering Spectra", Nuclear Instruments and Methods B9 344 (1985)

[3] L. R. Doolittle "A Semiautomatic Algorithm for Rutherford Backscattering Analysis", Nuclear Instruments and Methods B15 277 (1986)

[4] C. G. Levi, G. J. Abbaschian, and R. Mehrabian, "Interface Interactions During Fabrication of Aluminum Alloy—Alumina Fiber Composites", Met. Trans. A, 9A (5), 697–711 (1978)

[5] W. C. Oliver, C. J. McHargue, and S. J. Zinkle, "Thin Film Characterization Using a Mechanical Properties Microprobe", Thin Solid Films, 1987, 185–196

SYNTHESIS OF TITANIUM AND BORON CONTAINING POLYMERS: POTENTIAL PRECURSORS FOR ADVANCED CERAMICS

KENNETH E. GONSALVES AND K.T. KEMBAIYAN
Stevens Institute of Technology, Department of Chemistry and Chemical Engineering, Hoboken, New Jersey 07030

ABSTRACT

Titanium containing fibers have been synthesized via the modification of organic materials. Thermal processing of these chemically modified organic fibers in reactive/inert atmospheres, has produced materials containing Ti (C,O). Materials analysis was performed using SEM-EDS, ESCA and Auger spectroscopy. Novel monomeric titanium materials were synthesized for the above mentioned modification of organic materials. Boron containing monomers and polymers were also synthesized as possible precursors for B-N-C materials.

INTRODUCTION

Advanced ceramics (e.g. carbides, nitrides and borides) are materials having desirable properties for a variety of applications [1]. Their key properties are high hardness and strength along with resistance to heat, corrosion and wear. In our attempt toward such fibers we felt that if we utilized organic polymers (as a source of carbon) and suitably modified it with an organotitanium complex at the molecular level, we would have a system containing the right ingredients i.e. Ti and C. Such a polymeric system could conceivably be processed into a fiber [2,3]. $TiCl_4$ and alkyl cyanides [4] are reported to form complexes of the type $(R-C\equiv N)_2 \cdot TiCl_4$ and titanium amides $(R_2N)_4Ti$ also complex with the $-C\equiv N$ group [5]. Polyacrylonitrile (PAN) has pendant nitrile groups and therefore was selected as a prime candidate for modification by organotitanium complexes. PAN can also be easily fabricated into fibers and processed at elevated temperatures to produce "carbon" which can simultaneously undergo a "solid state" reaction with titanium. Under appropriate processing conditions it was envisaged that Ti-C would be the main constituent of the end product.

Thus far in this initial investigation, we have synthesized organometallic monomers and modified organic polymers. The fabrication of such modified organic polymers into fibers followed by their thermal processing into ceramics has been undertaken [6].

RESULTS AND DISCUSSION

Titanium containing monomers, tetrakis(dialkylamino)titanium $(R_2N)_4Ti$, $(R=C_2H_5, CH_3)$ were synthesized by a slight modification of the procedure reported in the literature [7,8]. In general, first the lithium dialkylamides (LDA) were obtained by reacting methyl lithium in diethyl ether with the appropriate dialkylamine at low temperatures (-20^oC). To the resulting LDA was then added $TiCl_4$ in benzene, the temperature being maintained below 10^oC. These reactions were conducted in an argon environment to exclude moisture and air. In both instances the titanium monomers were high boiling viscous fluids. The above sequence of reactions are presented below in Scheme 1.

Scheme 1

$$R_2NH + CH_3Li \xrightarrow[\text{Temp.} < -20°C]{\text{diethyl ether}} R_2NLi + CH_4$$

$$TiCl_4 + R_2NLi \xrightarrow[\text{Temp.} < 10°C]{\text{benzene}} (R_2N)_4Ti + LiCl$$

$$[R = CH_3, C_2H_5-]$$

Polyacrylonitrile (PAN) was prepared by known procedures [8]. Purified acrylonitrile was polymerized with a redox initiator system in water by precipitation polymerization. The intrinsic viscosity [η], was determined at 25°C in dimethylformamide (DMF) and the molecular weights calculated from the Mark-Houwink equation ranged between 18,000-20,000, suitable for fiber spinning.

Preliminary investigation regarding the feasibility of these monomers as potential sources of "active Ti" in solid state reactions were conducted prior to fiber fabrication and processing studies. The titanium monomer $[(CH_3)_2N]_4Ti$, on pyrolysis from ambient to 1200°, in an argon atmosphere left a black residue. On chemical analysis it was observed that the residue contained 61% titanium. The residue was also found to contain carbon, nitrogen and oxygen. In a parallel experiment, polyacrylonitrile powder (PAN) and the monomer $Ti[N(CH_3)_2]_4$ in a 10:1 ratio by weight, mixed under argon for 24h, were similarly pyrolyzed in argon. A black residue was obtained which contained 5.1% Ti, 87.5% C and 5.2% N, along with oxygen. It can be assumed that a mixture of Ti(C,N) and carbon were present. The total weight loss in both experiments was ca 40%. The above pyrolyses are summarized in Scheme 2.

Scheme 2

$$Ti(NR_2)_4 \xrightarrow[\text{ambient - 1200°C}]{\text{argon}} \begin{array}{c} Ti(C,N,O) \\ [Ti-60\%] \end{array}$$

$$Ti(NR_2)_4 + \underset{\underset{CN}{|}}{(CH_2-CH)} \xrightarrow[\text{ambient - 1200°C}]{\text{argon}} Ti(C,N,O) + C$$

From these initial experiments it was concluded that these monomers could provide titanium, a crucial source for reaction with carbon. The production of titanium with excess carbon in the latter experiment was critical as it was anticipated that an analogous solid state reaction would occur in the titanium modified PAN fibers. In another experiment, PAN was mixed with excess $TiCl_4$ and heated in an argon atmosphere for 48 hrs. at 60°C, followed by distilling off the excess $TiCl_4$ at 200°C. The brown residue was pyrolyzed in nitrogen till 1200°C. The residue after pyrolysis consisted of particles ranging from 0.1-0.5μ. The powder was subjected to X-ray diffraction analysis. Initial evaluation indicates the presence of a "TiC" phase with possibly "TiO" and graphite also present. SEM-EDS analysis, using a windowless detector to enable detection and quantification of elements till Z=5, indicates the presence of titanium and carbon in near stoichiometry, relative to the TiC phase. ESCA studies also support the above analyses, including the presence of TiN.

Fiber Processing

PAN powder was dissolved in DMF and the "dope" extruded into a chilled water bath to coagulate the polymer. The as-spun fibers were drawn in a hot water bath before drying. The drawn fibers were dried overnight in an oven at $60°C$, under tension.

Modification of the PAN fiber was done after drawing and drying. The fibers were immersed in solution of $Ti(NEt_2)_4$/benzene for 48h at $10°C$. The concentration of $Ti(NEt_2)_4$ in benzene was varied to determine the effect on fiber structure. The modified fibers were pyrolyzed under different conditions. The presence of a "skin" and an "inner core" along with substantial voids made the fibers extremely brittle. In the modification of PAN fibers by the monomer $Ti(NMe_2)_4$, no solvent was used as the monomer was less viscous and easier to handle. In the cross section of one such fiber pyrolysed in ammonia the voids, compared to fibers pyrolyzed in argon, were substantially reduced and the fiber was consequently less brittle. Modification of PAN fibers by immersion in $TiCl_4$ was also attempted and gave brittle fibers.

Materials Characterization

Scanning Auger electron spectroscopy (AES) was performed on a finely powdered single fiber mounted on a copper backed adhesive tape. The Auger transitions of the major elemental constituents of the fiber i.e. C and Ti were evident. Contributions from the carbon and oxygen of the adhesive are possible as copper signals were observed in the ESCA survey. From these initial AES and ESCA surveys, it appears that the fibers are composed of Ti(C, O). Semi quantitative elementalanalysis were performed on the fiber samples by SEM-EDAX. The Ti content on the surface ranged around 90-95% and in the interior from 65-90%, depending on the conditions of PAN fiber modification. The distribution averaged over different cross-sections was fairly uniform.

We are currently attempting to densify the fiber by conducting the pyrolysis in reducing hydrocarbon atmospheres and increasing the maximum processing temperature to $1700-2000°C$. Tensile testing of these fibers will be reported subsequently.

BORON CONTAINING POLYMERS

We have also synthesized a boron containing monomer, TITMB, and polymers PUB [10] & PBA which can possiblby be copolymerized with the titanium monomers described above, or blended with titanium containing polymers. These could be precursors for Ti-C-B-N fibers. Pyrolysis of PUB in ammonia to $1200°C$ gave an 80% char yield determined by XRD analysis to comprise of a graphite phase, a small amount of BN and other complex B-N-C phases.

Trimethyl-triisocyanate borazine (TITMB)

Poly(ureidoborazines) (PUB)

Poly(phenylboronamine) (PBA)

CONCLUSIONS

Developments in fiber-reinforced high temperature ceramic or metal matrix composites require the development of thermally stable ceramic fibers such as the Ti-C-N-B systems attempted in this work. Fiber coatings technology, via such novel organometallic precursors, also needs to be simultaneously developed to extend the capabilities of available and future ceramic fibers.

ACKNOWLEDGMENTS

Partial support of the National Science Foundation (Grant No. MSM-8612801) is gratefully acknowledged. Thanks to Surya Manukutla for synthesizing the PAN samples.

REFERENCES

[1] N.D. Corbin, T.M. Resetar and J.W. McCauley in Innovations in Materials Processing", Sagamore Army Materials Research Conference, G. Bruggerman and V. Weiss Eds. Vol. 30, Plenum, New York 1985.

[2] M.K. Jain and A.S. Abhiraman. J. Mater. Sci. 22, 278 (1987).

[3] W.J. Wynne and R.W. Rice Ann. Rev. Mater. Sci. 14, 297 (1984).

[4] H.J. Coerver and C. Curran, J. Am. Chem. Soc. 80, 3522 (1958).

[5] G. Chandra, T.A. George and M.F. Lappert Chem. Commun. 116 (1967); G. Chandra, A.D. Jenkins, M.F. Lappert and R.C. Srivastava J. Chem. Soc. (A) 2550 (1970).

[6] K. Gonsalves and R. Agarwal, J. Applied Polym. Sci. (in press).

[7] D.C. Bradley and I.M. Thomas J. Chem. Soc. 3857 (1960).

[8] A.D. Jenkins, M.F. Lappert and R.C. Srivastava, Polym. Journal 7, 289 (1971).

[9] D. Braun, H. Chedron and W. Kern. "Techniques of Polymer Synthesis and Characterization", Wiley, New York (1971) p. 132.

[10] K. Gonsalves and R. Agarwal, Applied Organomet. Chem. (in press).

PYROELECTRIC AND DIELECTRIC PROPERTIES OF POLYMER-CERAMIC COMPOSITES

D.K. DAS-GUPTA and M.J. ABDULLAH
School of Electronic Engineering Sciences,
University College of North Wales,
Bangor, Wales, U.K.

INTRODUCTION

Ferroelectricity in well-poled polyvinylidene fluoride (PVDF) is a phenomenon which is now well supported. However, the piezo- and pyroelectric responses in this polymer are significantly weaker than those of ceramic oxides. It would be attractive to design composite materials which will have the mechanical properties of polymers with the electro-active properties of ceramics. Such materials will be useful for diverse sensor applications, viz, acoustic emission detection, hydrophones, biomedical applications, thin film capacitors etc. One of the chief requirements of a capacitor is that a large capacitance to volume ratio is desirable. A high ratio requires a thin film, a high dielectric constant and an acceptable electrical breakdown strength. Such a material may be designed by a judicious incorporation of fine grain ceramics in a suitable polymer matrix.

Present work reports the result of an investigation of the dielectric and pyroelectric properties of composite films in which $BaTiO_3$ and PZT have been located in the matrix of PVDF.

EXPERIMENTAL

PVDF pellets (types A and B) and fine grain (1-2μm) PZT5 and PZT8 were obtained from Laporte Industries and Unilator Technical Ceramics Limited., respectively. Cookson plc. generously provided us with $BaTiO_3$ grains (~1μm). The composite hides were prepared with suitable concentrations in a rolling mill at 433K. A temperature controlled hydraulic press was subsequently used in the second stage to prepare films of minimum thickness of 200μm. The absorption current studies, the dielectric measurements and pyroelectric behaviour investigations were made as described in our previous work[1-4]. The different composite mixtures used in this work will be designated thus:

Composite A: PZT5/PVDF-A (Solef 11010): 50/50 (volume fraction)

Composite B: P2T8/PVDF-B (Solef 1008): 50/50 (volume fraction)

Composite C: $BaTiO_3$/PVDF-B (Solef 1008): 40/60 (volume fraction)

PVDF-A pellets (Solef 11010) of Composite A films were observed to be more opaque compared to PVDF-B (Solef 1008) of Composite B films. Above observations would suggest that the Solef 11010 pellets are more crystalline in nature than the Solef 1008 PVDF pellets.

RESULTS AND DISCUSSION

Figure 1 shows typical charging and discharging current transients $I_c(t)$ and $I_d(t)$ respectively for the composite films A, B and C for a charging field of 3.5 x 10^5 Vm^{-1} at 363K. It may be observed that composite C reaches a quasi-steady state value of $I_c(t)$

at ~10^3s while composites A and B seem to take a longer time (~10^4s) to reach such conduction levels. It has been shown that the steady state conduction mechanisms[1] in such composites is of ionic nature which is also true for ceramic materials. Furthermore, ionic conduction has also been observed in PVDF at high fields and high temperatures where the charge carriers hop through the defect sites along the chains[5]. The discharge current $I_d(t)$, behaves according to the well known expression:

$$I_d(t) = A(T) \, t^{-n} \quad \ldots (1)$$

where $A(T)$ is a temperature dependent factor, t the time after the removal of the charging voltage and $n \leq 1$. In this work n-value was found to be 0.6 - 0.7 for all three composites. A very broad distribution of relaxation times is indicated by $n \simeq 1$.

The observed dielectric behaviour (ϵ' and ϵ'') of the three composites are given in figure 2 in the frequency range of 10 Hz - 65 kHz. For composites A and B the ϵ' - values (real part of the dielectric constant) are very similar and higher than that of composite C. The permittivity ϵ of a composite system may be expressed thus[6]:

$$\epsilon = \epsilon_1 \left[1 + \frac{nq(\epsilon_2 - \epsilon_1)}{n\epsilon_1 + (\epsilon_2 - \epsilon_1)(1-q)} \right] \quad \ldots (2)$$

where ϵ_1 is the permittivity of the continuous medium (i.e. PVDF in our case), ϵ_2 permittivity of the ellipsoidal particles (ceramics), n a parameter attributed to the shape of the ellipsoidal particles. With n = 8, ϵ_1(PVDF) = 12, ϵ_2(PZT) = 1300 and ϵ_2 (BaTiO$_3$) = 1700, the calculated values of permittivities of composites A and C are 96 and 70 respectively which are in good agreement with the experimentally observed values, i.e. ϵ'composite A = 100, ϵ'composite B = 105 and ϵ'composite C = 80 at 1kHz at 363K.

The ϵ'' behaviour shows a broad relaxation (noticeably more for Composite C) at ~1kHz which is attributed to α_c-relaxation of the polymer (PVDF) phase. For composites A and B, the ϵ'' values increase significantly as the frequency is reduced and this may be attributed to an ionic conductivity or electrode polarization effect. Composite C also shows such a behaviour as the frequency goes below 10 Hz.

The behaviour of ϵ'' at 1kHz in the temperature range of 293-378K is shown in figure 3.

The observed high temperature peak at ~363K is in agreement with that in PVDF and it is associated with molecular motions in the crystalline region of the polymer[4] with added contributions from the ceramic phase. The increased dielectric loss with increasing temperature is usually ascribed to the ionic conductivity which is present in both the polymer and the ceramic phases.

Thermally stimulated discharge current (TSDC) plots of the three composites which were initially poled at a field of 7×10^6 Vm^{-1} for ~3 hours at 373K and cooled down to room temperature in presence of this field, is shown in figure 4. The TSDC spectra of composites A and B are similar and have higher values of thermally stimulated currents than that of composite C.

The pyroelectric coefficient p may be expressed thus:

$$p = \frac{1}{a} \left[I_p \bigg/ \frac{dT}{dt} \right] \quad \ldots (3)$$

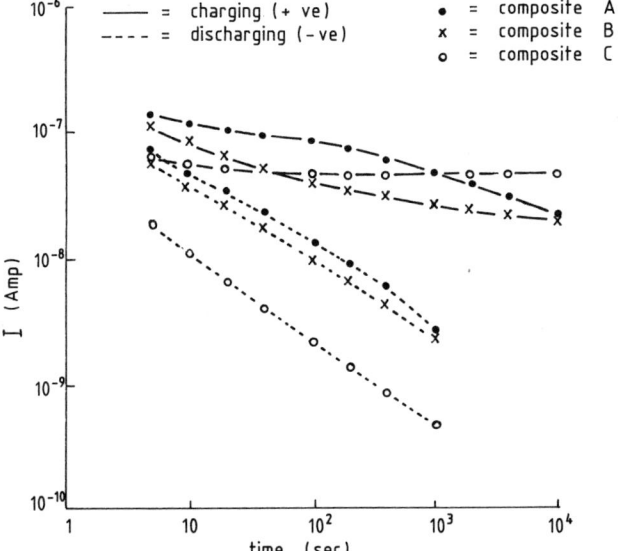

Figure 1: Charging and discharging currents in ceramic/polymer composites at 363 K and field 3.5×10^5 Vm^{-1}.

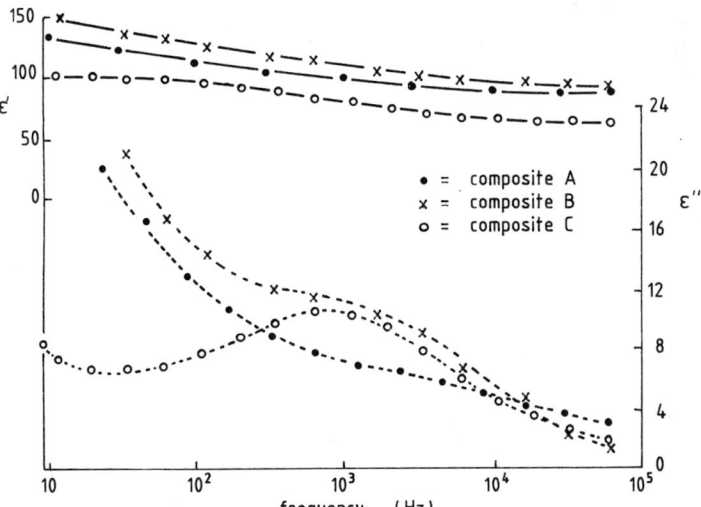

Figure 2: The behaviour of ε' and ε'' against frequency in ceramic/polymer composites at 363 K. ε' = ——, ε'' = ----

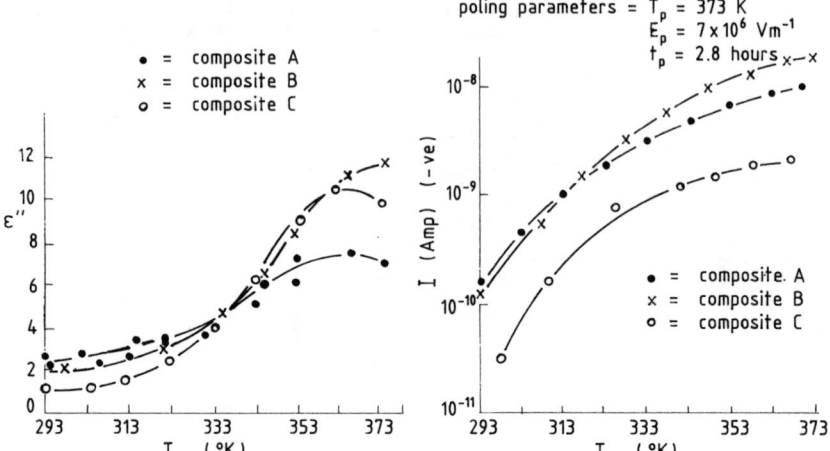

Figure 3 : Dielectric loss behaviour against temperature in ceramic/polymer composites at 1kHz.

Figure 4 : Thermally stimulated discharge current in ceramic/polymer composites.

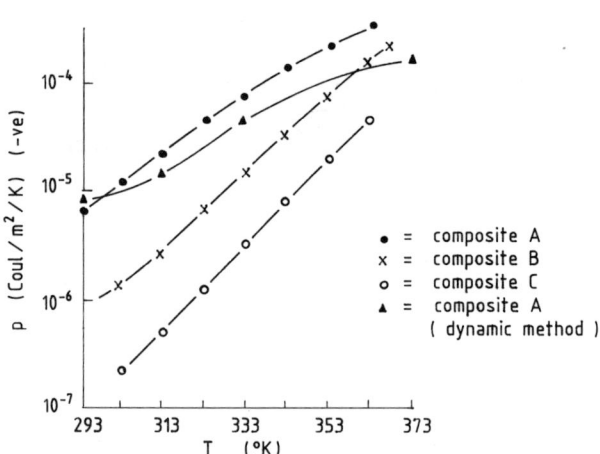

Figure 5: Pyroelectric coefficient against temperature in ceramic / polymer composites.

where I_p is the reversible pyroelectric current and dT/dt is the rate of rise in temperature in a TSDC run. This is the direct method of determining pyroelectric coefficient. For our measurements I_p values were identified from the third successive run of TSDC experiment thus ensuring that the trapped space charges have been eliminated and that I_p is truly reversible. The heating rate in the present TSDC run is 1°c/min and a is the electrode area of the samples. Figure 5 shows the behaviour of the pyroelectric coefficient p with temperature in the range of 303K - 373K from which it may be observed that Composite A has the highest pyroelectric coefficient as compared with the other two composites. It may also be noticed that p values of Composite A is greater than that of Composite B by approximately an order of magnitude at 303K. It is suggested that the observed higher dielectric losses (see figure 2) coupled with its lower elastic stiffness in the polymer phase may be responsible for its smaller p-values compared with that of Composite A which would be in agreement with Galgoci et al[7]. Tamura et al[8] also have observed increased piezoelectric coefficient with increasing elastic stiffness. The pyroelectric figure of merit p/ϵ', in Composite A (1.5×10^{-6} $Cm^{-2}K^{-1}$) is approximately 3.8 times greater than that of PZT (3.8×10^{-7} $Cm^{-2}K^{-1}$) at 343K.

Pyroelectric coefficient has also been measured in this work by a dynamic method in which p(T) may be obtained from the following relationship[9].

$$I_{p(max)} = Kp(T) \qquad \ldots (4)$$

where $\qquad K = \dfrac{F_o a}{Pc_p L} \theta^{\frac{\theta}{(1-\theta)}} \qquad \ldots (5)$

where F_o is the radiation power absorbed per unit area of the electroded sample, a the electrode area, P the density of the material, C_p the specific heat, L the sample thickness, $I_{p(max)}$ the peak pyroelectric current response to a stepwise radiation incident on the sample, and $\theta = \tau_E/\tau_T$ where τ_E and τ_T are the electrical and thermal time constants of the system.

$I_{p(max)}$ was measured at different temperatures using a stepwise thermal radiation from a tungsten filament lamp. The pyroelectric coefficients p(T) were calculated using the measured values of $I_{p(max)}$ of the dynamic method and equations 4 and 5 for composite A and are also shown in Figure 5. It may be observed that the p(T) values obtained by the direct method which is indeed the most accurate method, are somewhat higher than those obtained by the dynamic method, particularly at high temperatures. Further work is in progress in the study of electro-active properties of Composite A which appears to be the most attractive material of the three Composites investigated in the present work. Table 1 provides a summary of results obtained in this work.

ACKNOWLEDGEMENT

This work is financed by a research contract from the European Research Office of the U.S. Army in Great Britain. One of the Authors (M.J.A.) is also grateful to the Malaysian Government for a maintenance grant.

TABLE 1 - SUMMARY OF RESULTS

	$\sigma(\Omega^{-1}m^{-1})$ (90°C)	ϵ' (70°C) (1KHz)	ϵ'' (70°C) (1kHz)	Pyrelectric Coefficient p(T) (Coul.m^{-2}K^{-1}) (70°C)	Pyroelectric Figure of Merit p/ϵ'(70°C)
PZT	3x10^{-9}	1300	1.2	5x10^{-4}	3.8x10^{-7}
BaTiO$_3$	10^{-9}-10^{-8}	1700		7x10^{-4}	4.1x10^{-7}
PVDF	3x10^{-13}	12	1.0	9.0x10^{-6}	7.5x10^{-7}
Composite A [PZT5/PVDF(A)] 50:50	1.4x10^{-10}	95	6.0	1.4x10^{-4}	1.5x10^{-6}
Composite B [PZT8/PVDF(B)] 50:50	1.3x10^{-10}	96	6.5	3.3x10^{-5}	3.4x10^{-7}
Composite C [BaTiO$_3$/PVDF(B)] 40:60	3x10^{-10}	73	6.2	8.0x10^{-6}	1.1x10^{-7}

REFERENCES

1. Das-Gupta, D.K. and Abdullah, M.J., British Ceramic Proceedings "Novel Ceramic Fabrication Processes and Applications", Ed. R.W. Davidge, No. 38, 231 (1986).

2. Abdullah, M.J. and Das-Gupta, D.K., Ferroelectrics, 76, 393 (1987).

3. Das-Gupta, D.K. and Doughty, K., Thin Solid Films, to be published (1988).

4. Das-Gupta, D.K. and Abdullah, M.J., J. Mater. Sci. Lett., to be published (1988).

5. Das-Gupta, D.K., Doughty, K., and Brockley, R.S., J. Phys. D., 13, 2101 (1980).

6. Yamada, T., Ueda, T. and Kitayama, T., J. Appl. Phys., 53, 4328 (1982).

7. Galgoci, E.C., Schreffler, D.G., Devilin, B.P., & Runt, J., Ferroelectrics, 68, 109 (1986).

8. Tamura, M., Ogasawara, K., & Yoshimi, T., Ferroelectric, 10, 125 (1976).

9. Simhony, M., & Shaulov, A., J.Appl. Phys., 42, 3741 (1971).

PART IV

Ceramic Composite Mechanical Performance

THE MECHANICAL PERFORMANCE OF FIBER REINFORCED CERAMIC MATRIX COMPOSITES

A. G. EVANS* and D. B. MARSHALL**
*College of Engineering, University of California, Santa Barbara, CA 93106
**Rockwell International Science Center Thousand Oaks, CA 91360

ABSTRACT

This article evaluates the current understanding of relationships between microstructure and mechanical properties in ceramics reinforced with aligned fibers. Emphasis is placed on definition of the micromechanical properties of the interface that govern the composite toughness. Issues such as the debond and sliding resistance of the interface are discussed based on micromechanics calculations and experiments conducted on both model composites and actual composites.

INTRODUCTION

Practical ceramic matrix composites reinforced with continuous fibers exhibit important failure/damage behaviors in mode I, mode II and mixed mode I/II, as well as in compression. The failure sequence depends on whether the reinforcement is uniaxial or multiaxial and whether woven or laminated architectures are used. However, the underlying failure processes are fully illustrated by the behavior of uniaxially reinforced systems. The basic features are sketched in Fig. 1. The intent of the present article is to provide an assessment of relationships between the properties of the constitutents (fiber, matrix, interface) and the overall mechanical performance of the composite. At the outset, it is recognized that the composite properties are dominated by the interface, such that upper bounds must be placed on the interface debond and sliding resistance in order to have a composite with attractive mechanical properties. A major emphasis of the article thus concerns the definition of optimum properties for coatings and interphases between the fibers and the matrix, subject to high temperature stability and integrity. Residual stresses in the composite caused by thermal expansion differences are also very important and are confronted throughout.
The strong dependence of ceramic matrix composite properties on the mechanical properties of the interface generally demands consideration of fiber coatings and/or reaction product layers, at least for high temperature use. Thus, while composites fabricated using low temperature matrix infiltration procedures, such as chemical vapor infiltration (CVI), can create composites that exhibit limited interface bonding and, therefore, have acceptable ambient temperature properties, experience indicates that moderate temperature exposure causes diffusion, coupled with the ingress of O_2, N_2, etc., from the environment, resulting in chemical bonding across the interface. The resultant interphase consisting of oxides, nitrides, carbides (either separately or in combination) invariably have sufficiently high fracture resistance that desirable composite properties are not retained. Consequently, a major objective of this article and of continuing research on ceramic matrix composites is the identification of interphases that are both stable at high temperature and bond poorly to either the fiber or the matrix. Certain refractory metals and intermetallics seem to have these attributes, as elaborated in the following chapters.
The basic philosophy of this article is that the overall mechanical behavior is sufficiently complex and involves a sufficiently large number of independent variables that empiricism is an inefficient approach to microstructure optimization. Instead, optimization only becomes practical when each of the

important damage and failure modes has been described by a rigorous model, validated by experiment. The coupling between experiment and theory is thus a prevalent theme. It is also noted that this objective can only be realized if the models are based on homogenized properties that describe representative composite elements, while also taking into account the constitutent properties of the fibers, matrix and interface. Models that attempt to discretize microstructural details have little merit in the context of the above objective. In this regard, the present philosophy is analogous to that used successfully to describe process zone phenomena such as transformation and microcrack toughening,[1-5] as well as ductile fracture, [6,7] wherein the behavior of individual particles, dislocations, etc., provides input to the derivation of constitutive properties that describe the continuum behavior.

The behavior of the composite is intimately coupled to some basic features of crack propagation and sliding along interfaces. Indeed, the response of the composite can be simulated by studying interface responses in judiciously selected test specimens. The basic mechanics and the implications of tests used to study interface debonding and sliding are presented first. The characteristics of the damage and fracture processes that occur in each of the important modes depicted in Fig. 1 are then described. Finally, implications for the choice of matrices, fibers and coatings that provide good mechanical properties are discussed.

INTERFACE DEBONDING AND SLIDING

Mechanics of Interface Cracks

Interface debonding in ceramic matrix composites occurs both at the matrix crack front and in the crack wake (Fig. 2). Both processes are <u>mixed mode</u>. Furthermore, debond cracks typically occur between materials (fiber, matrix, coating) having quite different elastic properties. The requisite mechanics thus concern cracks on <u>bimaterial</u> interfaces, resulting in a more complex fracture mechanics formulism than the familiar stress intensity factors used in elastically homogeneous systems. The additional features that need to be introduced when considering interface cracks derive from the fact that the mixity of opening and shearing of the surface depends on the modulus mismatch and on the crack length, as well as on the mode of loading. Consequently, the debond resistance of a bimaterial interface must <u>always</u> be characterized by two parameters: the critical strain energy release rate, $\mathcal{G}_{ic}$, and the phase angle of loading, ψ (Fig. 3). The latter quantity is somewhat dependent on the choice of units, for reasons elaborated elsewhere.[8] However, this presents no difficulty, provided that a consistent choice is made. The strain energy release rate, $\mathcal{G}$, and the phase angle can be calculated for any problem of interest and can be expressed in terms of the applied loads, specimen dimensions and debond length.

The basic relationship used to calculate $\mathcal{G}$ and ψ can be defined with reference to Fig. 4. The phase angle is related to the angle of rotation ϕ of the crack surface as the crack opens by;[8,9]

$$\psi = \phi - (\ln r/2\pi)\ln[(1-b)/(1+b)] \qquad (1)$$

with b being one of the Dundurs' parameters[10] (a and b);

$$a = [G_1(1-\upsilon_2) - G_2(1-\upsilon_1)]/[G_1(1-\upsilon_2) + G_2(1-\upsilon_1)]$$

$$b = \left(\frac{1}{2}\right)[G_1(1-2\upsilon_2) - G_2(1-2\upsilon_1)]/[G_1(1-\upsilon_2) + G_2(1-\upsilon_1)] \qquad (2)$$

G is the shear modulus, υ is Poisson's ratio and r is the distance from the crack front. The energy release rate is related to the crack surface displacements (u,v) by;[8,9]

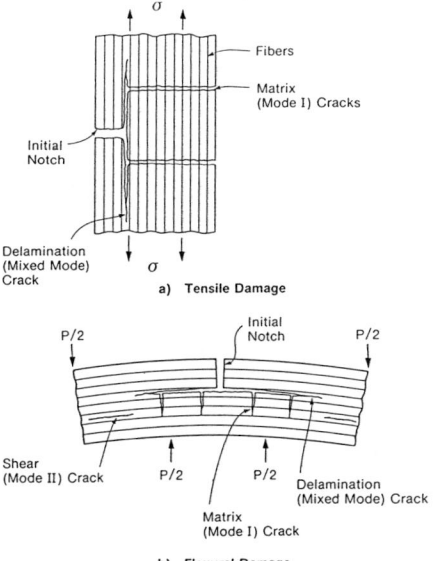

Fig. 1. A schematic illustrating the failure modes observed in high toughness uniaxially reinforced ceramic matrix composites; a) tension b) flexure

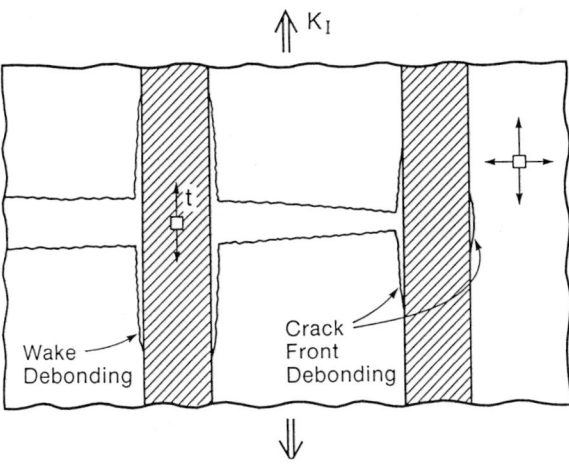

Fig. 2. A schematic illustrating the initial debonding of fibers at the crack front and fiber debonding in the crack wake

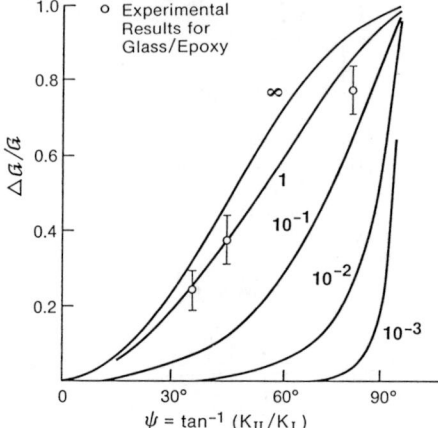

Fig. 3. Trends in critical strain energy release rate with the phase angle of loading: experimental results and predictions for a model involving contact at asperities on the crack surface

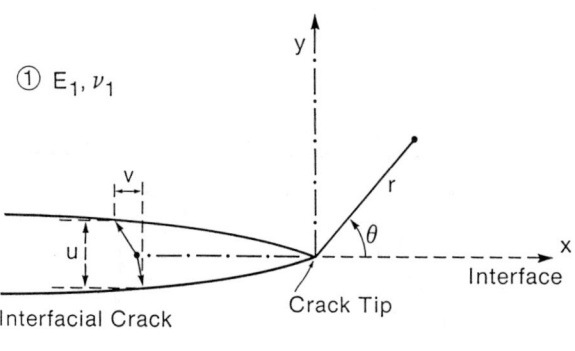

Fig. 4. The displacement of the surface of a crack at a bimaterial interface indicating the shear and opening displacement that accompnay most external loading conditions

$$G = \frac{\pi\left(1 + 4[\ln[(1-b)/(1+b)]/2\pi]^2\right)(u^2 + v^2)}{8r[(1-\upsilon_1)/G_1 + (1-\upsilon_2)/G_2]} \quad (3)$$

where the fiber and matrix have essentially the same elastic properties, G can be related to the stress intensity factor, K, in the usual manner, $K^2 = EG/(1-\nu^2)$.

Debonding Mechanics

Debonding solutions are required for axisymmetric loading, representative of the debonding of fibers (Fig. 2), as well as for planar cracks characteristic of macroscopic delamination (Fig. 1). In both cases, G and Ψ are strongly influenced by the <u>residual stress</u>. Furthermore, when the phase angle becomes large, $\Psi \to \pi/2$, frictional sliding and crack surface locking effects become important.[11] A comprehensive set of solutions that fully encompass the spectrum of residual stress and of frictional sliding relevant to composites does not yet exist. The known solutions are described below.

<u>Axisymmetric</u> solutions exist for composites with interfaces subject to residual radial <u>tension</u>, wherein a net crack opening exists for the full range of applied loads, elastic moduli and fiber volume fractions.[12] All solutions have the general features that G is small, but non-zero, when the debond length is zero and increases to a steady-state value G_{ss} when the debond length d exceeds $\sim R$ (Fig. 5a). Such behavior indicates the insightful bound that the debond <u>must</u> extend without limit when G_{ss} exceeds G_{ic} at the appropriate Ψ (see Fig. 3). The basic trends in G_{ss} and Ψ relevant to wake debonding, determined using finite elements, are summarized in Figs. 5b,c. The variables in the analysis are: the ratio Σ of Young's modulus for the fiber, E^f, the test of the matrix, E^m, the fiber volume fraction, f, the stress-free (residual) strain ϵ, the stress imposed on the fiber, t, and the Poisson's ratios ν^m and ν^f. Note that the phase angles are typically large, indicative of a large ratio of shear to opening.

Rigorous axisymmetric solutions for interfaces subject to residual radial compression have not been derived. However, some approximate solutions based on a modified shear lag approach are insightful.[13] This approach has merit when the friction coefficient μ is small ($\mu \lesssim 0.2$). For this case, crack opening does not occur until t reaches a critical value t_c given by;

$$t_c/E^f\epsilon = 1/\nu^f \quad (4)$$

For $t > t_c$, steady-state obtains for long debonds and the solutions given in Fig. 5 are directly applicable. For $t < t_c$, the debond crack is subject to normal compression and G diminishes with increase in debond length, d, representative of λ <u>stable</u> debonds,

$$G/E^f R \epsilon^2 \approx F^2 + 2F - \frac{4\mu d(1-f)(1+F)(1-\upsilon F)}{R[1+f+(1-f)(1-2\upsilon)]} \quad (5)$$

where

$$F = (t-p)/E^f\epsilon$$

$$\xi = (1-f)/[(1-f)(1-2\nu) + \Sigma(1-f)]$$

for $\nu^f \approx \nu^m \approx \nu$ and p is the axial residual stress in the matrix as governed by ϵ, f and Σ. The G in this instance is strictly mode II and debonding should thus be predicted by equating G to G_{ic} at $\Psi = \pi/2$. Such predictions have not been attempted. However, it is insightful to note that, for "weak" interfaces ($G_{ic} \ll G_{fc}$), the debond length and the slip length, ℓ, are likely to be closely related, with ℓ given by;[13]

$$\ell/R \approx t/2E^f\epsilon\mu\xi(1-\nu^f F) \quad (6)$$

Fig. 5. Trends in energy release rate and phase angle for loads exerted on a fiber in the crack wake

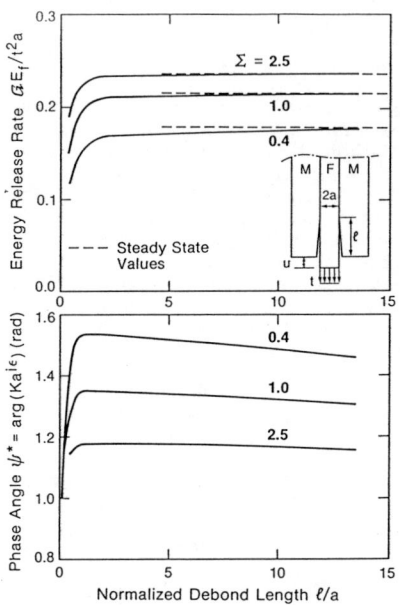

a) effects of debond length

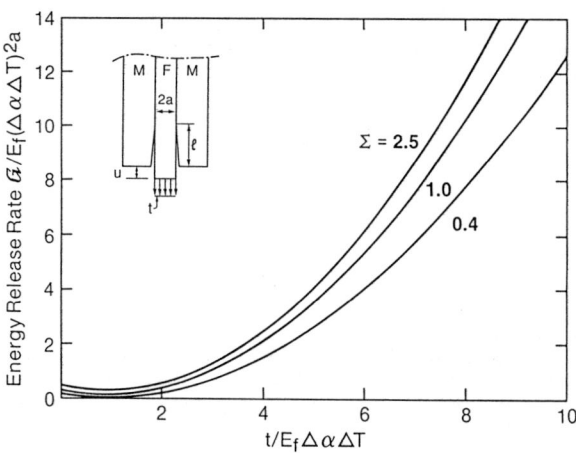

b) effects of applied stress t on steady-state, $\mathcal{G}_{ss}$

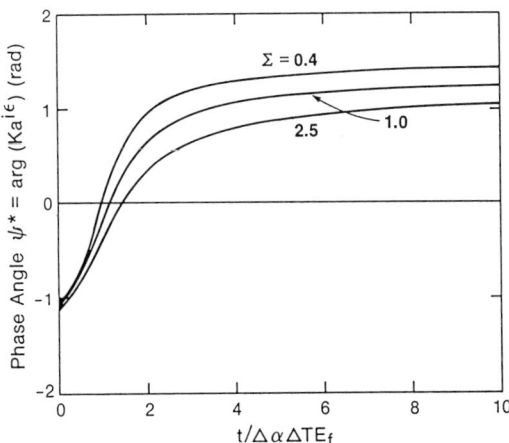

c) effects of applied stress t on phase angle Ψ in steady-state regime

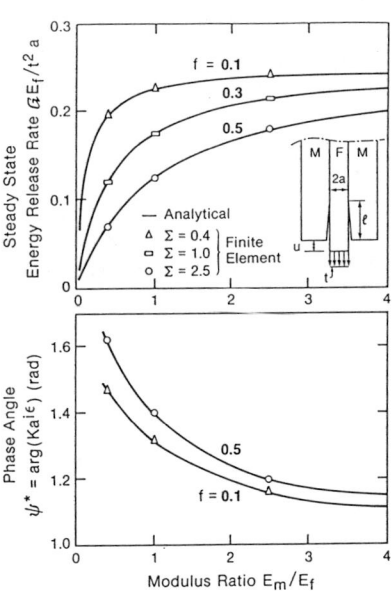

d) effects of elastic modulus ratio on $\mathcal{G}_{ss}$ and Ψ

For the plane delamination problem depicted in Fig. 1, a comprehensive analysis exists,[14] expressible in terms of imposed axial forces and bending moments. The solution having greatest relevance to problems in ceramic matrix composites involves the four-point bending of a bimaterial beam with debond cracks between the inner loading points.[9,14] The general form of the solution (Fig. 6a) indicates that $\mathcal{G}$ has a maximum upon initial debonding and then decays to a steady-state level when d exceeds the thickness of the upper (matrix) layer by ~ 3. Such behavior suggests that initial debonding occurs stably, but then becomes unstable. Trends in the steady-state value $\mathcal{G}_{ss}$ are summarized in Fig. 6b. The corresponding non-dimensional phase angle, Ψ^* is ~ 0.68 for all Σ, when $\varepsilon = 0$. Clearly, the elastic properties have strong influences on both $\mathcal{G}_{ss}$ and Ψ.

The preceding results also indicate that initial debonding along the interface is expected, provided that $\mathcal{G}_{ic}$ at $\Psi \simeq \pi/4$ is less than the critical strain energy release rate for the fiber, $\mathcal{G}_{fc}$, by a ratio that depends on the elastic properties for fiber and matrix. For the elastically homogeneous case, debonding occurs in preference to fiber failure when

$$\mathcal{G}_{ic}/\mathcal{G}_{fc} \lesssim 1/4 \qquad (7)$$

Solutions for fibers and matrix having different elastic properties ($\Sigma \neq 1$) do not yet exist. Further crack front debonding is not addressed by this solution. Useful insights concerning this debonding problem can be gained by interpolating between the above mixed mode $\mathcal{G}(\Psi)$ solution and $\mathcal{G}_{ic}(O)$ for long, cylindrical debonds in the crack tip field[15] (Fig. 7). The latter solution indicates that debond lengths substantially larger than the fiber diameter result in very small values of $\mathcal{G}_{ic}(O)$ at the interface compared with that of the matrix crack front, $\mathcal{G}_{mc}$, as given approximately by;

$$\mathcal{G}_{ic}(O)/\mathcal{G}_c \approx 0.1 R/d \qquad (8)$$

Consequently, it is surmised that the $\mathcal{G}_{ic}(O)$ required for debonding decreases rapidly with debond length, as sketched in Fig. 7. Extensive crack front debonding thus appears unlikely in the absence of residual stress, even when $\mathcal{G}_{ic}$ is quite small.

This conclusion about crack front debonding is substantially changed when residual stress exists. Residual radial tension results in opening mode steady-state interface cracking, characterized by;[12]

$$E^f \mathcal{G}_{ss} \approx \pi q^2 R/2 \qquad (9)$$

where q is the residual stress normal to the interface, as governed by the stress-free strain, ε, the fiber volume fraction and the elastic properties.[12,15] Superimposing opening mode K's from the residual and applied fields provides a full debonding solution. For elastically homogeneous composites,

$$\mathcal{G}/\mathcal{G}_m \approx \left[0.3\sqrt{\frac{R}{d}} + q\sqrt{\frac{\pi R}{2E\mathcal{G}_m}} \right]^2 \qquad (10)$$

Equating $\mathcal{G}$ to the critical value for the interface, the debond length becomes

$$d/R \approx \frac{0.1 E \mathcal{G}_M/q^2 R}{\left[\sqrt{E\mathcal{G}_i/q^2R} - \sqrt{\pi/2}\right]^2} \qquad (11)$$

As expected, once the debond initiation condition has been satisfied, residual tension encourages further debonding. Indeed, when $q \gtrsim K_i \sqrt{\pi/2R}$, then $d \to \infty$ at

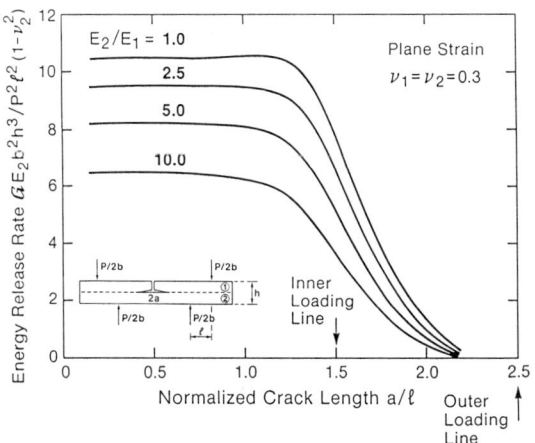

a) trends with crack length for an elastically homogeneous system

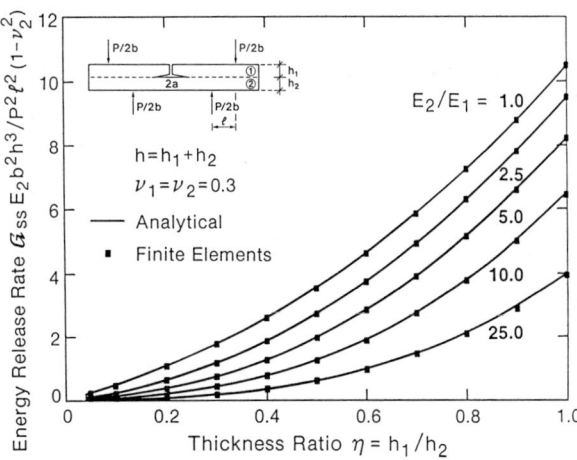

b) trends in steady-state energy release rate, G_{ss} with modulus and thickness ratios

Fig. 6. Energy release rates for a bimaterial beam tested in flexure

the crack front. For this condition, partial debonding is likely upon cooling[16] (See Fig. 10).

Initiation of debonding is a necessary but not sufficient condition for good composite properties. It is also required that the debond crack remain in the interface and not kink into the fiber to cause premature fiber rupture, either along the crack front or in the crack wake. Analysis of this problem[17] (Fig. 8) indicates that kinking out of the interface is strongly influenced both by the elastic properties of fiber and matrix and by the phase angle. In particular, debond deviation into the fiber is less likely when the fiber has a high modulus: an important rationale for choosing high modulus fibers. Furthermore, even when the fiber has a relatively low modulus, kinking out of the interface is only likely whenever $\mathcal{G}_i(\Psi) \gtrsim 0.4\, \mathcal{G}_f$.

The Interface Fracture Resistance

The preceding mechanics provide the essential background needed for the measurement of debond resistances relevant to composite performance, as summarized in Fig. 9. Three basic test methods have been identified:[9,11,18] compact tension tests, flexural tests and pull-out tests. The former provides data for $\Psi \simeq 0$, the latter for $\Psi \simeq \pi/2$ and the flexural test for $\Psi \simeq \pi/4$. Critical aspects of interface fracture testing concern the initial introduction of a well-defined debond crack and measurement of the residual stress.[18] Another important testing issue concerns friction at the loading points.[19] A procedure that takes frictional effects into account, based on measurements of the hysteresis in loading and unloading compliance has been developed and validated.[19] These rigorous demands on the testing needed to generate valid $\mathcal{G}_i(\Psi)$ data, have limited the extent of available results. Preliminary results indicate that $\mathcal{G}_i$ tends to increase with increase in Ψ, especially as $\Psi \to \pi/2$, and furthermore, that the rate of increase depends on the morphology of the fracture interface.[11,18] Specifically, rough fracture interfaces cause $\mathcal{G}_i$ to increase more rapidly with increase in Ψ (Fig. 3). Analysis of this phenomenon[11] has attributed this trend to the sliding and locking of crack surface asperities that make contact at large phase angles (Fig. 3). The material parameter that governs the magnitude of the above effect is;[11]

$$\chi = EH^2 / \mathcal{G}_o L \tag{12}$$

where H is the amplitude and L the wavelength of undulations on the fracture interface and $\mathcal{G}_o$ is the intrinsic fracture resistance of the interface. Specifically, large χ results in the greatest effects on $\mathcal{G}_i(\Psi)$. The quantity χ is a measure of the length of the contact zone, which increases as either H increases or $\mathcal{G}_o$ decreases.

The magnitude of $\mathcal{G}_o$ is clearly influenced by the presence of interphases, the atomistic structure of the interface, etc. However, as yet, residual stress and morphological influences have not been sufficiently decoupled to explore these basic relationships. Nevertheless, preliminary measurements reveal that $\mathcal{G}_o$ is typically quite small for oxides bonded to refractory metals (Nb), to intermetallics (TiAl) and to noble metals (Au,Pt), as well as for oxides bonded with inorganic glasses and for carbides and nitrides having graphite and boron nitride interlayers.

Debonding During Composite Fracture

The preceding debonding results, while still incomplete, are consistent with the following sequence of events during matrix crack propagation. Initial debonding occurs along the interface at the crack front, provided that $\mathcal{G}_f/\mathcal{G}_i$ (at $\Psi \simeq \pi/4$) $\gtrsim 4$. The extent of debonding is typically small when residual compression exists at the interface, but can be extensive when the interface is in residual tension. However, more importantly, further debonding is invariably

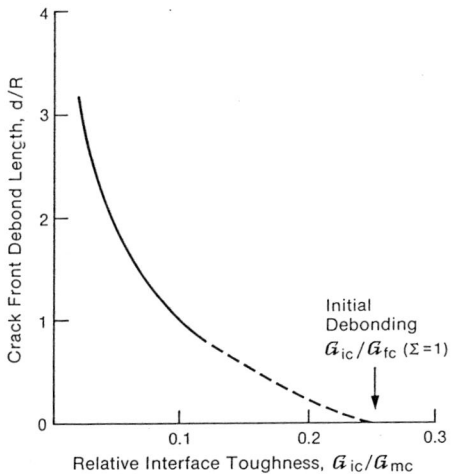

Fig. 7. The energy release rate for crack front debonding

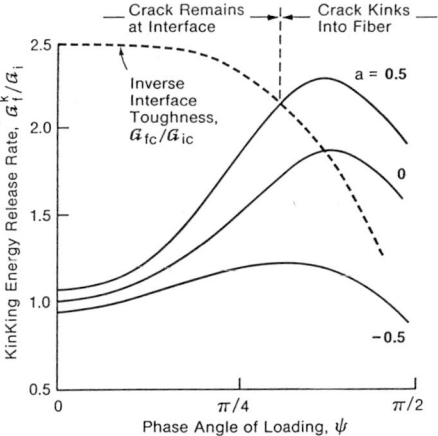

Fig. 8. Energy release rates for kinking of an interface crack into a fiber as a function of the phase angle of loading. Also shown is a typical critical energy release rate for an interface crack: a is one of the Dundurs' parameters (Eqn. 2). Kinking occurs when the critical values $\mathcal{G}_{fc}/\mathcal{G}_{ic}$ fall below the kinking level, $\mathcal{G}_f^k/\mathcal{G}_i$

induced in the crack wake.[12] The extent of debonding is again governed largely by the residual field. Residual radial tension results in unstable conditions and encourages extensive debonding.[12] Residual compression causes stable debonding,[13] with extent determined by the friction coefficient and morphology of the debonded interface. Fracture of the fiber by kinking of the debond crack into the fiber is unlikely,[16] especially when the fiber modulus is relatively large. Fiber failure and pull-out toughening thus appear to be strongly influenced by the statistics of fiber failure.[20]

The above sequence indicates that, while debonding is a prerequisite for high toughness, the properties of the composite are not otherwise limited by the extent of debonding. The <u>basic fracture requirement</u> for the interface is thus;

$$\mathcal{G}_i/\mathcal{G}_f \lesssim 1/4$$

at $\Psi \simeq \pi/4$ when $\Sigma = 1$. Trends in this prerequisite with Σ remain to be characterized. Subject to this requirement, the <u>sliding resistance</u> of the debonded interface is the most important interface property, as elaborated in subsequent sections.

One additional issue concerns thermal debonding in <u>composites</u> subject to residual radial tension, as governed by the inequality,[12] $q \gtrsim \sqrt{\pi E\, \mathcal{G}_{ic}/2R}$. When this inequality is satisfied, the incidence of thermal debonding is influenced by the interface flaw statistics, such that the debond probability increases as q becomes large. The thermal debonds typically extend about 2/3 around the circumference (see Fig. 10) before arresting.[17] When thermal debonds exist, fiber failure at the crack front is inhibited and the requirements on $\mathcal{G}_i$ needed for good composite toughness are less stringent than those expressed above.

There is no direct experimental validation of the preceding hypothesis for ceramic matrix composites. However, various observations of crack interactions with fibers and whiskers are supportive of the general features.[17,21] In particular, experiments on LAS/SiC composites reveal that as-processed materials with a C interlayer debond readily and demonstrate extensive pull-out (Fig. 10a), whereas composites heat treated in air to create a continuous SiO_2 layer between the matrix and fiber exhibit matrix crack extension through the fiber without debonding (Fig. 10b). Furthemore, composites with a thin interface layer of SiO_2 having a partial circumferential gap exhibit intermediate pull-out characteristics (Fig. 10c). The associated constituent properties are summarized in Table I. Based on these properties, the preceding arguments would indicate that crack front debonding should <u>not</u> occur when a complete SiO_2 layer exists at the interface; whereas, appreciable crack front debonding should obtain when the C layer is present, in complete accordance with the observations.[17,21] The composites with only a partial SiO_2 interface layer are also interesting. For these materials, $\mathcal{G}_i$ is related to that fraction of the circumference that bonds the fiber to the matrix: typically 1/3 (Fig. 10c). Reference to Table I and to the initial debonding requirement (Eqn. 7) would thus indicate that debonding, while marginal, is certainly possible.

A second pair of examples concerns SiC whiskers in Al_2O_3 and Si_3N_4 matrices with an amorphous silicate interphase.[22] For these materials, the properties indicated in Table II indicate that crack front debonding conditions are marginally satisfied, consistent with the variability in debonding evident from toughness data.

<u>Pull-Out</u>

An integral aspect of the analysis of composite fracture involves consideration of fiber failure and of subsequent pull-out. As noted above, fiber failure can usually be described using concepts of weakest link statistics based on the fiber strength parameters S_0 and m,[20] in accordance with the frequency distribution,

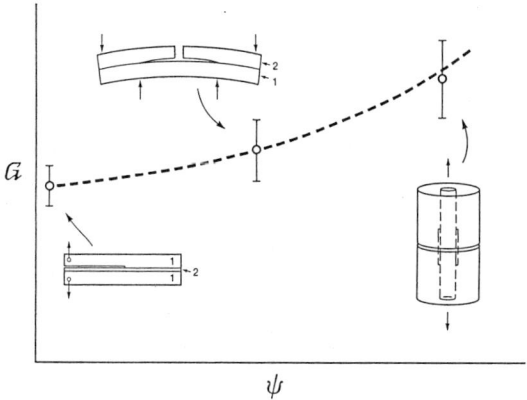

Mixed Mode Interface Fracture Resistance Tests

Fig. 9. Some test specimens used to determine the fracture resistance of interfaces as a function of phase angle

TABLE I

Constituent Properties of LAS/Nicalon Composites

	E(GPa)	$\mathcal{G}_c$ (Jm^{-2})	α (K^{-1})
Fiber (Nicalon)	200	4 – 8$^{\neq}$	4×10^{-6}
Matrix (LAS)	85	40	1×10^{-6}
Interface Amorphous C Amorphous SiO$_2$	——— 80	< 1 * 8	——— 1×10^{-6}

$\neq$ Determined from fracture mirror radii

* Determined by indentation: takes into account initial thermal debonding (Fig. 10)

TABLE II

Constituent Properties of Whisker–Reinforced Ceramics

	E(GPa)	G_c (Jm^{-2})	α (K^{-1})
Al$_2$O$_3$	400	20 – 30	7 x 10^{-6}
Si$_3$N$_4$	300	60 – 80	3 x 10^{-6}
SiC	400	15 – 20	4 x 10^{-6}
Amorphous interface	70	4 – 8	——

TABLE III

Effect of Residual Stress on Toughness

STRESS/DISPLACEMENT LAW		RUPTURE CONDITION	RESIDUAL STRESS IN FIBER	
			COMPRESSION	TENSION
Linear		Stress	Decrease	Decrease
		Displacement	Increase	Decrease
Frictional With Pull-Out	Surface Roughness	Stress	————	Negligible
	Coulomb Friction	Stress	Decrease	————

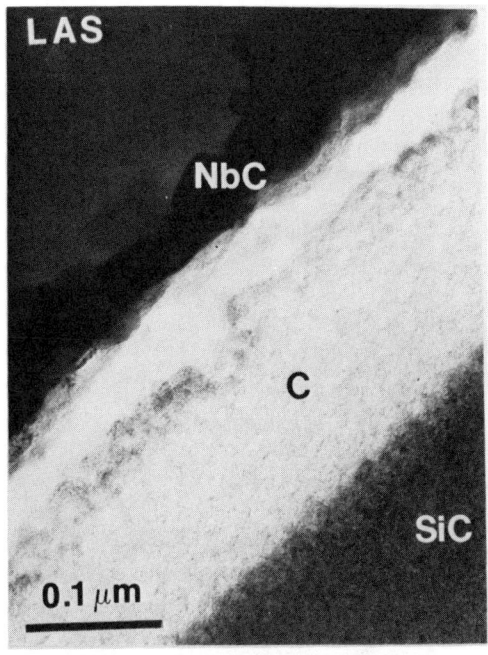

a) As-processed indicating C interlayer and extensive pull-out

Fig. 10. Interfaces and pull-out in a composite consisting of LAS matrix and SiC (Nicalon) fibers

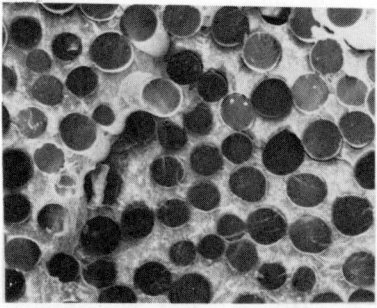

b) heat treated in air for 16 hours at 800°C indicating a complete SiO₂ layer and no pull-out

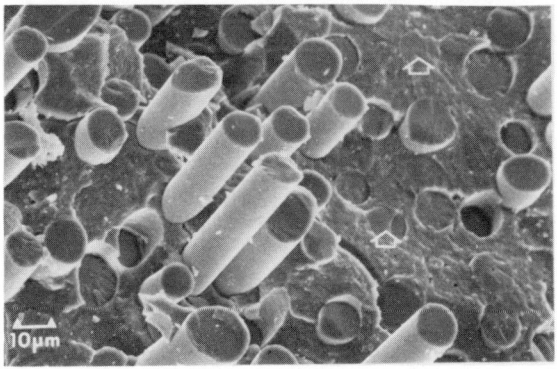

c) heat treated in air for 4 hours at 800°C indicating a partial SiO$_2$ layer-with gap-and variable pull-out

$$\int g(S)\,dS = (S/S_o)^m \tag{13}$$

The locations of fiber failure that govern the pull-out distributions can, in principle, be determined from the stresses on the fibers ahead of the matrix crack and in the crack wake. The former analysis has not been attempted, partly because the problem is complex and partly because of a perception that fiber failures close to the crack plane that cause pull-out are most likely to occur in the crack wake, following the debond extension process. Indeed, such behavior has been observed in glass reinforced plastics.[23] However, there is no direct evidence that fiber failure ahead of the matrix crack can be neglected in ceramic matrix composites.

While it is important to be aware of the above uncertainties, it is nevertheless insightful to fully analyze the wake failure phenomenon. Comparisons with experimental fiber pull-out data then allow assessment of the hypothesis. The fiber failure analysis commences with the basic weakest link description of failure. Then, by incorporating an axial fiber stress distribution, the fiber failure locations can be derived.[20] Such analysis has been performed for composites with negligible residual stress and having debonded interfaces subject to a constant sliding stress τ. For this purpose, the fundamental probability density function is[20]

$$\phi(z,t) = \frac{2\pi R m}{S_o^m}(t - 2z\tau/R)^{m-1} \exp\left[-(t/T)^{m+1}\right] \tag{14}$$

where

$$T = \left[\frac{S_o^m \tau (m+1)}{2\pi R^2}\right]^{\frac{1}{(m+1)}}$$

and z is the distance from the crack plane. Then, the cumulative probability that the pull-out length will be $\tilde{<}h$ is ($h < t R/2\tau$),[21]

$$\Phi(h) = 2\int_o^h \int_o^{\infty} \phi(z,t)\,dz\,dt \tag{15}$$

Trends in cumulative probability are plotted on Fig. 11, indicating that the pull-out lengths tend to increase as m decreases. The effects of residual strain on Φ (h) are expected to be substantial. Preliminary estimates suggest that the pull-out length usually decreases as the residual strain ε increases, when the residual stress at the interface is compressive. However, specific trends are sensitive to m, as well as to the friction coefficient μ.

Experimental results concerning trends in the pull-out distribution with interface properties[21] have been obtained for heat treated LAS/SiC composites, having the features described in the preceding section (Fig. 10). The results reveal that as the gap caused by C removed is filled with SiO_2, the pull-out distribution gradually changes (Fig. 12). In particular, the median length decreases and that proportion of fibers that actually pull out exhibit length distributions consistent with those predicted by the above weakest link fiber failure analysis (Fig. 12) such that the interface τ increases by about an order of magnitude when a partial SiO_2 layer replaces C. This change in τ causes a dramatic change in the mechanical properties of the composite, as elaborated below.

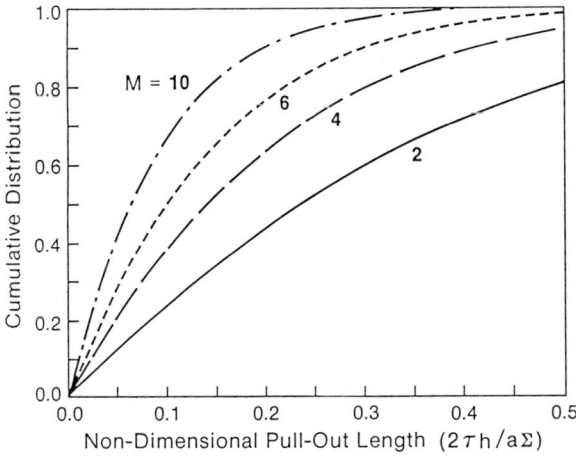

Fig. 11. The cumulative pull-out distribution for several values of the shape parameter, m

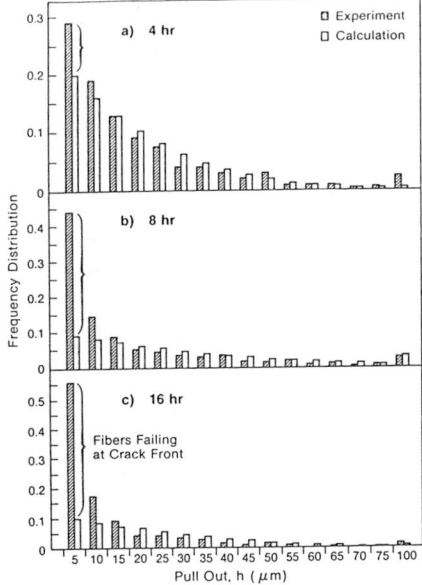

Fig. 12. Histograms indicating trends in pull-out length with heat-treatment

TENSILE PROPERTIES: MODE I FAILURE

Axial Stress-Strain Behavior

The axial tensile properties of ceramic matrix composites are strongly influenced by the relative debond resistance $G_i(\Psi)/G_f$, by the friction coefficient along the debonded interface μ and by the residual strain, ε, as intimated in the preceding section. When the interface is in residual tension and when G_i/G_f and μ are both small, as exemplified by C interlayers between fiber and matrix, experience[21,24] has indicated that the tensile stress/strain behavior illustrated in Fig. 13a obtains. Three features of this curve are important: matrix cracking at a stress σ_o, fiber bundle failure at σ_u and the pull-out stress. Increases in either G_i/G_f or μ cause the stress-strain curve to become linear[21] (Fig. 13b). Furthermore, the ultimate strength then coincides with the propagation of a single dominant crack (albeit, sometimes with a desirable "tail" caused by delamination, as discussed in the Mixed Mode Failure section). Composites having this macroscopic characteristic exhibit properties governed by a fracture resistance curve. The individual properties within each of these two regimes are discussed below, as well as criteria that dictate the transition between regimes.

Matrix Cracking

The stress σ_o at which matrix cracking occurs has been the most extensively studied behavior in ceramic matrix composites.[15,24-26] For composites in which q is tensile and the interface properties can be effectively represented by an unique sliding stress, τ, the matrix cracking stress is given by;[15]

$$\frac{\sigma_o}{E} = \left(\frac{6 f^2 E^f \tau G_M}{(1-f) E (E^M)^2 R} \right)^{\frac{1}{3}} - \frac{p}{E^M} \qquad (16)$$

This result is <u>independent</u> of the matrix crack length because the crack is bridged by fibers. Experiments conducted on a number of ceramic matrix composites have validated Eqn. (16). When the interface is subject to residual compression, τ depends on the applied stress and the solution for σ_o is more complex. However, to first order, τ may be simply replaced by μq. At σ_o, multiple matrix cracking is expected[25] and observed[26] with a saturation crack spacing D in the range;

$$\sigma_o R/2 f \tau < D < \sigma_o R/f \tau \qquad (17)$$

Experimental observations[24] have again confirmed this feature of matrix cracking.

The most crucial aspects of the above interpretation of steady-state cracking and of behavior prediction concern determination of τ and q for actual composite system. Both are difficult to measure. Two basic approaches have been used to measure the <u>sliding resistance</u> τ: indentation[27,28] and crack opening hysteresis.[24] Both approaches are readily applicable when G_i and τ are small. The former method is most insightful when used with a nanoindenter system, whereupon τ can be obtained on single fibers either from a push through force on thin sections or from the hysteresis in the loading/unloading cycle on thick sections (Fig. 14a). However, this method has the obvious disadvantage that the fiber is in axial compression such that the debond interface is also compressed during the test, with attendant changes in τ. Matrix cracking followed by measurement of the crack opening hysteresis (Fig. 14b) is more desirable, when feasible, because the fibers are subject to axial tensile loading. However, when

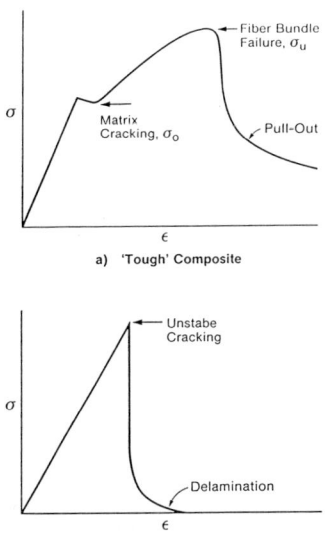

Fig. 13. Tensile stress-strain curves for ceramic matrix composites; a) small G_{ic}, small μ, b) small G_{ic}, large μ

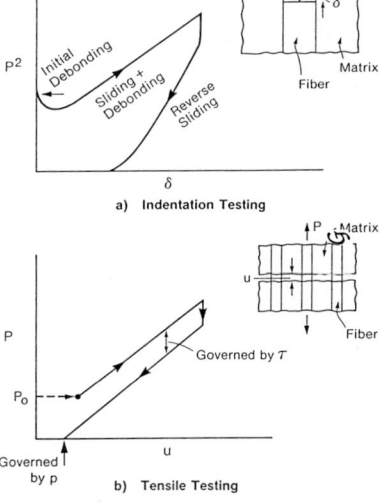

Fig. 14. a) A load/unload cycle for nanoindentation of a fiber b) Crack opening hysteresis for a composite with intact fibers revealing the trends in both sliding resistance and residual stress

appreciable fiber failure accompanies matrix cracking, erroneous results also obtain for this method. Both approaches give about the same value of τ for composites having the following characteristics: tensile residual strain exists at the interface,[24] G_i/G_f is small and τ is small (<10MPa). Otherwise, both approaches are problematic. Consequently, other approaches applicable to composites having larger τ are being investigated. One of these is discussed in the following section.

The Ultimate Strength

Following multiple matrix cracking, the fibers are subject to an oscillating stress field with maximum equal to σ_∞/f between the crack surfaces where σ_∞ is the applied stress. The probability of fiber failure within such a stress field subject to weakest link statistics can be readily derived.[29] However, derivation of a load maximum requires that the stress redistribution caused by the fractured fibers be modeled. Such an analysis has not been attempted. Nevertheless, a lower bound for the load can be derived by simply allowing failed fibers to have no load bearing ability. Then, a modified bundle failure analysis allows the ultimate strength to be;[34]

$$\sigma_u = f\hat{S} \exp\left[-\frac{\left[1-(1-\tau D/R\hat{S})^{m+1}\right]}{(m+1)\left[1-(1-\tau D/R\hat{S})^m\right]}\right] \quad (18)$$

where S_o and m are the statistical parameters that represent the fiber strength distribution and

$$(R\hat{S}/\tau D)^{m+1} = (1/2\pi RL)(RS_o/\tau D)^m\left[1-(1-\tau D/R\hat{S})^m\right]^{-1} \quad (19)$$

with L being the gauge length. In the one composite system for which analysis of the ultimate strength has been performed (LAS/SiC),[21] surprisingly Eqn.(18) agrees quite well with measured values.

The ultimate strength anticipated from the above logic is expected to be influenced by the residual stress. Specifically, in systems for which the fiber is subject to residual compression, the axial compression should suppress fiber failure and elevate the ultimate strength to a level in excess of that predicted by Eqn.(18). This effect may be estimated by regarding the matrix as clamping onto the fiber and thus, simply superposing the residual stress onto $\hat{S}$.

Resistance Curves

When mode I failure is dominated by propagation of a single dominant matrix crack, accompanied by fiber failure and pull-out, the mechanical properties are characterized by a resistance curve. Analysis of this phenomenon utilizes the distribution of fiber failure sites determined in the pull-out analysis (Fig. 11). From such analysis, the mean failure length of all fibers that fail at stress t acting on the fiber between the crack surfaces is first evaluated.[20] Then, by taking into account the reduced stress caused by fiber failure and knowing the associated crack opening, u, the total stress on the fibers between the crack at a fixed crack opening may be determined,[20] as plotted on Fig. 15. Several features of the t(u) curve are notable. The initial, rising position is dominated by intact fibers, the peak is dominated by multiple fiber failures, analogous to bundle failure, and the tail is governed by pull-out. The role of the shape parameter, m, on these features is particularly interesting. As m decreases, corresponding to a broader distribution of fiber strengths, more fibers fail further from the matrix crack (Fig. 11) causing the extent of pull-out to substantially increase.

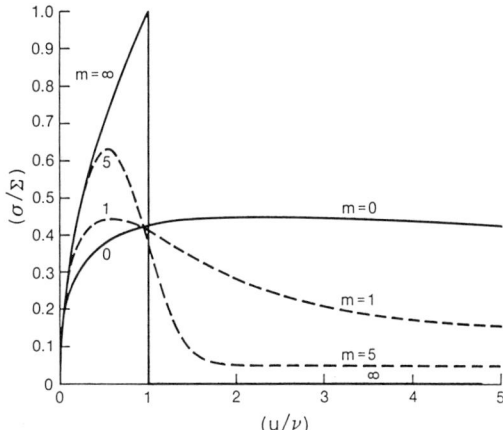

Fig. 15. Trends in non-dimensional crack openings stress t, with opening u for several values of the shape parameter, m

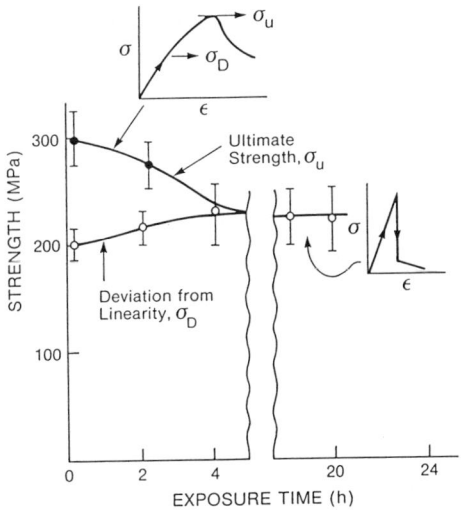

Fig. 16. Effects of heat treatment on the stress-strain behavior of LAS/SiC composites

The trends in t(u) directly associate with the two most relevant features of the fracture resistance curves: the asymptotic toughness and the slope (or tearing modulus), respectively. The <u>asymptotic toughening</u> can be simply derived from the J integral result,

$$\Delta \mathcal{G}_{ss} = 2f \int_0^{\tilde{u}} t(u) du \qquad (20)$$

The expressions that govern the trends in $\Delta \mathcal{G}_{ss}$ with material properties are unwieldy in form.[20] However, inspection reveals that $\Delta \mathcal{G}_{ss}$ always increases as the scale parameter S_0 incease, thereby establishing that high fiber strengths are invariably desirable. However, the dependence on τ and R is ambivalent. The essential details are highlighted by considering separately the bridging and pull-out contributions to the toughness integral. The <u>bridging</u> component

$$\Delta \mathcal{G}_b = 4fTU/(m+1) \qquad (21)$$

where $U = T^2R/4E^f\tau(1 + \zeta)$ is proportional to $[R^{m-5}/\tau^{m-2}]^{1/(m-1)}$. A notable feature is the inversion in the trend with that τ occurs at $m = 2$, and with R at $m = 5$. The corresponding pull-out contribution can be examined by recognizing that the toughening has the form

$$\Delta \mathcal{G}_b \sim \langle h \rangle^2 (\tau/R) \qquad (22)$$

which, with Eqn.(15) indicates a toughness proportional to $[R^{m-3}S_0^{2m}/\tau^{m-1}]^{\frac{1}{m+1}}$. The toughness thus increases with increasing R when $m > 5$, and decreases when $m < 3$. Conversely, it increases with increasing τ when m is very small (≤ 1), and decreases when $m > 2$. These limits arise because of the competing importance of the contribution to toughness from the intact bridging fibers and the failed fibers that experience pull-out. Knowledge of the magnitude of the statistical shape parameter, m, for the fibers *within* the composite is therefore a prerequisite to optimizing the shear properties of the interface for high toughness.

The slope of the resistance curve has not yet been evaluated, because numerical methods are needed to determine the upper limit of Eqn.(2), as dictated by the crack opening at the end of the bridging zone. Yet, this shape has a major influence upon the ultimate strength of the composite. Further research concerning this phenomenon is a major priority.

Property Transition

Non-linear macroscopic mechanical behavior in tension is most desirable for structural purposes and thus, analysis of the transition between this regime and the linear regime is important. A useful preamble involves comparison of the basic trends in the steady-state matrix cracking stress, σ_0 (Eqn.16), and in the asymptotic fracture resistance, $\Delta \mathcal{G}_{ss}$ (Eqn.20). At the simplest level (these trends apply except at small m), these trends are

$$\sigma_0 \sim (\tau/R)^{1/3}$$

$$\Delta \mathcal{G}_{ss} \sim RS_0^2/\tau \qquad (23)$$

Most significantly, σ_0 increases but $\Delta \mathcal{G}_{ss}$ decreases as τ increases. These opposing trends with τ suggest the existence of an <u>optimum</u> τ that permits good matrix cracking resistance while still allowing high toughness.

More specifically, a property transition is expected when the matrix cracking stress attains the stress needed for fiber bundle failure. One bound on the property transition can be obtained by simply allowing σ_0 to exceed the ultimate strength.[21,31] Then, the parameter B which governs the transition when τ is small is;

$$B = \tau E \mathcal{G}_M / S_o^3 R \qquad (24)$$

Specifically, when B exceeds a critical value, linear behavior initiates. Experiments on heat treated LAS/SiC composites[21] have examined the conditions associated with this transition (Fig. 16).

Residual Stress

Large mismatches in thermal expansion between fiber and matrix are clearly undesirable. In particular, relatively large matrix expansions, $\alpha_m/\alpha_f >> 1$, cause premature matrix cracking (Eqn.16). Such behavior is not necessarily structurally detrimental, but concerns regarding thermal fatigue, the ingress of environmental fluids, etc. have discouraged the development of materials having these characteristics. Conversely, very small matrix expansions, $\alpha^f/\alpha^m >> 1$, thermally debond the fiber from the matrix. When sufficiently extensive, such debonding results in axial separations that negate the influence of the fibers. Consequently, values of α_f/α_m close to unity are required. Indeed, mode I axial properties subject to an interface that easily debonds and slides freely along the debond involve an optimum residual stress, with a maximum matrix cracking stress, when ε is positive, given by,[15]

$$\sigma_o / E = (2/3) \left[f \mu \mathcal{G}_M / \lambda_2 E_M R \right]^{\frac{1}{2}} \qquad (25)$$

where

$$\lambda_1 = \frac{1 - (1 - E/E_f)(1 - v_f)/2 + (1 - f)(v_m - v_f)/2 - (E/E_f)[v_f + (v_m - v_f) f E_f / E]^2}{(1 - v_m)\Delta},$$

$$\lambda_2 = \frac{[1 - (1 - E/E_f)/2](1 + v_f) + (1 + f)(v_m - v_f)/2}{\Delta},$$

and

$$\Delta = 1 + v_f + (v_m - v_f) f E_f / E$$

When $\alpha_f/\alpha_m >> 1$, such that ε is negative, asperities on the debond surface may provide a discrete sliding stress, τ, that depends on such features as the asperity amplitude. For such cases, the optimum residual strain has not been determined. It is noted, however, that good properties have been demonstrated for LAS/SiC composites having an expansion mismatch, $\Delta\alpha \sim 3 \times 10^{-6}$ C^{-1}.

The above remarks concerning residual strains are qualified by debonding requirements. While the condition $\mathcal{G}_i/\mathcal{G}_f < 1/4$ ($\Sigma = 1$) is an adequate prerequisite for debond initiation of the matrix crack front, it is not clear at this stage whether smaller values are needed when ε is positive, in order to ensure that debonding occurs to a sufficient extent that the pull-out distribution is not diminished by fiber failure from the debond.

The fracture resistance is also influenced by the residual stress. However, the sign and magnitude of the change in toughness induced by residual stress depends on the mechanisms of interface sliding and fiber failure, as summarized in Table III. Subject to adequate debonding, the salient result for ceramics reinforced with brittle fibers is that $\Delta \mathcal{G}_{ss}$ is unaffected when ε is negative and the interface is characterized by an unique τ, whereas $\Delta \mathcal{G}_{ss}$ usually decreases with increase in ε when ε is positive because the pull-out lengths decrease, as apparent when τ in Fig. 11 is equated to μq.

Residual stresses in composites are difficult to measure. Even when the composite is fully elastic, such that no interface debonding or sliding occur on cooling, the residual stress at the surfaces are still complex. Consequently, methods such as X-ray diffraction that probe thin surface layers are difficult to interpret. Neutron diffraction, which typically averages over a much larger volume of material, is usually more satisfactory. Measurement difficulties are exacerbated when debonding and sliding occur on cooling. These processes initiate preferentially at the surface and spread into the body along the interface, thereby alleviating the residual stress over the debond/slip length. For small τ and $\mathcal{G}_i$, these lengths are large (many multiples of the fiber diameter).[24] Consequently, a valid measure of the residual stress can only be obtained using processes that penetrate well into the material. One independent approach for measuring q that has merit in some cases involves use of the same crack opening measurements described by Fig. 14b. Specifically, the residual axial stress in the matrix is related directly to the stress at which crack closure occurs.[24] The method is, however, restricted to materials for which matrix cracking is not accompanied by extensive fiber failure.

Transverse Failure

The transverse strength of high toughness composites are generally very low. There have been no systematic studies of this property. However, experimental studies on composite laminates[32] indicate that the transverse cracks typically propagate along the interface layer and through the matrix between neighboring fibers. Furthermore, because the interfaces have sufficiently small $\mathcal{G}_{ic}$ to allow debonding, overall failure is preceded by interface failure. This process is assumed to occur at a critical stress, σ_c, which can be determined in a manner analogous to that for the steady-state cracking of thin films,[33] to give,

$$\sigma^c \approx \sqrt{2E\mathcal{G}_i/\pi R} - q \qquad (26)$$

In some cases, q is sufficiently large that $\sigma^c < 0$ and the interfaces debond upon cooling (Fig. 10). <u>Overall failure</u> involves the coalescence of interface cracks. This occurs at another critical stress, which is related to the solution for an array of parallel cracks.

MIXED MODE FAILURE

Mode II Failure Mechanisms

Flexural tests performed on uniaxial composites reveal that a shear damage mechanism exists (Fig. 11)[24,32,34] and that such damage often initiates at quite low shear stresses, e.g., 20MPa in LAS/SiC. The damage consists of an echelon matrix microcracks inclined at about $\pi/4$ to the fiber axis (Fig. 17). With further loading, the microcracks coalesce, causing matrix material to be ejected and resulting in the formation of a discrete mode II crack. The crack is defined by the planar zone of ejected matrix. The crack also has a microcrack damage zone similar to that present upon crack initiation.

The microcracks that govern mode II failure are presumably caused by stress concentrations in the matrix and form normal to the local principal tensile stress, but then deflect parallel to the mode II plane and coalesce. An adequate model that incorporates the above features has not been developed. Consequently, the underlying phenomena are briefly noted without elaboration. The stress concentrations in the matrix have magnitude governed by the elastic properties, the fiber spacing and the interface strength. The growth and coalescence of the microcracks scales with the matrix toughness $\mathcal{G}_{mc}$. The shear strength seemingly decreases as the mode I toughness increases.

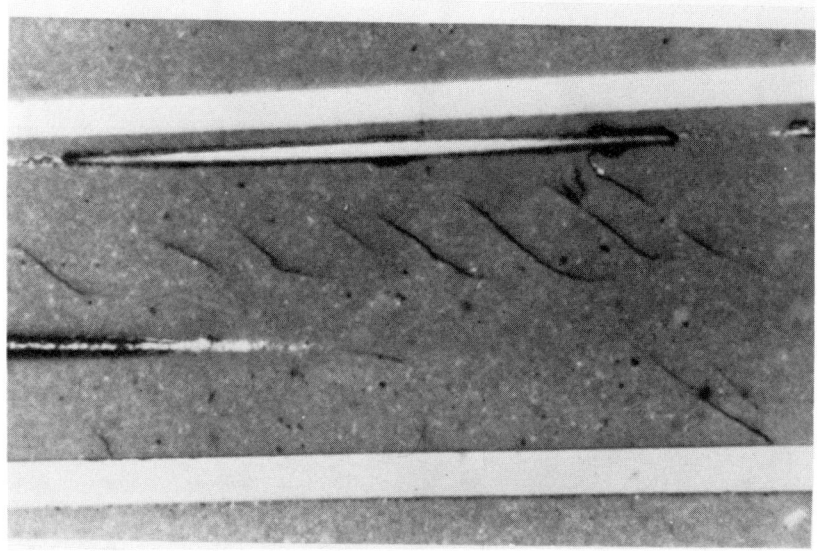

Fig. 17. Matrix microcracks observed in the zone of shear failure

Delamination Cracking

Delamination is a common damage mode in the presence of notches[32] (Fig. 11). Delamination cracks nucleate near the notch base and extend stably (Fig. 18). Analysis of such data is based on the solutions used for mixed mode interface cracking in beams,[9] modified to take account of the elastic anisotropy. The fracture resistance is found to increase with crack extension and, because of the large phase angle, the fracture mechanism is essentially identical to that noted for mode II failure, involving matrix microcracking and spalling. The existence of a resistance curve is attributed to intact fibers within the crack that resist the displacement of the crack surfaces and thus shield the crack tip in a manner analogous to fiber bridging in mode I. However, explicit analysis has yet to be conducted.

MICROSTRUCTURE DESIGN

Many of the microstructural parameters that control the overall mechanical properties of ceramic matrix composites are now known and validated, as elaborated in the preceding sections. Consequently, various general remarks about microstructure design can be made. However, important aspects of damage and failure are incompletely understood because there have been few organized studies of failure in mode II, mixed mode and transverse mode I. The remarks made in this section thus refer primarily to axial mode I behavior with no special regard to attendant problems in other loading modes.

The basic microstructural parameters that govern mode I failure are the relative fiber/matrix interface debond toughness, $\mathcal{G}_i/\mathcal{G}_f$, the residual strain, ε, the friction coefficient of the debonded interface, μ, the statistical parameters that characterizes the fiber strength, So and m, the matrix toughness, $\mathcal{G}_m$, and the fiber volume fracture f. The prerequisite for high toughness is that $\mathcal{G}_i/\mathcal{G}_{t} \lesssim 1/4$ ($\Sigma = 1$). Subject to this requirement, the residual strain must be small ($\Delta\alpha \lesssim 4 \times 10^{-6} C^{-1}$) and negative such that the interface is in tension. Furthermore, the friction coefficient along the debonded interface should be small ($\mu \lesssim 0.1$). The ideal fiber properties are those that encourage large pull-out lengths, as manifest in an optimum combination of a high median strength (layer S_o) and large variability (small m).

The above conditions can be satisfied, in principal, by creating interphases between the fiber and matrix, either by fiber coating or, in-situ, by segregation. The most common approach is the use of a dual coating: the inner coating satisfies the above debonding and sliding requirements, while the outer coating provides protection against the matrix during processing. However, the principal challenge is to identify an inner coating that has the requisite mechanical properties while also being thermodynamically stable in air at elevated temperatures. Most existing materials have either C or BN as the debond layer. However, both materials are prone to degradation in air at elevated temperatures. More stable alternatives have been proposed (e.g., Nb,Mo,Pt,NbAl) but have not been evaluated.

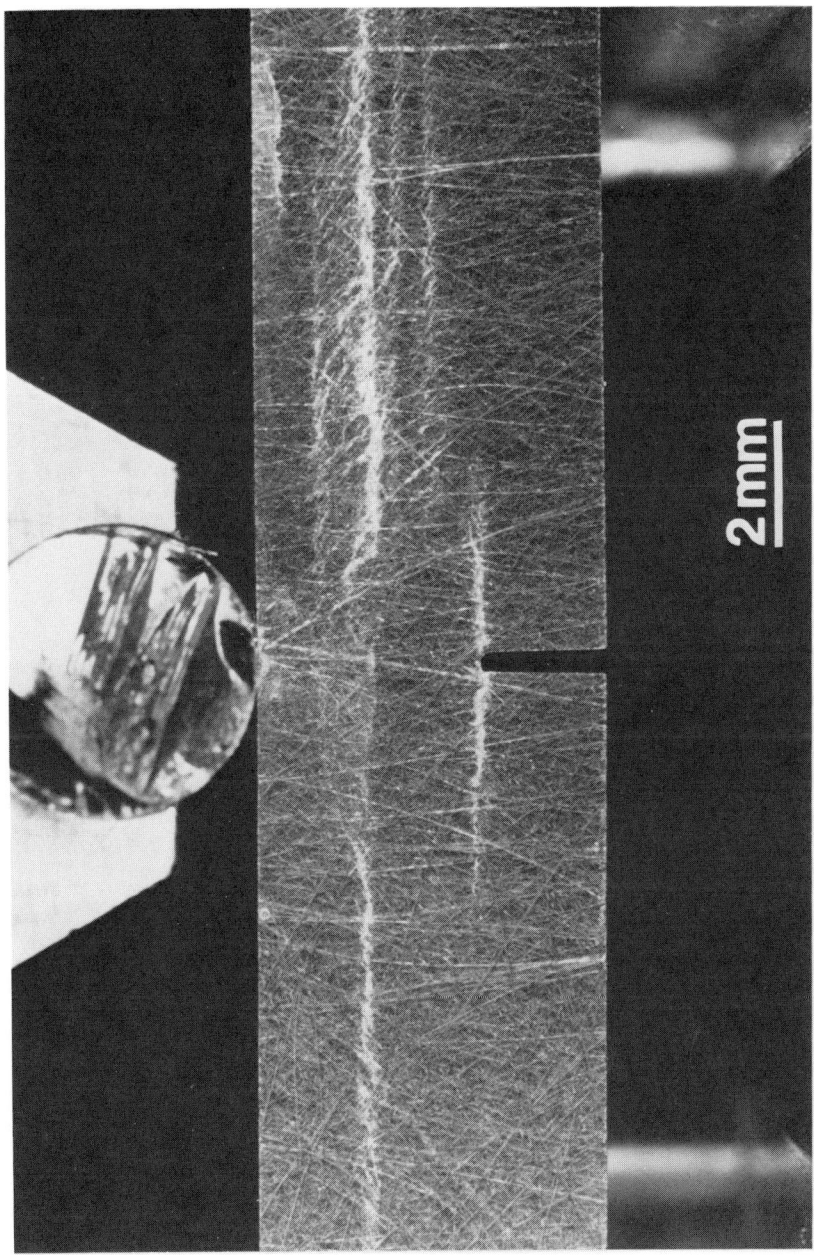

Fig. 18. Delamination cracking in notched flexure tests

NOTATION

α	linear thermal expansion coefficient
ε	stress–free strain ($\Delta\alpha\Delta T$): positive refers to residual compression normal to the interface
μ	friction coefficient
ν	Poisson's ratio of composite
ξ	coefficient = $(1-f)[(1-f)(1-\nu+\nu\Sigma)+\Sigma(1-f)]$
Σ	ratio of Young's modulus of fiber to matrix, E^f/E^m.
σ_o	matrix cracking stress
σ_u	ultimate strength
σ_∞	applied stress
σ^c	stress for transverse interface failure
τ	shear resistance of interface after debonding
$\Phi(h)$	cumulative pull-out distribution
ϕ	crack surface shear angle
$\phi(z, t)$	probability density function for fiber failure
χ	interface fracture parameter = $EH^2/\mathcal{G}_o L$
ψ	phase angle of loading
a	Dundurs' parameter = $\dfrac{G_1(1-\nu_2)-G_2(1-\nu_1)}{G_1(1-\nu_2)+G_2(1-\nu_1)}$
B	Transition parameter

b	Dundurs' parameter	$= \dfrac{G_1(1-2v_2) - G_2(1-2v_1)}{2[G_1(1-v_2) + G_2(1-v_1)]}$
D	matrix crack spacing	
d	debond length	
E	Young's modulus of composite	
F	non-dimensional stress	$= (t-p)/E^f \varepsilon$
f	fiber volume fraction	
G	shear modulus	
$\mathcal{G}$	strain energy release rate	
$\mathcal{G}_{ic}$	critical strain energy release rate for interface	
$\mathcal{G}_{ss}$	steady-state strain energy release rate	
$\mathcal{G}_{fc}$	critical strain energy release rate for the fiber	
$\mathcal{G}_{Mc}$	critical strain energy release rate for the composite	
$\mathcal{G}_{\infty}$	intrinsic critical strain energy release rate for the interface	
$\mathcal{G}_R(\Delta a)$	increase in critical strain energy release rate with increase in crack length, Δa.	
H	amplitude of interface roughness	
h	pull-out length	
ℓ	slip length	
L	gauge length	
m	shape parameter for fiber strength distribution	
p	residual axial stress in the matrix	$= \dfrac{\lambda_2}{\lambda_1}\left(\dfrac{E^f}{E}\right)\dfrac{f\varepsilon E^M}{1-v^M}$
q	residual compression normal to interface	$= (1-f)E^M \varepsilon / 2\lambda_1(1-v^M)$
R	fiber radius	

r	distance from crack front
S	fiber strength
S_o	scale parameter for fiber strength distribution
T	pull-out parameters
t	stress acting on fiber between crack surfaces
U	pull-out parameter = $T^2 R/4E^f \tau(1+\xi)$
u	crack opening displacement
v	crack shear displacement
z	distance from crack plane

REFERENCES

[1] A. G. Evans and R. M. Cannon, Acta Met. 34, (1986) 761.

[2] J. W. Hutchinson, Acta Met, 35, (1987) 1605.

[3] M. Rühle, A. G. Evans, R. M. McMeeking and J. W. Hutchinson, Acta Met. 35, (1987) 2701.

[4] A. G. Evans and K. T. Faber, J. Am. Ceram. Soc. 67, (1984) 255.

[5] B. Budiansky, J. W. Hutchinson and J. Lambropolous, Intl. Jnl. Solids and Structures, 19, (1983) 337.

[6] J. W. Hutchinson, Non-Linear Fracture Mechanics, Tech. Univ. Denmark (1979).

[7] J. R. Rice, Fracture, Vol. 11 (ed., H. Liebowitz), Academic Press, NY (1968) p. 191.

[8] J. R. Rice, J. Appl. Mech., in press.

[9] P. G. Charalambides, R. M. McMeeking and A. G. Evans, J. Appl. Mech., in press.

[10] J. Dundurs, Mathematical Theory of Dislocations, ASME, NY (1969) p. 70.

[11] A. G. Evans and J. W. Hutchinson, Acta Met., in press.

[12] P. G. Charalambides and A. G. Evans, Advanced Ceramic Materials, in press.

[13] L. S. Sigl and A. G. Evans, to be published.

[14] Z. Suo and J. W. Hutchinson, Harvard University Report MECH, 118 (1988), Intl. J. Frac., in press.

[15] B. Budiansky, J. W. Hutchinson and A. G. Evans, J. Mech. Phys. Solids, 34, (1986) 167.

[16] E. Bischoff, O. Sbaizero, M. Rühle and A. G. Evans, Advanced Ceramic Materials, in press.

[17] M. Y. He and J. W. Hutchinson, Harvard University Report MECH, 113 (1988), J. Appl. Mech., in press.

[18] H. C. Cao and A. G. Evans, to be published.

[19] P. G. Charalambides, H. C. Cao, J. Lund and A. G. Evans, to be published.

[20] M. D. Thouless and A. G. Evans, Acta Met. 36, (1988) 517.

[21] M. D. Thouless, O. Sbaizero, L. S. Sigl and A. G. Evans, Advanced Ceramic Materials, in press.

[22] M. Rühle, B. J. Dalgleish and A. G. Evans, Scripta Met. 21, (1987) 681.

[23] J. K. Wells, Ph.D. dissertation, Cambridge Univ. (1982).

[24] D. B. Marshall and A. G. Evans, J. Am. Ceram. Soc. 68, (1985) 225.

[25] J. Aveston, G. A. Cooper and A. Kelly, The Properties of Fiber Composites, JPC Science Technology (1971) p. 15.

[26] D. B. Marshall, B. N. Cox and A. G. Evans, Acta Met. 33, (1985) 2013.

[27] D. B. Marshall and W. Oliver, J. Am. Ceram. Soc., 70, (1987) 542.

[28] T. Wiehs and W. D. Nix, MRS Symposium, Reno, NV, Spring 1988.

[29] D. Johnson Walls, M.S. Thesis, University of California, Berkeley (1986).

[30] D. B. Marshall and A. G. Evans, Materials Forum, in press.

[31] E. Y. Luh and A. G. Evans, J. Am. Ceram. Soc. 70, (1987) 466.

[32] O. Sbaizero and A. G. Evans, J. Am. Ceram. Soc. 69, (1986) 481.

[33] M. S. Hu and A. G. Evans, Acta Met., in press.

[34] A. G. Evans, Mat. Sci. Eng. 71, (1985) 3.

THE FRICTIONAL RESISTANCE TO SLIDING OF A SiC FIBER IN A BRITTLE MATIX

T. P. WEIHS, C. M. DICK and W. D. NIX
Dept. of Materials Science and Engineering, Stanford University, Stanford, Ca 94305

ABSTRACT

The frictional resistance to sliding of a SiC fiber in a brittle, ceramic matrix has been measured with two different experimental techniques. Both techniques utilize a load-controlled indentation instrument. In the first technique, the ends of individual fibers are displaced down into the matrix. The frictional resistance to sliding, τ, was calculated using the elastic model of Marshall and Oliver and the load-displacement data. Alternatively, fibers have been displaced along their complete lengths through thin sections of the matrix. The critical force for complete slip and the sample geometry determined τ for a given fiber. For this technique slip over the complete length of a fiber was verified by the protrusion of that fiber from the bottom of the sample. By inverting the sample and loading the protruding fiber, the frictional resistance to reverse sliding was also measured. The results obtained from the two complementary techniques are in general agreement.

INTRODUCTION

A major goal in the design of ceramic based fiber composites has been to optimize the toughening achieved by fiber pull-out during crack propagation. This toughening is directly related to the frictional work associated with fiber pull-out[1-3]. Therefore, it depends strongly on how easily a fiber can slide in a brittle matrix. If the frictional resistance to sliding is too high, the fibers will fracture in unison with the matrix as a crack passes. In this case no pull-out will occur. If the friction is too low, little work is required to pull the fibers out of their matrix and minimal toughening is achieved. Consequently, characterization of the frictional stresses at the interfaces between fibers and matrix is necessary to understand and to optimize the toughening of these systems. In an attempt to characterize the frictional stresses for a model system, experiments have been performed using two different experimental techniques[4-5]. The frictional resistances to sliding of a SiC fiber in a brittle matrix obtained by these two techniques are presented, and both the testing techniques and the results are compared.

THEORY

The frictional resistance to sliding of a stiff fiber in a brittle matrix can be measured using two different but complementary methods. Consider a fiber loaded as shown in Figure 1. Given sufficient force and a sufficiently weak interface, part of the fiber will debond from the matrix and slide relative to the matrix as it is compressed. Marshall and Oliver (M&O)[4] have demonstrated that the sliding length, L, is determined by the frictional resistance to slip, τ, the applied force, F, and the fiber radius, R (Figure 1). If the interface is homogeneous and the frictional resistance is constant along the interface, the debond (or slip) length should increase linearly with the applied force according to

$$L = F/2\pi R\tau \qquad (1)$$

As the load on the fiber is increased, the fiber will continue to debond and to compress elastically. M&O have also shown that by modeling the compression of the fiber, and by measuring the load on the fiber and the displacement of the top of the fiber, one can determine the frictional resistance, τ.

An alternative experimental technique has been reported recently by Weihs and Nix[5]. They have shown that if the sample in Figure 1 is thin enough, a fiber can be forced to slide along its entire length. Such a technique yields a more direct measurement of τ. For a sample of thickness t, the critical force, F_c, for complete slip is given by

$$F_c = 2\pi R t \tau \qquad (2)$$

If a load-controlled instrument (such as a Nanoindenter [6,7]), applies the critical force, F_c, to a fiber, it will continue to slide until the indenter contacts the matrix surrounding the fiber. The length of slip is determined by the shape of the indenter and the radius of the fiber. For a blunt indenter, like the one used here, a typical fiber will slide 1 to 2 microns. Such a protrusion of the fiber from the bottom of the sample can be seen easily to verify slip along the complete length of the fiber. An additional benefit to the critical force technique is achieved by inverting a sample and loading a protruding fiber: the frictional resistance to reverse sliding can be measured.

PROCEDURE

The model composite tested consisted of SiC fibers (Nicalon) in a lithium-alumino-silicate ceramic matrix, SiC/LAS III.* The composite was hot-pressed at 1250°C and then heated (ceramed) in Argon at 1135°C to crystallize the matrix. The SiC fibers tested varied in diameter from 9 to 24 μm while the average diameter was 15.4 μm.

To prepare samples for testing, thin rectangular sections (5.0 x 3.0 mm with thicknesss ranging from 0.3 to 0.7 mm) were cut from a larger rectangular bar. The samples were cut so that one half of the fibers were perpendicular to the face of the sample and one half were parallel to it. Next, both faces of the samples were polished to a 0.25 μm finish with diamond paste. However, for the thinnest sample (0.3 mm prior to polishing), fibers were found to protrude up to 0.05 mm from one face of the sample after it was cut from the rectangular bar. The formation of these protrusions could not be avoided during the cutting procedure. Subsequent polishing of the rough face of the thinnest sample forced protruding fibers to slide down into the matrix and out the back side of the sample. The significance of such sliding prior to testing will be discussed later. One point to note here is that the protrusion of the fibers from the bottom of the thinnest sample prevented the study of reverse sliding in this specimen. When the sample was inverted after an initial test, the fibers that had slipped could not be distinguished from the other fibers since all fibers were protruding from the bottom of the sample.

The thicknesses of the two samples tested (to be labeled 1 and 2) were measured after polishing to be 0.696 mm and 0.128 mm, respectively. Once measured, the thick sample was mounted on two closely spaced glass plates to form a simple beam. For the thinner, more fragile sample, washers were bonded to the top and bottom of the faces of the sample to reduce the chance of cracking during handling and testing.

Using a Nanoindenter[6,7], a load-controlled indentation instrument, the ends of fibers were displaced into the matrix as shown schematically in Figure 1. The loading rate was controlled to displace the indenter tip at a constant velocity between 3 and 6 nm/sec. Once a force of 120 mN was applied to the fiber (the maximum loading capability of the instrument), the load was removed at a constant rate. During each test the force applied by the Nanoindenter and the displacement of the indenter tip were continuously recorded. In the case of the thicker sample, the specimen was inverted in order to load protruding fibers in the reverse direction. All tests were performed at room temperature.

To determine the displacement of the fiber, the penetration of the indenter tip into the fiber, u_I, had to be subtracted from the total displacement of the indenter tip, (Figure 1). Following the procedure used by M&O[4], a separate sample was annealed in air at 1000°C for 2 hrs. This heat treatment permitted strong bonds to form at the fiber/matrix interface that prevented the displacement of fibers upon testing. The lack of displacement was verified by SEM analysis and by cyclic loading of the fibers. The load-depth curve for an indentation into a SiC fiber is presented in Figure 2. An equation fitted to the curve in Figure 2 was used to determine the average indentation depth into a SiC fiber for any applied load. The indentation depth was then subtracted from the total displacement to obtain the fiber displacement in the

---
* United Technologies Research Center, East Hartford, Ct.

above experiments.

RESULTS

Figure 3 presents a plot of the square of the force versus fiber displacement (F^2 vs u) for a fiber in Sample 1. Based on the elastic analysis presented by M&O [4], if the frictional stress is

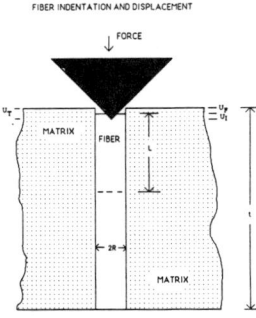

Fig. 1. Schematic of Test Configuration

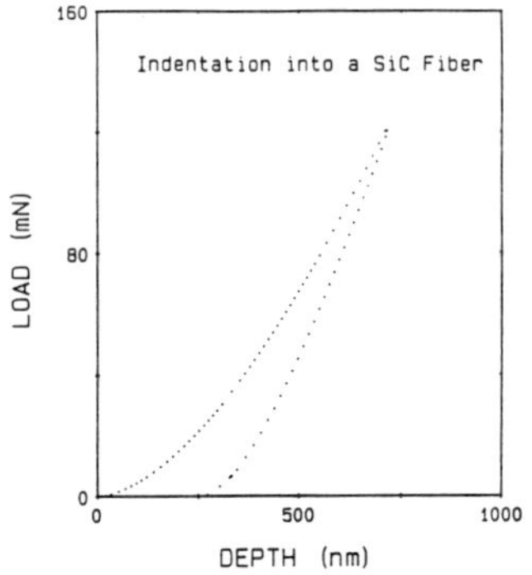

Fig. 2. Load/Depth Curve for an Indentation into a SiC Fiber

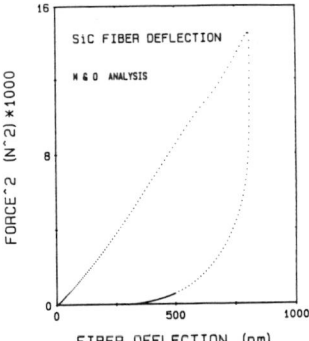

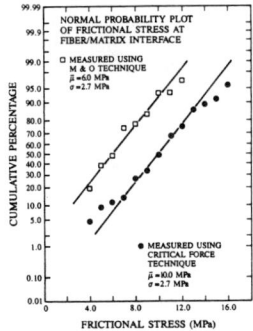

Fig. 3. Load Squared/Displacement Curve for a Fiber in Sample 1.

Fig. 4. Normal Probability Plot of Frictional Resistance to Sliding

constant along the fiber/matrix interface and if the interface is homogeneous, then the loading part of the curve should be linear and its slope should determine τ for that fiber. Results similar to the one in Figure 3 were obtained for 30 other fibers in Sample 1. These tests were used to determine the distribution of the values of τ based on the M&O technique. In determining this distribution, the results from tests on fibers in Sample 2 were not considered because they showed unusually low (< 1.0 MPa) values of τ compared to the other sample and to previous results [4]. The unreasonably low values are partially attributed to the large compliance of the thin sample. Based on simple elastic beam and circular plate[8,9] theory, Sample 2 could be expected to deflect by amounts which are smaller than, but similar in magnitude, to the deflections of the individual fibers being tested. Since deflections of the whole sample are not considered in the analysis of the load-deflection data, this large compliance could significantly lower the slope of the loading curves (Figure 3) and thereby lower the calculated values of τ as seen. Consequently, the M&O analysis has been applied only to tests on the thicker sample which is expected to have only negligible bending deflections.

The results from a M&O analysis of the tests from Sample 1 are plotted on normal probability paper in Figure 4. This particular type of plot was used to present the data for three different reasons. First, it allows one to qualitatively judge whether or not the quantity of interest, namely τ, is normally distributed[10,11]. A normal distribution is characterized by a linear spread of data between 16 and 84 cumulative percent. The plotted points below and above these percentages typically show more deviation as the number of data points in these regions is small. Secondly, given a normal distribution, which is expected for the frictional resistance[10], the mean and standard deviation of the quantity can be graphically determined[10,11]. The mean is given by the intersection of the 50% horizontal line and the line fitted to the data between 16 and 84 cumulative percent. Similarly, the standard deviation is given by the difference between the mean and the quantity given by the intersection of the 16% horizontal line and the fitted line. Lastly, a normal probability plot provides an informative means for graphically comparing the results obtained by the two techniques.

With regard to the data from the M&O analysis (Figure 4), the values of τ fall close to the fitted line within the percentages of 16 and 84. This linear behavior suggests that the frictional stress follows a normal distribution. As mentioned earlier, such behavior is expected since the frictional stresses are thought to be determined by the additive effects of random events[10]. The small deviations from linearity at high percentages are often seen in normal distributions and do not contradict the linearity of the bulk of the data. The fitted curve in Figure 4 gives a value of 6.0 MPa for the mean and a standard deviation of 2.7 MPa.

Figure 5(a) presents the displacement of a fiber that was reported earlier[5]. (This fiber, called Fiber A, was from Sample 1.) During the initial loading, Fiber A was steadily compressed into the matrix. However, once the critical load was reached, the fiber moved rapidly through the matrix at a velocity greater than 2-4 µm/sec and protruded from the bottom of the sample. During the reverse loading, Fiber A showed similar behavior and slipped at about the same critical force

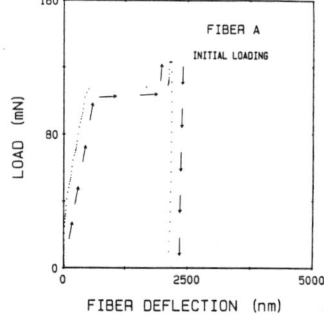

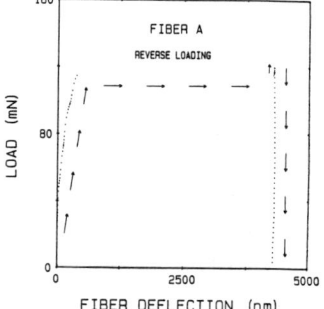

Fig. 5(a). Load/Displacement Curve for the Initial Loading of Fiber A

Fig. 5(b). Load/Displacement Curve for the Reverse Loading of Fiber A

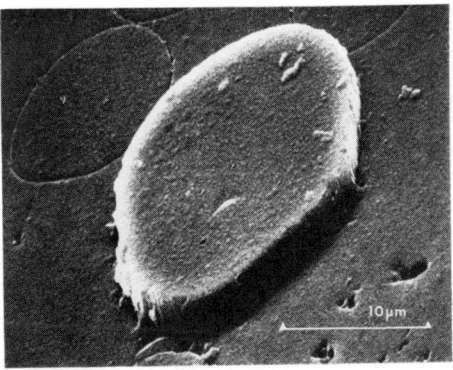

Fig. 6. SEM Micrograph of SiC Fiber A (rad = 11 μm) Protruding from the Sample after Reverse Loading

as for the initial loading (Figure 5(a) & (b)). An SEM micrograph of Fiber A protruding from the matix after a reverse loading is presented in Figure 6. One can notice that an indentation from the initial loading marks the top face of the fiber. Results similar to those for Fiber A were obtained for 2 of the other 30 fibers tested in Sample 1 and 44 of the 45 fibers tested in Sample 2.

To analyze the distribution of the frictional stresses based on the critical force technique, only the results from tests on Sample 2 were utilized. For this specimen almost all of the fibers (44 of 45) were able to slide at loads less than the maximum load of 120 mN. Thus, the sampling of fibers is not biased towards those fibers with small frictional stresses. (Such a bias is present for the critical force results from the thicker sample in which approximately 10% of the fibers slid over their complete length.) The results from Sample 2 are presented in Figure 4 along with those from the M&O analysis. Considering the cumulative percentage range between 16 and 84, the deviations from the fitted line are small. Thus, the critical force data also appears to follow a normal distribution of τ. However, the calculated mean value is higher than for the M&O analysis. Following the procedure outlined earlier, the mean and standard deviation of the frictional stress were determined to be 10.0 MPa and 2.7 MPa, respectively.

DISCUSSION

The two techniques presented here yield similar distributions of the frictional resistance to sliding of a SiC fiber in a brittle matrix. Both distributions are normal distributions and the standard deviations are small and equal. However, the mean values are different. To determine which technique may offer the more accurate mean value of τ, the three fibers which slid over their complete lengths in Sample 1 should be considered. The frictional stresses for these three fibers were reported to be 2.27, 3.09, and 2.79 MPa using the critical force technique[5]. (The M&O analysis for the same tests yielded similar results.) This indicates that at least 10% of the 31 fibers tested in Sample 1 had frictional stresses less than 3.1 MPa. Such a cumulative percentage agrees with the normal distribution plot in Figure 6 for the M&O analysis, but it does not agree with the data obtained by the critical force technique. Measurements using the latter technique suggest that less than 1.0% of the fibers should have frictional stresses under 3.1 MPa. This disparity implies that the data obtained from Sample 2 using the critical force technique may be in error.

To examine reasons why τ may be over estimated by the critical force technique, several questions should be addressed. First, does the sliding of the fibers during the preparation of Sample 2 distort the values of τ that are measured? From an earlier work [5] the answer appears to be no for two reasons. First, since the resistance for reverse sliding was equal to that for forward sliding [5], the sliding of the fibers during polishing of the sample should not reduce the friction at the interface. Secondly, although the sliding of the fibers during polishing may induce residual stresses at the fiber/matrix interface, these stresses will be removed by the sliding of the fiber which occurs before F_c is reached.

The second question to be addressed concerns the bending of the sample during the loading of individual fibers. As also reported earlier [5], such bending is negligible for Sample 1. However, for Sample 2, which is approximately 5 times thinner than Sample 1, bending can be significant. Two detrimental effects are produced by this bending. First, a M&O analysis of the data produces unusually low values of τ because the analysis does not account for the additional deflections. (This effect has already been described.) Secondly, stresses in the sample that result from the bending may impede the sliding of fibers and thereby raise the value of τ measured by the critical force technique. Using elastic theory for a circular plate [8], the maximum radial stresses could be as high as 3.0 MPa for a 120 mN load. In future work this effect will be reduced by minimizing the support length for the sample.

CONCLUSION

The frictional resistance to sliding of a SiC fiber in a brittle, ceramic matrix has been measured using two different experimental techniques. Both techniques yield normal distributions of the frictional stress with similar standard deviations. However, the technique developed by Marshall and Oliver yields a mean value of τ which is 39% smaller than the value of τ measured directly using the critical force technique developed by Weihs and Nix. Although the reason for the difference in mean values is still unclear, the present data suggests that the elastic bending stresses in the thin sample may account for the higher frictional stresses measured using the latter technique.

ACKNOWLEGEMENTS

The authors would like to Dr. J. Brennan from United Technologies Research Center for supplying the test material. This work was financially supported by the Defense Advanced Research Projects Agency through the University Research Initiative at UCSB under ONR contract N00014-86-k-0753.

REFERENCES

1) D. C. Phillips, J. Mat. Sci., 9, 1847, (1974).
2) K. M. Prewo and J. J. Brennan, J. Mat. Sci., 15, 463, (1980).
3) D. B. Marshall and A. G. Evans, J. Am. Cer. Soc., 68, 225, (1985).
4) D. B. Marshall and W. C. Oliver, J. Am. Cer. Soc., 70, 542, (1987).
5) T.P. Weihs and W.D. Nix, Scripta Met., 22, 271, (1988).
6) M. F. Doerner and W. D. Nix, J. Mat. Res., 1, 601, (1986).
7) W. C. Oliver, R. Hutchings, and J. B. Pethica, "Measurements of Hardness at Indentation Depths as Low as 20 Nanometers," in Microindentation Techniques in Materials Science and Engineering, Ed. P. J. Blau and B. R. Lawn (ASTM, STP 889, 1985).
8) S.P. Timoshenko and J.H. Goodier, Theory of Elasticity, (McGraw-Hill, San Francisco, 1970), pp. 385-8.
9) A.C. Ugural, Stresses in Plates and Shells, (McGraw-Hill, San Francisco, 1981), pp. 37-8.
10) L. Blank, Statistical Procedures for Engineering, Management, and Science, (McGraw-Hill, San Francisco, 1980), pp. 428-35.
11) L. Sachs, Applied Statistics - A Handbook of Techniques, (Springer-Verlag, N.Y., 1984), pp. 107-8.

EFFECT OF THERMAL EXPANSION MISMATCH ON FIBER PULL-OUT IN GLASS MATRIX COMPOSITES

U. V. Deshmukh*, A. Kanei*, S. W. Freiman** and D. C. Cranmer**
*Department of Materials Engineering, Drexel University, Philadelphia, PA 19104
**Ceramics Division, National Bureau of Standards, Gaithersburg, MD 20899

ABSTRACT

Single fiber pull-out tests can be used to directly measure the fiber-matrix interfacial shear stress in glass matrix composites. The system under investigation consisted of a soda-lime-silica glass matrix containing SiC monofilaments with a carbon-rich surface. The presence of the carbon-rich layer on the surface of these fibers makes them non-wetting to most glasses; hence the fibers are held in the matrix only by frictional forces acting at the interface. The mechanical gripping responsible for this force can be changed by manipulating the glass matrix/fiber thermal expansion coefficient mismatch. Frictional stresses (τ) and friction coefficients (μ) obtained for SiC monofilaments in a soda-lime-silica glass matrix were compared with previously obtained data on a borosilicate glass matrix (τ = 2-3 MPa, μ = 0.72 ± 0.36). For the soda-lime-silica system, τ's of 4-20 MPa and μ of 0.10 ± 0.03 were obtained. τ in the soda-lime-silica system is higher due to the larger difference in thermal expansion mismatch between the fiber and matrix. The differences in μ may be due to lubrication effects caused by water at the fiber-matrix interface.

INTRODUCTION

Strengths of fiber-matrix interfaces have been measured by several different techniques, but single fiber pull-out tests give the most direct measure of the interface strength [1-3]. Depending on the nature of the interaction (chemical and/or mechanical) between the fiber and the matrix, a single fiber pull-out test gives information about both the debonding and pull-out processes occurring in composites. However, single fiber pull-out tests have been used for glass matrix systems only recently [4-6].

The stress analysis of the test is based on the shear-lag theory of Cox [7], as modified by Greszczuk [8] and Lawrence [9] for elastically loaded brittle matrices. A good review of recent developments in this area is given by Gray [10]. In the absence of a chemical bond across the interface, or if an existing bond is broken, the resistance to fiber pull-out is mainly from interfacial frictional stresses, τ. If the matrix shrinks more than the fiber (due to thermal expansion), a residual compressive stress acting at the interface is produced, which depends on the elastic properties of the fiber and matrix, the difference in thermal expansion between the fiber and matrix, and the temperature difference between the glass strain point (T_s) and room temperature (T_{rt}). When $\alpha_{glass} > \alpha_{SiC}$, a larger thermal expansion coefficient difference should lead to a larger compressive stress, and thus to a higher resistance to pull-out. To confirm this hypothesis, experimental data were collected on a soda-lime-silica glass/SiC fiber system for comparison with data [4] on a borosilicate glass/SiC fiber system.

BACKGROUND

In a previous study [4] of a borosilicate glass/SiC monofilament system, a simple model was used where τ was estimated as the maximum pull-out force divided by the contact area. The contact area was taken to be $2\pi rL$, where r is the fiber radius and L is the initial embedded length. This value of τ is probably related to the debonding of the fiber from the matrix. This model is inaccurate since it does not account for Poisson contraction of the fiber radius as a result of the tensile pull-out force nor does it account for shear stresses occurring where the fiber emerges from between the glass plates. This latter effect is assumed to be a fixed quantity which cannot yet be accounted for. In the present paper, we have used the model of Takaku and Arridge [11], which takes into account the effect of Poisson's contraction. Their expression for the axial stress acting on the fiber is:

$$\sigma_f = \sigma_0/k \, [1 - \exp(-2\mu k L/r)] \qquad (1)$$

where σ_0 is the normal compressive stress acting at the interface in the unstressed material, k is a constant determined by the elastic properties of the fiber and matrix [$= E_m \nu_f / E_f (1+\nu_m)$; E is Young's modulus, ν is Poisson's ratio], and μ is the friction coefficient. By measuring σ_f as a function of embedded length, in principle we can obtain values for σ_0 and μ. As demonstrated in reference 4, however, due to scatter in the data for σ_f, a reliable value for σ_0 could not be obtained in this way. We therefore determined σ_0 using the expression of Vedula et. al. [12] for the residual stress developed in a composite due to anisotropic thermal expansion of fiber and matrix:

$$\sigma_0 \equiv \sigma_{rf} = (\sigma_1 \, \Delta\alpha_\ell + \sigma_2 \, \Delta\alpha_r) \, \Delta T \qquad (2)$$

where σ_1 and σ_2 are constants, $\Delta\alpha_\ell$ and $\Delta\alpha_r$ are the differences in thermal expansion coefficients between the fiber and matrix in the longitudinal and radial directions, respectively, and ΔT is the cooling range viz. $T_{rt} - T_s$.

EXPERIMENTAL PROCEDURE

Soda-lime-silica glass has a thermal expansion coefficient of 82 x 10^{-7}/°C compared to 32 x 10^{-7}/°C for borosilicate glass. The SiC monofilament longitudinal expansion is 26 x 10^{-7}/°C, while the radial expansion is 25.3 x 10^{-7}/°C [6]. Samples were fabricated by sandwiching SiC monofilaments[*] between soda-lime-silica glass plates[+] and heating under dead weight loading corresponding to 14-21 KPa pressure as shown in Figure 1. Molybdenum sheets were placed between the dead weight and the glass plates to prevent adhesion during fabrication. The entire assembly was immersed in graphite powder to prevent oxidation of the fibers. Industrial grade argon was kept flowing through the furnace. The samples were held at temperature (725-760°C) for different times (30-90 min) to obtain good flow of the glass around the monofilament. A schematic of the finished samples is shown in Figure 1. Different embedded lengths were obtained by varying the length of the monofilament between the plates. Individual samples were loaded in uniaxial tension on a universal test machine[#]. The samples were gripped using swivel hooks attached to ball joints to facilitate alignment of the fiber with the stress axis. Samples were pulled at a rate of 0.05 cm/min. The embedded length was determined from the force-displacement curves (Figure 2) since accurate optical measurements could not be made due to uncertainty in where the monofilament emerged from between the glass plates.

---

[*] SCS-6 SiC monofilaments, AVCO Corp., Lowell, MA. Trade names and companies are identified in order to adequately specify the materials used. In no case does such identification imply that the products are necessarily the best available for the purpose.

[+] Fisher Brand, Fisher Scientific, Pittsburgh, PA.

[#] Instron Corp., Canton, MA.

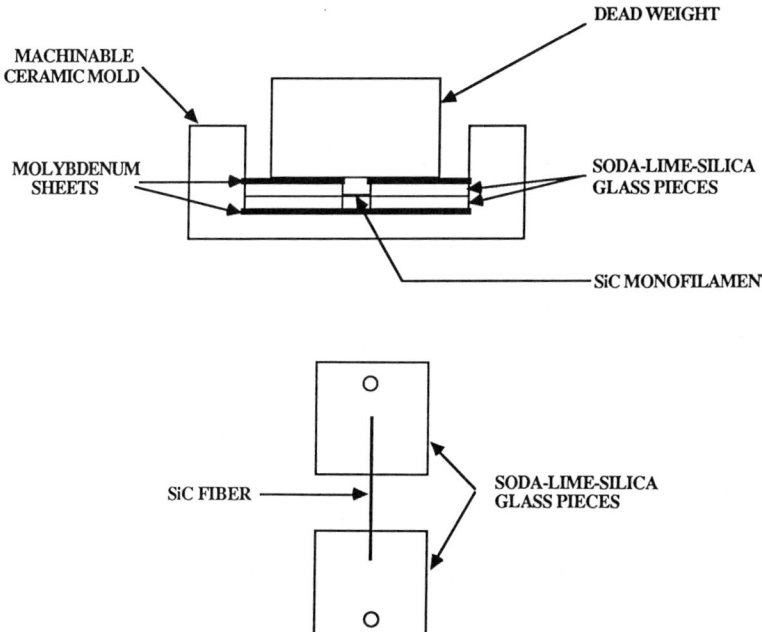

Figure 1. Schematic of the sample-making mold assembly and individual sample.

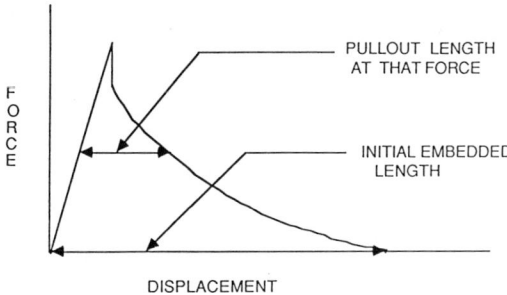

Figure 2. Measurement of pull-out and initial embedded length from a force-displacement curve.

RESULTS AND DISCUSSION

Two types of force-displacement curve were observed in the soda-lime-silica glass/SiC fiber system as shown in Figure 3. In both cases, the load increased linearly with displacement until the fiber debonded from the matrix. In the first case, there was a sharp drop in load, followed by pull-out of the fiber. In the second case, the load decreased gradually from the maximum while the fiber pulled out. The incidence of "stick-slip" behavior during pull-out was much less frequent than in the borosilicate glass/SiC system [4], suggesting the presence of a lubricating layer between the fiber and the soda-lime-silica matrix.

The maximum force required to initiate pull-out of the fiber is plotted vs embedded length in Figure 4. Although there is considerable scatter in the data, it can be seen that the pull-out force increases with increasing embedded length as long as the fiber strength is not exceeded. Some of the scatter may be due to non-uniform flow of glass around the fiber, leading to an air gap at the interface, and therefore reduced contact area and pull-out force. Also plotted in Figure 4 are the loads at which only frictional forces are acting on the interface. The difference between the maximum load and the frictional load is not constant. To confirm that the first load drop is due to debonding, several samples were unloaded following the initial load drop, then reloaded. Upon reloading, pull-out was observed at the same load where unloading had occurred, showing that the fiber-matrix bonds were in fact broken initially and that only frictional stress was operating at the interface. Occasionally, after reloading, a very small load drop was observed which is attributable to the difference between the static and dynamic coefficients of friction. The stability of τ during pull-out over a wide range of instantaneous embedded length is demonstrated in Figure 5.

Table 1 shows τ calculated from Equation 1 for both the soda-lime-silica and borosilicate glass systems. The large standard deviation for the soda-lime-silica system is believed to be due to variations in processing conditions. After making appropriate substitutions in Equation 2 for $\Delta\alpha_\ell$, $\Delta\alpha_r$, ΔT, σ_1, and σ_2 (see Table 1), the frictional stress, σ_0, for the soda-lime-silica/SiC system is estimated to be about 150 MPa. Similar calculations for borosilicate glass/SiC give a value for σ_0 of about 16 MPa. By substituting values for σ_0 in Equation 1, the friction coefficient, μ, can be obtained. For the soda-lime-silica system, μ was calculated to be 0.10 ± 0.03; for the borosilicate system, μ is 0.72 ± 0.36. The frictional stress acting at the interface may now be calculated as the product of μ and σ_n, where $\sigma_n = \sigma_0 - k\sigma_f$, σ_f being the axial stress corresponding to the frictional force and $k\sigma_f$ representing the reduction in σ_n due to Poisson's contraction. For the soda-lime-silica system, τ was calculated to range from 4-20 MPa while for the borosilicate system, it was in the range from 2-3 MPa.

SUMMARY

(1) Single fiber pull-out tests were used successfully to evaluate the frictional shear stress in glass matrix composites.

(2) For soda-lime-silica glass/SiC monofilament, τ of 4-20 MPa and μ of 0.10 ± 0.03 were obtained. These values may be significantly affected by the glass/fiber/air surface stress discontinuity.

(3) Compared to borosilicate glass system, the frictional stresses in the soda-lime-silica system are higher due to the higher normal compressive stress at the fiber-matrix interface. These higher stresses are the result of the larger difference in thermal expansion coefficients between the soda-lime-silica glass and SiC fiber than between the borosilicate glass and SiC fiber.

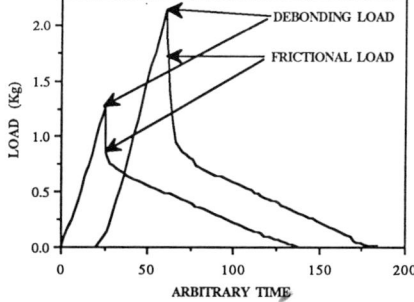

Figure 3. Typical force-time curves observed in soda-lime-silica/SCS-6 system.

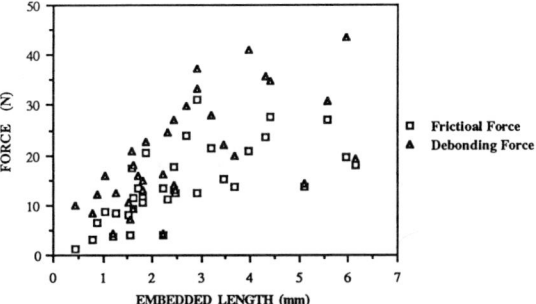

Figure 4. Plot of maximum pull-out force vs embedded length for soda-lime-silica glass/SiC monofilament. Debond force is based on the maximum load observed during the test; frictional force is based on a lower load for which only frictional forces are operating on the interface.

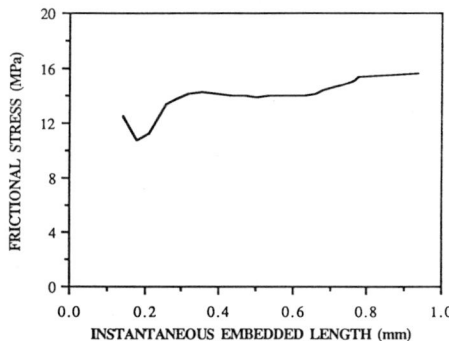

Figure 5. Plot of frictional stress throughout a single pull-out test as a function of instantaneous embedded length.

Table 1
Frictional Shear Stress and Friction Coefficients
Calculated from Equation 1
(Numbers in parenthesis represent number of samples averaged)

Matrix Material	Borosilicate Glass	Soda-Lime-Silica Glass
$\Delta\alpha_\ell$ (x 10^6/°C)	0.6	5.6
$\Delta\alpha_r$ (x 10^6/°C)	0.67	5.67
ΔT (°C)	495	440
σ_1 (GPa)	6.6	7.5
σ_2 (GPa)	47.0	52.7
σ_0 (MPa)	16.2	153.8
μ	0.72 ± 0.36 (9)	0.10 ± 0.03 (27)
τ (MPa)	3.6 ± 0.7 (9)	13.9 ± 4.1 (27)

(4) The low friction coefficients obtained in the soda-lime-silica system suggest the presence of a lubricating layer between the fiber and the matrix, perhaps due to processing.

ACKNOWLEDGEMENTS

The authors would like to thank Profs. M. J. Koczak and M. Barsoum for helpful discussions. We are grateful to Dr. S. G. Fishman for support through SDIO/IST contract N00014-86-F-0096.

REFERENCES

1) B. Harris, J. Morley, and D. C. Philips, J. Mat. Sci., 10 2050 (1975).

2) M. R. Piggott, A. Sanadi, P. S. Chua, and D. Andison in *Composite Interfaces*, H. Ishida and J. L. Koenig, Elsevier, 1986, p. 109.

3) L. S. Penn and S. M. Lee, Fibre Sci. Tech., 17 19 (1982).

4) U. V. Deshmukh and T. W. Coyle, to be published in Cer. Eng. Sci. Proc.

5) C. W. Griffin, S. Y. Limaye, D. W. Richerson, and D. K. Shetty, *ibid*.

6) R. W. Goettler and K. T. Faber, *ibid*.

7) H. L. Cox, Brit. J. Appl. Phys., 3 72 (1952).

8) L. B. Greszczuk in *Interfaces in Composites*, ASTM STP 452, 1969, p. 43.

9) P. Lawrence, J. Mat. Sci., 7 1 (1972).

10) R. J. Gray, J. Mat. Sci., 19 861 (1984).

11) A. Takaku and R. G. C. Arridge, J. Phys. D, 6 2038 (1973).

12) M. Vedula, R. N. Pangborn, and R. A. Queeney, Composites, 19 55 (1988).

ROLE OF FIBER-MATRIX INTERFACIAL SHEAR STRESS ON THE
TOUGHNESS OF REINFORCED OXIDE MATRIX COMPOSITES

RAJ N. SINGH
General Electric Company, Corporate Research and Development, P.O. Box 8, Schenectady, NY 12301

ABSTRACT

Mullite and zircon ceramic matrix composites uniaxially reinforced with as-supplied and BN-coated silicon carbide filaments were fabricated. The filament-matrix interfacial shear stresses were measured by a modified indentation technique and mechanical properties were determined in flexure mode. Significant improvements in strength and toughness were observed for the composites in comparison to monolithic mullite and zircon ceramics. Lower filament-matrix interfacial shear stresses were measured in composites with BN-coated filaments than in composites containing as-supplied filaments. The influence of filament-matrix interfacial shear stress on the critical stress for first matrix cracking, maximum composite strength, and toughness of the composites were studied.

I. INTRODUCTION

Mechanical properties of a fiber-reinforced ceramic matrix composite (CMC) are influenced by the properties of the fiber and the matrix materials, processing conditions, and the fiber-matrix interfacial shear stress. In particular, the toughness of CMC's is dependent on the fiber-matrix interfacial shear stress. In a tightly bonded fiber-matrix system little or no fiber pullout is expected and as a consequence brittle ceramic-like behavior is observed. In contrast, an optimum level of the interfacial shear stress can lead to a significant amount of toughness due to the fiber pullout mechanism. There are a number of factors which influence the fiber-matrix interfacial shear stress such as fiber-matrix interaction[1], thermal expansion mismatch between the fiber and matrix[2], and composite processing conditions. In addition, reinforcement coatings can modify fiber-matrix reactions and consequently interfacial shear strengths[1]. In this investigation uncoated and BN-coated silicon carbide monofilaments (AVCO SCS-6) were incorporated into mullite and zircon matrices and fabricated into uniaxially reinforced composites. The fiber-matrix interfacial shear stress was measured by the modified indentation technique[2] and the influence of measured interfacial shear stress on the critical stress for first matrix cracking, maximum composite strength, and toughness of the composites were studied.

II. EXPERIMENTAL

Silicon carbide monofilaments (AVCO-SCS-6) were used as reinforcement in the mullite and zircon matrices. Two types of fiber-matrix interfaces were created in the fully consolidated composites. In one case as-supplied filaments containing the carbon-rich surface were used, and in another case a boron nitride coating of about 1 μm thickness was deposited by a technique described elsewhere [3]. The interface between as-supplied filaments and mullite or zircon matrix is designated as interface A and that between BN-coated filaments and mullite or zircon matrix is designated as interface B.
    The composites were fabricated by uniaxially aligning the filaments and incorporating the mullite or zircon matrix around each of the filaments using a proprietary process. The final consolidation of the matrix was done by hot-pressing between 1500 and 1650°C. This produced fully consolidated composites with little porosity (<1%). A typical fiber loading of 25% by volume was used in all the composites of this study.

Mechanical testing of the uniaxially reinforced composite was performed in 3-point flexure mode using a Universal Testing Machine at a crosshead speed of 0.0127 cm/min (0.005 in/min)[4]. Load-deflection data were obtained for monolithic (mullite and zircon) samples and fiber-reinforced composites until complete failure.

The interfacial shear stress between filament and oxide matrix was determined by the modified indentation technique as described by Brun and Singh[2]. A similar technique was used by Laughner et al.[5] to measure interfacial shear stress between AVCO SiC filaments and an RBSN matrix. In this technique we have the ability to turn the sample over and repeat the measurement. The force to move the filament for the second time was invariably lower because of the debonding of the filament-matrix interface. The difference between the first and subsequent measurements made it possible to separate the bonding and frictional components of the contribution to the interfacial shear stress. Typically 10 to 15 filaments were pushed in each composite to determine the average value of the interfacial shear stress.

III. RESULTS AND DISCUSSION

(A) Interfacial Shear Stress

The interfacial shear stress was measured for mullite-silicon carbide and zircon-silicon carbide composites at room temperature. An average value of 64.3 MPa was obtained for the first push in mullite-silicon carbide composite containing as-supplied SiC filaments. The second push of these filaments needed only 16.4 MPa of stress which is significantly lower than the first push. Similarly processed mullite-SiC composites containing BN-coated SiC filaments showed somewhat lower value of 54.7 MPa for the interfacial shear stress. The second push of these filaments needed an average stress of 31 MPa. These values for the BN-coated filaments in mullite matrix are comparable to an earlier measurements in which values of 85 MPa and 29 MPa for the first push and subsequent push, respectively, were measured[1]. However, similar measurements on uncoated AVCO filaments showed that it was tightly bonded and required stresses beyond 200 MPa[1,2]. The composites of the earlier study had only a few AVCO SiC filaments i.e., very low fiber loading whereas the composites of the present study had 25 % filaments in the mullite matrix. The composites of this study showed microcracking of the mullite matrix in the fully consolidated samples. In contrast, the sample of the earlier study did not develop microcracking. The microcracking was caused by the somewhat higher coefficient of thermal expansion of mullite than the AVCO SiC filaments which also resulted in lower values of the interfacial shear stress because of stress relief due to microcracking.

The zircon-SiC composites containing SiC filaments do not show any evidence of matrix microcracking in the fully consolidated samples. This is because of closer match of filament-matrix thermal expansion coefficients. The measured interfacial shear stress values of 96 MPa and 15 MPa were obtained for the first push and the subsequent push, repectively, in composites containing uncoated filaments. In contrast, values of 12 MPa and 9 MPa for the first and subsequent push, respectively, were obtained for samples with BN-coated SiC filaments in the zircon matrix. These values for the BN-coated filaments are significantly lower than the values for the uncoated filaments; and lower yet than the corresponding values for BN-coated filaments in the mullite matrix which is indicative of the smaller contribution of compressive hoop stress on the interfacial shear stress measurements[2]. These observations on interfacial shear stress measurements provide an opportunity for altering the toughness in ceramic matrix composites via proper selection of the reinforcement-matrix system and also the reinforcement coatings.

(B) Mechanical Properties

The load-deflection data were obtained at room temperature in 3-point flexure mode for monolithic and composite samples containing SiC filaments in mullite and zircon matrices. The composites were fabricated with SiC reinforcements in the as-supplied condition as well as those with the BN coating.

The results for the mullite-SiC and zircon-SiC systems are shown in Figs.1 and 2, respectively. The monolithic mullite and zircon samples show elastic loading up to the onset of brittle failure, which is typical of most ceramic materials. In contrast, the composites containing SiC filaments show elastic response in the initial stages followed by an extended regime of inelastic behavior as the load increases. The slope of the curve progressively decreases as the specimen deforms because of filament failures. The load reaches a maximum value followed by two types of behaviors, depending on the nature of the filament-matrix interface. In the case of composites fabricated with filaments in the as-supplied state (interface A) the load-deflection data show a significant drop in load after the point of maximum load. In contrast, composites fabricated with the BN-coated filaments (interface B) show a gradual decrease in load after the point of maximum load as shown in Figs. 1 and 2. In addition to these features the zircon-SiC composites show clear evidence of first matrix cracking stress which is not seen in the mullite-SiC composites because of matrix microcracking. These behaviors are consistent with the lower value of interfacial shear stress for composites containing BN-coated filaments as compared to samples containing as-supplied filaments.

A summary of interfacial shear stress and mechanical property data for mullite-SiC and zircon-SiC composites is given in Table 1. The mullite-SiC composite samples show ultimate strengths between 709 and 777 MPa which are significantly higher than the strength of monolithic mullite. The toughness values (a measure of the work of pullout) were calculated from the area under the stress-deflection curve in 3-point flexure. The toughness for a mullite-SiC composite containing as-supplied filaments was found to be 289 KJ/m^2 and for the sample containing BN-coated filaments it was 565 KJ/m^2. These values are significantly higher than the value of about 10 KJ/m^2 for monolithic mullite which was calculated in a similar manner.

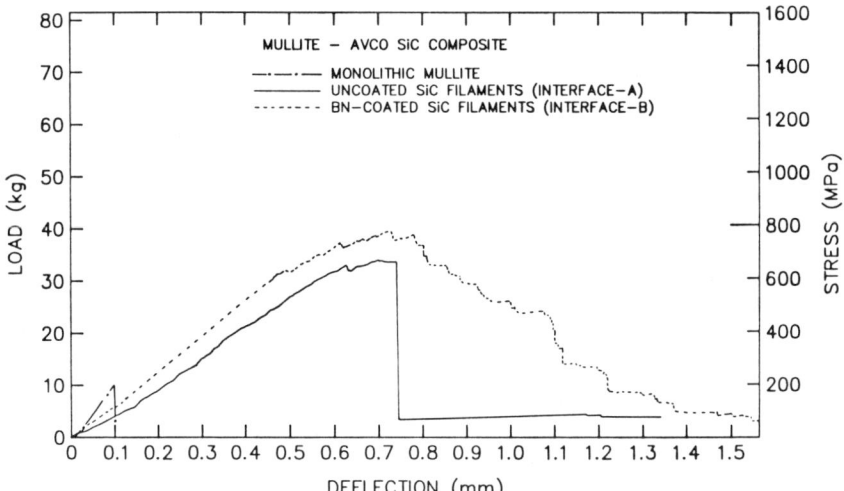

Fig. 1. Load-deflection behaviors for monolithic mullite and mullite-SiC composites containing as-supplied and BN-coated filaments.

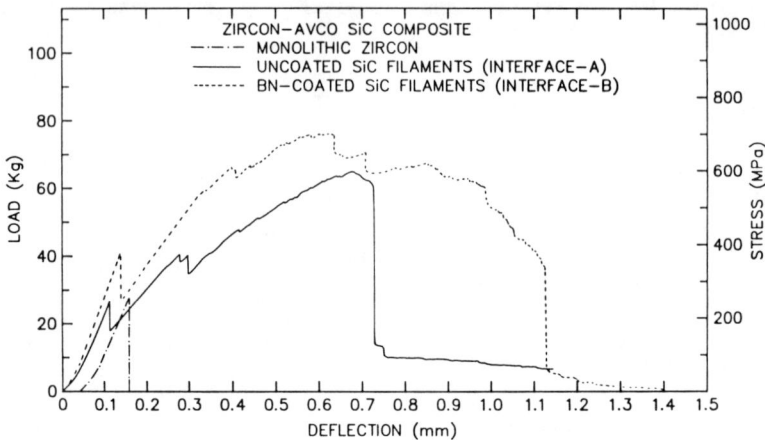

Fig. 2. Load-deflection behaviors for monolithic zircon and zircon-SiC composites containing as-supplied and Bn-coated filaments.

Table I. A Summary of Interfacial Shear Stress and Mechanical Property Data for Mullite-SiC and Zircon-SiC Composites.

Sample No.	Reinforcement/ coating/matrix	Interfacial Shear Stress		Mechanical Properties		
		First Push (Mpa)	Second Push (MPa)	σ_{cr} (MPa)	σ_m (MPa)	Toughness (KJ/m^2)
1	AVCO SiC/None/Mullite	64.3	16.4	-	709	289
2	AVCO SiC/BN/Mullite	54.7	30.9	-	777	565
3	AVCO SiC/None/Zircon	96.0	15.0	273	634	311
4	AVCO SiC/BN/Zircon	12.0	9.0	349	712	545

Interfacial shear stress and mechanical property data for zircon-SiC composites are also given in Table 1. The critical stress for the first matrix cracking of 273 MPa was obtained for composite containing as-supplied filaments. The ultimate strength of 634 MPa and the toughness value of 311 KJ/m^2 were also obtained for this composite. Similarly fabricated composites containing BN-coated filaments gave the first matrix cracking stress of 349 MPa, the ultimate strength of 712 MPa, and the toughness of 545 KJ/m^2. These values can be compared with the strength of 250 MPa and the toughness of 13 KJ/m^2 for monolithic zircon.

The role of filament-matrix interfacial shear stress on the mechanical properties of composites can be discussed with the data for zircon-SiC composites as given in Table 1. The first matrix cracking stress is related to the interfacial shear stress according to the following relation[6]

$$\sigma_{cr} = E \ (6 \ V_f^2 \ E_f/V_m \ E_m \ E)^{1/3} \ (\tau \ \gamma_m/a \ E_m)^{1/3}$$

where σ_{cr} is the first matrix cracking stress, E_f, E_m, and E are the moduli of filament, matrix, and composite, V_f and V_m are the filament and matrix volume fractions, τ is the fiber-matrix interfacial shear stress, γ_m is matrix Poission's ratio, and a is the filament diameter.

The calculated values of 331 and 179 MPa were obtained as critical stress for first matrix cracking using the measured values of 96 and 15 MPa for interfacial shear stress in zircon-SiC composites containing as-supplied filaments. The first value as determined from the first push is close to the measured critical stress of 273 MPa. Similarly σ_{cr} values of 166 and 151 MPa were calculated for composites containing BN-coated filaments using values of 12 and 9 MPa for the interfacial shear stress. Again these calculated σ_{cr} values are significantly lower than the measured values of 349 MPa. The reasons for such a disagreement are not apparent. The experimental results were obtained in 3-point flexure but a true tension test may be more appropriate for comparison with the model. Also the effect of residual compressive stresses due to thermal expansion mismatch must be considered in the calculations. Nonetheless the effect of lower interfacial shear stress in increasing the toughness of composites containing BN-coated filaments is quite apparent from this study.

IV. CONCLUSIONS

Uniaxially reinforced composites containing as-supplied and BN-coated SiC monofilaments in mullite and zircon matrices were fabricated. The important observations on the filament-matrix interfacial shear stress and mechanical properties are summarized below.

1. The interfacial shear stress measured for the first push was always higher than the values for the subsequent push because of the filament-matrix bonding. Lower interfacial shear stress values were also measured for BN-coated filaments in comparison to the as-supplied filaments in mullite and zircon matrices.

2. The strength and toughness of the first matrix cracking composites were significantly higher than the monolithic samples. The toughness was also higher in composites with BN-coated filaments as compared to composites containing as-supplied filaments because of lower interfacial shear stress for BN-coated filaments.

ACKNOWLEDGMENTS

The author wishes to thank A.R. Gaddipati and W.A. Morrison for help in sample preparation, M. Sutcu for helpful discussions and P.M. Breslin for manuscript preparation.

REFERENCES

1. R.N. Singh and M.K. Brun, "Effect of Boron Nitride Coatings On Fiber-Matrix Interactions," Ceramic Eng. and Science Proc. Vol. 8, No. 7-8, 636 (1987).

2. M.K. Brun and R.N. Singh, "Effect of Thermal Expansion Mismatch and Fiber Coating On The Fiber/Matrix Interfacial Shear Stress in CMC's," to be published in the J. Am. Ceram. Soc.

3. R.N. Singh,"LPCVD of Boron Nitride From Beta-trichloroborazine," Proc. of the tenth international conference on chemical vapor deposition, p.543 (1987); edited by G.W.Cullen, vol.87-8, The Electrochemical Society.

4. R.N. Singh and A.R. Gaddipati, "Mechanical Properties of a Uniaxially Reinforced Mullite-Silicon Carbide Composite," J. Am. Ceram. Soc., 71(2), C-100 (1988).

5. J.W. Laughner, N.J. Shaw, R.T. Bhatt, and J.A. DiCarlo, "Simple Indentation Method For Measurement Of Interfacial Shear Stress In SiC/Si3N4 Composites," Ceramic Eng. and Science Proc. 7(7-8), 932(1986).

6. J. Aveston, G.A. Cooper, and A. Kelly, "Single and Multiple Fracture," pp 15-26, in the Properties of Fiber Composites, Conference Proceeding of the National Physical Laboratory, IPC Science and Technology Press Ltd., Surrey, England, 1971.

THE MECHANICAL PROPERTIES AT HIGH TEMPERATURES OF SiC WHISKER-REINFORCED ALUMINA

KENONG XIA AND TERENCE G. LANGDON
Departments of Materials Science and Mechanical Engineering, University of Southern California, Los Angeles, CA 90089-1453

ABSTRACT

Creep tests were conducted at temperatures from 1673 to 1823 K on samples of Al_2O_3 containing 9.3, 18 and 30 vol % SiC whiskers. The results show that the strains to failure tend to decrease and the stress exponents for the steady-state creep rates tend to increase with increasing volume percentages of SiC whiskers.

INTRODUCTION

Ceramic materials have many potential applications at high temperatures because of their high melting points. However, this potential is usually not fully utilized because the conventional ceramics are inherently brittle and lack toughness. Two methods are available for improving the toughness: (i) in transformation toughening, a second phase is introduced which undergoes a stress-induced transformation near the tip of the crack, and (ii) in ceramic composites, fibers, whiskers or particles are added to inhibit crack opening or cause crack deflection.

This paper describes the high temperature creep properties of three different alumina composites reinforced with SiC whiskers. When SiC whiskers are incorporated into an Al_2O_3 matrix, it has been shown that the fracture toughness at room temperature is increased due to both a deflection of the crack at the whiskers and by a whisker pull out mechanism [1].

EXPERIMENTAL MATERIALS AND PROCEDURES

The experiments were conducted using samples of high purity Al_2O_3 containing either 7.5, 15 or 25 wt % SiC whiskers: this is equivalent to 9.3, 18 and 30 vol % SiC (w) and the samples are so designated in this report, with (w) denoting that the SiC is in the form of whiskers. The whiskers were single crystals, hexagonal in cross-section, with diameters of $\sim$0.5-1.0 μm and lengths from $\sim$20 μm to $\sim$60 μm. The room temperature values of Young's modulus are $\sim$7.0 × 10^5 and $\sim$3.86 × 10^5 MPa for the SiC whiskers and the Al_2O_3 matrix, respectively, thereby giving an elastic modulus mismatch of $\sim$1.8 [2].

Inspection of the microstructures of the as-received samples revealed a tendency for the whiskers to be densely packed in agglomerates, and the tendency for agglomerate formation increased with increasing SiC content. Thus, the whiskers were reasonably uniformly distributed in the material containing 9.3 vol % SiC, but whisker agglomerates were visible in the 18 vol % SiC samples and the agglomerates were well-defined in the 30 vol % SiC material. Figure 1 shows the intermediate situation for an 18 vol % SiC sample: the photomicrographs were taken on a side section of the sample after etching, using a scanning electron miscroscope at 0° tilt, and they show (a) the formation of agglomerates separated by channels which are reasonably whisker-free and (b) a close-up view of the whiskers and a channel. The hot-pressing direction is vertical in Fig. 1, and it is apparent from Fig. 1(a) that the agglomerates are elongated essentially perpendicular to the hot pressing direction. The maximum agglomerate length was <50 μm in the 18

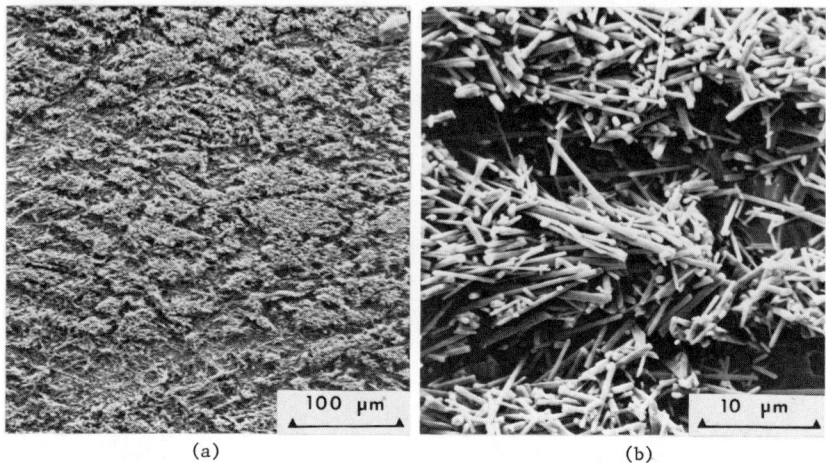

Fig. 1 Microstructure of the as-received 18 vol % SiC material: (a) the formation of whisker agglomerates, and (b) close-up view of the whiskers. The hot pressing direction is vertical.

vol % SiC material but it was >100 μm in the 30 vol % SiC samples. The average grain size of the matrix material was of the order of 1-2 μm for all three sets of samples.

The creep tests were conducted in air in four-point bending using test samples with a length of ∼26 mm and a cross-section of 2.0 × 3.0 mm. The specimens were held between four pivot points made of high purity sapphire and a constant load was applied through alumina loading rams. The loading assembly was contained within a vertical split furnace and the displacement of the loading ram was continuously recorded throughout each test on a strip-chart recorder. A schematic illustration of the testing assembly is given elsewhere [3]. Each specimen was held at the test temperature for at least 30 minutes prior to application of the load, and the creep tests were conducted at temperatures from 1673 to 1823 K with the testing temperature controlled to within ±2 K. The applied stress and the strain at the outer fiber were calculated using established procedures for four-point bending [4].

EXPERIMENTAL RESULTS AND DISCUSSION

A set of typical creep curves is shown in Fig. 2 in the form of strain, ε, versus time, t, for samples tested at an absolute temperature, T, of 1723 K and over a range of levels of the applied stress, σ: the volume percentages of SiC whiskers are (a) 9.3, (b) 18 and (c) 30%, respectively. Many of the samples were tested to failure and the failure point is denoted by a cross at the end of the creep curve.

Inspection of the curves in Fig. 2 shows that, for each material, there are the normal three stages of creep behavior: a primary stage where the creep rate decreases with time, a steady-state region where the creep rate is reasonably constant, and then a tertiary stage of accelerating creep leading to failure of the sample. The tertiary stage occurs abruptly in the 30 vol % material, possibly due to the formation and rapid propagation of a crack, but this final stage is less abrupt with 9.3 vol % SiC. The total strains exhibited by all specimens are fairly low, but there is a clear tendency for the strain at failure to increase when the volume percentage of SiC whiskers is reduced. For example, the total strains at failure are of the order of ∿6-8% for the 9.3 vol % SiC material, but this is reduced to ∿3% in the 18 vol % specimens and there is a further reduction to <2% for the 30 vol % material.

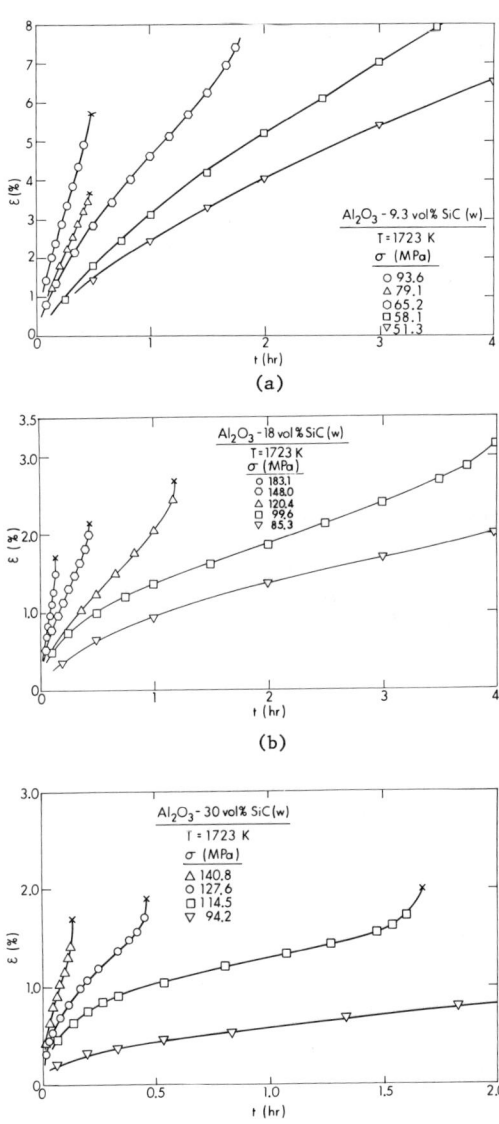

Fig. 2

Creep curves at 1723 K for (a) 9.3, (b) 18 and (c) 30 vol % SiC, respectively.

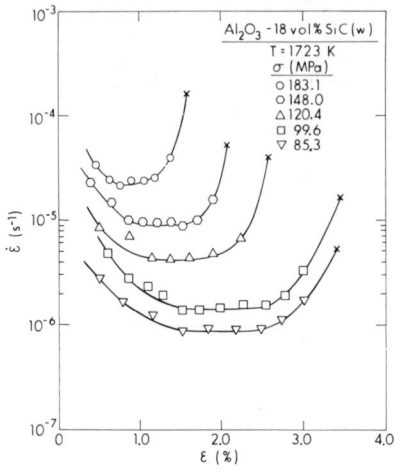

Fig. 3

Instantaneous creep rate versus total strain for Al_2O_3 containing 18 vol % SiC at 1723 K.

There is also a tendency for the region of steady-state creep to extend over longer intervals of strain, and therefore to be more clearly defined, when the level of the applied stress is reduced. This trend is apparent in Fig. 3 where the instantaneous creep rate is plotted as a function of the total strain for the five creep curves shown in Fig. 2(b) for the alumina containing 18 vol % SiC. For this material, there is a reasonably well-defined steady-state flow at the lower experimental stress levels but at the highest stress of 183.1 MPa the creep rate passes through a minimum and then accelerates into the tertiary stage.

It is well established in metals that the steady-state creep rate, $\dot{\varepsilon}$, may be expressed by a relationship of the form

$$\dot{\varepsilon} = \frac{ADGb}{kT} \left(\frac{b}{d} \right)^p \left(\frac{\sigma}{G} \right)^n \tag{1}$$

where D is the appropriate diffusion coefficient, G is the shear modulus, b is the Burgers vector, k is Boltzmann's constant, d is the grain size, and A, p and n are constants. Furthermore, equation (1) applies to ceramic materials, and it has been demonstrated that the experimental values of the stress exponent, n, for ceramics tend to be close to 1, 3 or 5, respectively [5]. Although there are many theoretical mechanisms of creep predicting different values for n [6], the exponents of 1, 3 and 5 in ceramics have been attributed to diffusion creep, dislocation climb from Bardeen-Herring sources and dislocation climb in a fully ductile material, respectively [5].

In order to check the values of n in the present materials, the steady-state creep rates (or minimum creep rates if there was no extended steady-state region) were logarithmically plotted against the applied stress, as shown in Fig. 4 for (a) 18 and (b) 30 vol % SiC, respectively. The stress exponents were ∼3.8-3.9 for the two materials containing the lower volumes of SiC whiskers, but the value of n increased to ∼6.3 for alumina with 30 vol % SiC. However, despite these differences in the values of n, inspection of Fig. 4 shows that the steady-state creep rates for these two materials are very similar. For example, at $\sigma = 10^2$ MPa and T = 1773 K, the values of $\dot{\varepsilon}$ are ∼7 × 10^{-6} and ∼4 × 10^{-6} s^{-1} for 18 vol % and 30 vol % SiC, respectively.

In polycrystalline materials, there is generally a transition from diffusion creep with n = 1 to an intragranular dislocation mechanism with n > 1 as the stress level is increased [5]. Thus, the present results are consistent with the occurrence of a dislocation process within the matrix.

It is appropriate to compare these results both with single phase Al_2O_3 and with Al_2O_3 composites reinforced with SiC whiskers.

In unreinforced single phase Al_2O_3, the stress exponent at high stresses is generally close to 3. However, these results refer to polycrystalline samples having grain sizes which are usually larger than in the composite materials: for example, for polycrystalline Al_2O_3 at 1803 K with a grain size of ∼13 μm [7]. For smaller grain sizes of the order of ∼1.6-2.0 μm, which is comparable to the composite materials, the value of n in single phase Al_2O_3 is typically ∼1.6-1.9 [8-10]. Thus, the stress exponents obtained in this investigation are significantly larger than in single phase Al_2O_3 with a similar grain size.

Two sets of creep data are available for Al_2O_3 reinforced with SiC whiskers. For Al_2O_3 with 15 wt % (18 vol %) SiC, a stress exponent of n = 5.2 was reported for a testing temperature of 1773 K [10]. For Al_2O_3 samples containing either 5, 15 or 20 vol % SiC, an exponent of n = 5 was reported at a temperature of 1773 K [11]. For the former material, which was obtained from the same source as in the present investigation, the steady-state creep rate was ∼9 × 10^{-6} s^{-1} at σ = 10^2 MPa and T = 1773 K [10]; whereas for the latter three materials, which were produced using a pressure-filtering technique, the steady-state creep rates were similar and of the order of

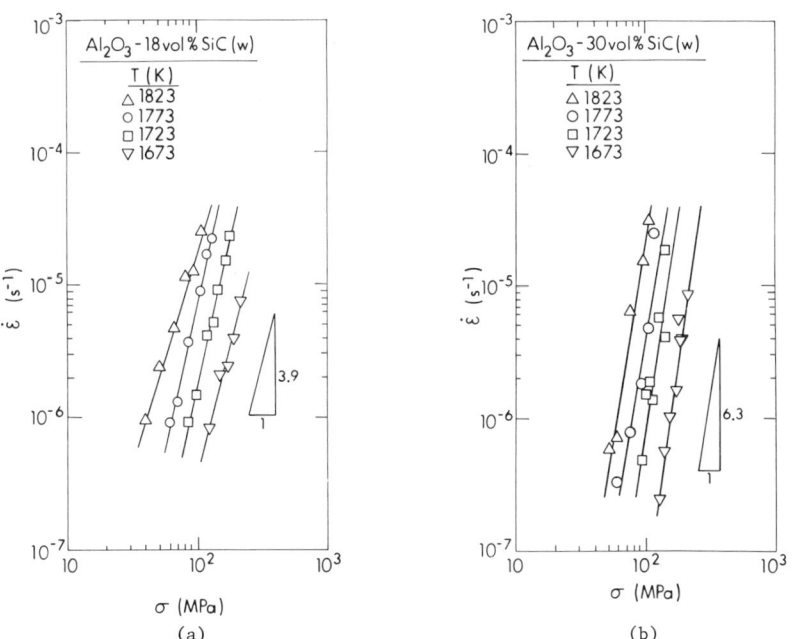

Fig. 4 Steady-state creep rate versus applied stress for (a) 18 and (b) 30 vol % SiC, respectively.

∼3 × 10^{-5} s^{-1} under the same conditions of stress and temperature [11]. Thus, the steady-state creep rates obtained in this investigation tend to be slower than in the earlier reports.

Finally, two additional points should be noted. First, there is a significant increase in the value of n when the SiC content is increased to 30 vol % [Fig. 4(b)]. Second, whereas the earlier creep data of $\dot{\epsilon}$ versus σ for Al_2O_3 with SiC whiskers referred to a single testing temperature of 1773 K [10,11], the present results reveal a consistent creep behavior, and similar values of n, over temperatures from 1673 to 1823 K.

SUMMARY AND CONCLUSIONS

1. Creep tests were performed on samples of Al_2O_3 containing 9.3, 18 and 30 vol % SiC whiskers.

2. At testing temperatures from 1673 to 1823 K, all three materials exhibit creep curves showing the normal three stages of creep behavior.

3. The total strains to failure are low (<10%) but they tend to increase when the volume percentage of SiC whiskers is reduced.

4. The stress exponent for the steady-state creep rate is independent of temperature within the range covered experimentally, and it increases from ∼3.8-3.9 for the materials containing 9.3 and 18 vol % SiC to ∼6.3 for the material with 30 vol % SiC.

ACKNOWLEDGEMENTS

We are grateful to Dr. J. F. Rhodes of the Advanced Composite Materials Corporation for supplying the three materials used in this investigation. This work was supported in part by a Lockheed Independent Research Grant.

REFERENCES

1. P.F. Becher, T.N. Tiegs, J.C. Ogle and W.H. Warwick, in Fracture Mechanics of Ceramics, edited by R.C. Bradt, A.G. Evans, D.P.H. Hasselman and F.F. Lange (Plenum, New York, 1986), vol. 7, p. 61.
2. J. Homeny, W.L. Vaughn and M.K. Ferber, Am. Ceram. Soc. Bull. 67, 333 (1987).
3. K. Xia and T.G. Langdon, in Proceedings of the IX Inter-American Conference on Materials Technology (Universidad de Chile, Santiago, Chile, 1987), p. 253.
4. G.W. Hollenberg, G.R. Terwilliger and R.S. Gordon, J. Am. Ceram. Soc. 54, 196 (1971).
5. W.R. Cannon and T.G. Langdon, J. Mater. Sci. 23, 1 (1988).
6. W.R. Cannon and T.G. Langdon, J. Mater. Sci. 18, 1 (1983).
7. G. Engelhardt and F. Thümmler, Ber. Deut. Keram. Ges. 47, 571 (1970).
8. J.R. Porter, W. Blumenthal and A.G. Evans, Acta Met. 29, 1899 (1981).
9. A.H. Chokshi and J.R. Porter, J. Mater. Sci. 21, 705 (1986).
10. A.H. Chokshi and J.R. Porter, J. Am. Ceram. Soc. 68, C-144 (1985).
11. J.R. Porter, F.F. Lange and A.H. Chokshi, Am. Ceram. Soc. Bull. 66, 343 (1987).

CREEP BEHAVIOR OF AN AL_2O_3-SIC COMPOSITE

P. LIPETZKY[*1], S.R. NUTT[*1], AND P.F. BECHER[**]
[*]Brown University, Engineering Division, Providence, RI 02912
[**]Oak Ridge National Laboratory, Metals and Ceramics Division, Oak Ridge, TN 37831

ABSTRACT

The addition of SiC whiskers to Al_2O_3 causes significant improvement in mechanical properties, including fracture toughness, thermal shock resistance, and creep resistance. The creep response of a whisker-reinforced alumina composite has been measured using four-point flexural loading at temperatures of 1200 and 1300C. Composites were fabricated by hot-pressing a blend of alumina powder with 33 volume percent SiC whiskers. The creep data showed a stress-dependent stress exponent equal to 1 at low stress levels and ranging from 4-6 at higher stresses. The applied stress at which the transition occurred was temperature dependent and ranged from 50-125 MPa. Mechanisms of creep deformation were determined from TEM observations of specimens prepared from interrupted creep tests. Voids were observed at grain boundary-interface junctions in tensile regions and whiskers within the composite were sometimes oxidized where voids had formed. TEM observations from specific stages of steady state creep reached under different applied loads are presented, and the relative contributions of different deformation mechanisms are discussed.

INTRODUCTION

Current interest in the high-temperature mechanical behavior of ceramic composites derives from the potential use of these materials in high-temperature structural applications, such as heat engines. The mechanical properties of composites at elevated temperatures depend on the properties of the constituents, the distribution and morphology of these phases, and the strength of the interface between the two phases. Creep resistance is perhaps the most critical high-temperature property of ceramic composites, and it is often the major limitation for high-temperature structural applications. Under creep conditions, the mechanical properties of the composite are strongly influenced by the presence of the reinforcing phase and the strength of the interface bond between the matrix and reinforcement. Creep of the composite can result in microstructural damage in the form of cavities and cracks that reduce load-bearing capability and eventually cause component failure. Before ceramic composites can be used reliably in structural applications at elevated temperatures, the creep response must be understood, including the mechanisms of deformation and damage that limit component lifetime and lead to failure. The objective of the present study is to measure the creep response of a model ceramic composite and determine the mechanisms of deformation and damage using TEM observations of crept specimens.

One of the most widely studied ceramic composites consists of an aluminum oxide matrix reinforced with silicon carbide whiskers. The addition of SiC reinforcements causes substantial improvements in room-temperature toughness [1] and increased resistance to erosion and thermal shock [2]. Commercial applications of these materials include cutting tools, valve and pump components, and extrusion dies. The composites are typically fabricated by hot-pressing a high-purity alumina powder mixed with short single-crystal fibers of SiC in volume fractions ranging from 15-35%. Whiskers tend to lie perpendicular to the hot-pressing axis (HPA), and aspect ratios are typically 5-10. While most research has focused on the room-temperature properties of these composites, interest in high-temperature behavior is increasing. Early work by Chokshi and Porter demonstrated a substantial improvement in creep resistance when alumina was reinforced with SiC whiskers [3]. Furthermore, they concluded that the applied creep stress was carried largely by the whisker reinforcements, and that the creep response of the whiskers controlled the creep of the composite. Their work demonstrated a need to better understand creep deformation mechanisms in composites and to determine the role of microstructural parameters during creep.

[1] Support from the Office of Naval Research (N00014-86-K-0125) is gratefully acknowledged.

EXPERIMENTAL PROCEDURE

The composite material selected for this study consisted of a polycrystalline alumina matrix reinforced with SiC whiskers and was obtained from the Greenleaf Corporation (WG-300). Microstructural characterization of as-fabricated composites revealed an average matrix grain size of 1-2 microns, and a SiC whisker volume fraction of 0.33. Additives such as yttria and magnesia are often used to facilitate the densification of alumina ceramics, and these additives generally result in residual pockets of glassy phase, particularly at triple grain junctions and (in composites) at grain boundary-interface junctions. However, the as-fabricated composite material was exceptionally "clean" in this regard, and in general, accumulations of glass phase were not detected at boundary junctions.

Creep tests were conducted in air at temperatures of 1200 and 1300C. Specimens measuring 0.1 x 0.2 x 2.0 inches were loaded in a four-point bend fixture in which the load points and support points were separated by 0.75 in. and 1.5 in., respectively. The displacement of the center of the specimen relative to the load pins was monitored by a linear variable differential transducer and recorded. Before the load was applied, specimens were heated to the test temperature and held for 30 min. to allow the apparatus to equilibrate. Temperature was monitored by a Pt-Rh thermocouple attached to the test fixture.

The applied stresses and resultant strains were determined using the method described by Hollenberg et al [4]. Consistent with Hollenberg's work, it was assumed that the deflection of the specimen was sufficiently small so that the stresses were nearly equal to the stresses predicted from linear elastic beam theory. Because ceramic materials creep at much different rates under tension and compression, the position of the neutral strain axis is sensitive to the amount of strain. To be certain that the stress gradient remained linear, the maximum strain was kept to less than 0.3 percent in most cases. This precaution resulted in a nearly constant stress on the tensile surface, and stresses ranging from 30 to 250 MPa were achieved by varying the applied load.

TEM specimens were prepared from crept material by cutting sections perpendicular to the load train on both tensile and compressive sides. Creep tests were interrupted during steady state and cooled under load before sectioning. Disks were cut and mechanically thinned before thinning to perforation by ion-milling. Most of the disks examined were cut from the tensile side of the specimen, and specimens were examined in an analytical TEM operated at 120 kV and equipped with a double-tilt specimen holder.

RESULTS AND DISCUSSION

Creep Data

The creep data for a range of stress levels at 1200 and 1300C are summarized in Figure 1a, a log-log plot of the steady-state creep rate versus applied stress. Both sets of data are bilinear and exhibit slopes of approximately 1 for low stress levels and strain rates (regime I) and 5 for higher stress (regime II) levels. Assuming a power law relation between the steady-state creep rate and the applied stress (creep rate proportional to applied stress raised to an exponent n), the slopes in Figure 1a give the value of the stress exponent n. The marked change in slope reflects the onset of a different deformation mechanism associated with higher stress levels. For a given stress level, an increase in temperature results in an increase in creep rate, although a given increase in stress causes the same increase in strain rate at both temperatures. The point of inflection shifts to a lower stress level at 1300C, implying that the deformation mechanism responsible for the change in stress exponent is thermally activated.

For fine-grain alumina, stress exponents in the range 1-2 have generally been attributed to diffusional creep mechanisms [5], and it is not unreasonable to suppose that the composite tested here deforms by a similar mechanism at low stress levels. However, specific creep mechanisms cannot be reliably inferred from higher values of n, particularly in multiphase materials such as composites. Chokshi and Porter reported a stress exponent of 5.2 for similar composites tested at temperatures of 1500-1600C, and concluded that the operative creep mechanism was intragranular glide of dislocations [3]. While the regime II stress exponent (n=5) measured in our tests is nearly identical to the value measured by Chokshi, our TEM observations (following section) indicate that dislocation glide plays only a minor role in creep at 1200 and 1300C.

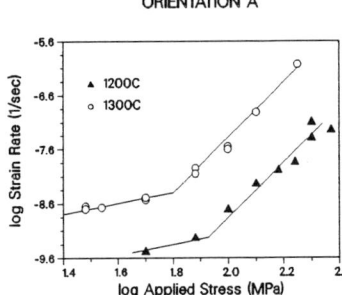 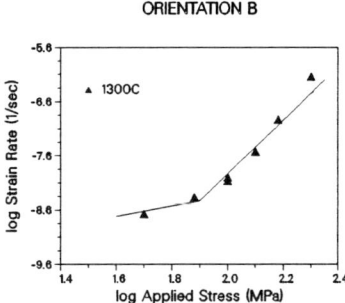

Figure 1. Strain rate vs stress relations for creep deformation of SiC-reinforced alumina composite. (a) Bilinear behavior yielding stress exponents of 1 and 5 for orientation in which the HPA and the load train are coincident. (b) Orientation B, in which the HPA and the load train are perpendicular.

Two different specimen orientations were used in the creep tests: type A in which the hot-pressing axis (HPA) was coincident with the four-point flexure load train, and type B, where the load train and the tensile axis were both perpendicular to the HPA. (During hot-pressing, the whiskers tend to be aligned in planes perpendicular to the HPA.) The two orientations yielded significantly different creep rates, as shown in Figure 1b. The type B orientation yielded a lower steady-state creep rate in Regime II than the type A orientation at 1300C. Because the type B orientation required a higher applied stress to change creep mechanisms, the onset of Regime II creep was delayed. The observed orientation dependence of the creep behavior is not well understood at present, and further experimentation is needed.

The measured steady state creep rates for two different stress levels are plotted against inverse temperature in Figure 2. The slopes of these plots provide a measure of the activation energy for steady state creep. Data from Regime I ($n=1$) give an activation energy of 450 kJ/mol, while data from regime II ($n=5$) give 500 kJ/mol, implying different creep deformation mechanisms for the two regimes. The activation energy for steady state creep of unreinforced alumina with a comparable grain size is 430kJ/mol [6]. Also plotted in Figure 2 are creep data from Chokshi and Porter for 15 % SiC tested at 1500C under a stress of 80 MPa [3]. Their data yielded an activation energy for steady state creep of 450 kJ/mol, which is similar to the measured activation energy for our material. Extrapolating their data to the temperatures used in our measurements would predict significantly higher creep rates than we observed. This difference can be attributed to the higher volume fraction of SiC whiskers in our material (33 % compared to 15 %), which effectively carry more of the applied load, and to different processing parameters used in fabrication. The effect of whisker reinforcements on the creep behavior of alumina has been compared to the effect of increasing the grain size, as reported by Jakus and Nair [7]. Cannon and co-workers showed that increasing the grain size from 1.2 to 2.7 microns decreased the creep rate of unreinforced alumina from 2.3 E-5 to 1.0 E-6 sec-1 at 1300C and 100 MPa [6]. Their data also indicated that unreinforced alumina with an average grain size of 15 microns would creep at about the same rate as what we observe for the whisker-reinforced composite, in which the grain size is 1-2 microns. However, in addition to bonding adjacent grains together, the whiskers also carry some of the applied load, as evidenced by the effects of volume fraction and orientation on creep response.

TEM Observations

The primary processing defect observed in the as-fabricated composite was a local variation in whisker volume fraction caused by inhomogeneous mixing of the matrix powder and whiskers prior to hot-pressing. Whisker clusters and whisker-free regions were only rarely observed, and the distribution of

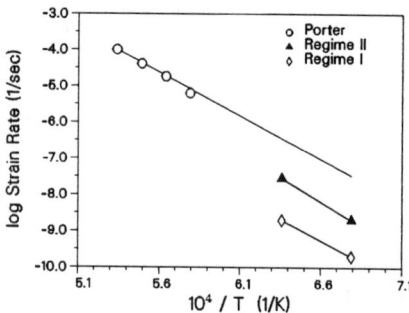

Figure 2. Arrhenius plot of steady-state creep rate vs inverse temperature for Al_2O_3-SiC composite. Regimes I and II correspond to n=1 and n=2 respectively. Data from Porter and Chokshi [10] are also shown.

whiskers was generally homogeneous. However, points of contact between whiskers were sometimes associated with small voids and residual amounts of yttrium-rich glass phase. As mentioned previously, the as-fabricated composite material contained relatively little glassy phase. During hot-pressing, additives [such as yttria and magnesia], tend to be squeezed out of the material or into the voids. Consequently, triple grain junctions and whisker-matrix interfaces were typically free of residual glass, although occasional pockets of yttrium-rich glass were sometimes observed, particularly at points of contact between whiskers. The observed dislocation density in the as-fabricated material was very low.

Specimens that had reached steady state creep in the first stress regime (n=1) showed significant microstructural changes. The most salient feature of the crept microstructures was the presence of silicon-rich glass phase pockets at grain boundary-interface junctions (Fig. 3). Compositional analysis by x-ray spectroscopy showed primarily silicon with small amounts of calcium, from which it was concluded that the phase was a silica glass. (Calcium is one of the primary impurities present in as-grown whiskers [8], and some of the calcium in the glass may come from the whiskers.) Similar glass pockets were also observed at triple grain junctions, although with less frequency, although calcium was generally not present in these pockets. In some cases the glass devitrified during cooling, forming mullite and other silicate structures. The dislocation density was negligible, supporting the previous assertion of a diffusional creep deformation mechanism in this stress regime.

Crept specimens from Regime II showed significant differences from the Regime I specimens, the most important of which was the appearance of cavities at grain boundary-interface junctions. Figure 4a shows a typical cavity formed at a whisker interface intersected by a grain boundary. The cavities were typically on the order of several hundred nanometers, which is consistent with the size of cavity predicted from the diffusional creep model developed by Cannon et al [6]. The cavities appeared triangular and were generally lined with glassy material composed primarily of silica. Incipient cavities, consisting of glass pockets containing small bubbles, were also observed along whisker interfaces, from which it was concluded that cavitation was generally preceded by the accumulation of glass phase. Cavities were also observed at triple grain junctions and along grain boundaries, as shown in Figure 4b, and grain boundary glass films were often detected. The viscosity of silica glass is about 10 P at temperatures of 1200-1300C, and the viscosity is further reduced by the presence of impurities such as calcium [9]. Consequently, the viscous flow of glass phase during composite creep is an important deformation mechanism, and it qualitatively resembles lubrication, facilitating deformation by intergranular and interfacial sliding.

In addition to cavitation at interfaces and boundaries, other types of defects contributed to the creep deformation in Regime II. Intragranular dislocations were occasionally observed, although the overall dislocation density was very low. The relative importance of dislocation glide is probably small, although the total strains in the specimens examined was generally no more than 0.25 %. Low-angle grain boundaries were frequently pinned by small intragranular particles and fibers of SiC and by intergranular whiskers (Fig. 5).

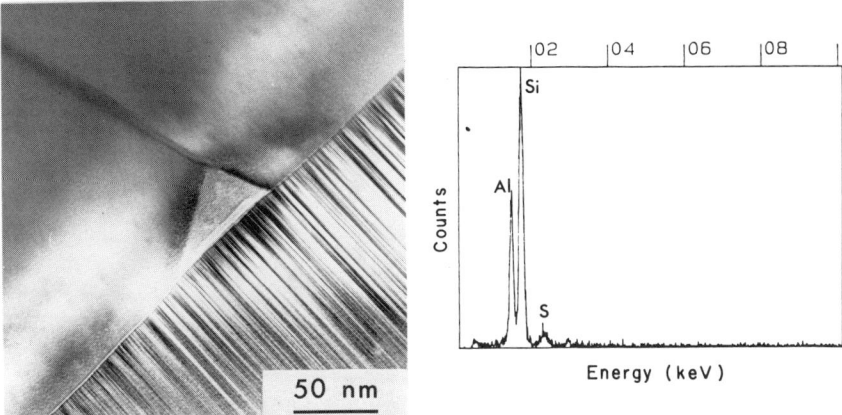

Figure 3. Accumulation of noncrystalline silica phase at grain boundary-interface junction after creep in Regime I. X-ray spectra shows primarily silicon.

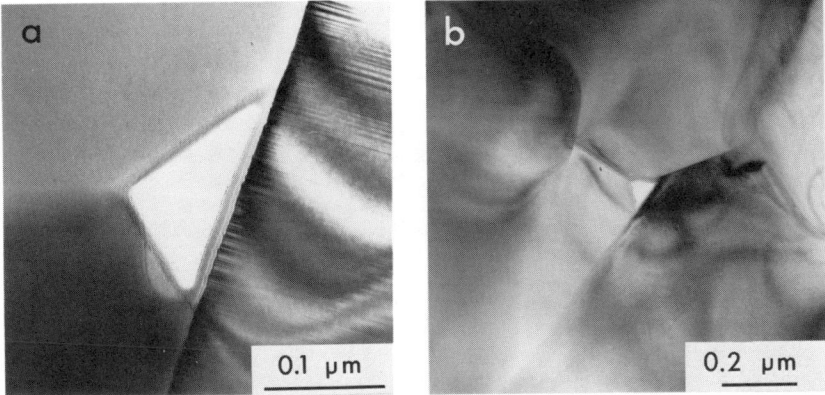

Figure 4. Cavity formation in composite after steady-state creep at 1300C, (a) at grain boundary-interface junction, and (b) at triple grain junction.

The marked change in stress exponent associated with higher stress levels (shown in Figure 1) is caused by the onset of cavitation, and the cavitation process is preceded by the accumulation of glass phase at interface junctions. Consequently, the presence of glass phase is critical to the creep response of this composite, and the mechanism by which glass is formed will now be considered. Although alumina and SiC show no evidence of chemical reaction at temperatures well above 1300C, it is well-established that SiC will oxidize when exposed to air at temperatures below 1300C, forming SiO_2 glass. Porter reported that alumina-SiC composites undergo a topochemical thermal oxidation reaction in which the silica glass formed by oxidation of SiC subsequently reacts with alumina to form a mullite surface scale [10]. We propose that most of the silica glass which accumulates at interfaces prior to creep cavitation is a result of the oxidation of SiC whiskers both at the specimen surface and internally. Evidence of oxidation of internal (subsurface) SiC whiskers is shown in Figure 6, a TEM image of part of a void (V) at a whisker side. The indented profile of the whisker side results from thermal oxidation of SiC at the cavity, producing silica glass.

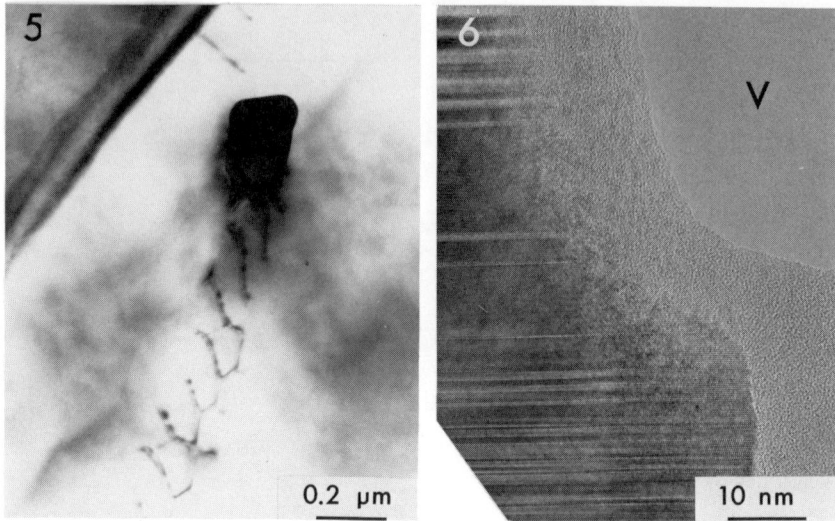

Figure 5. Dislocations pinned by small intragranular SiC inclusions in composite after creep at 1200C.
Figure 6. Oxidation of internal SiC whisker at interface cavity after creep at 1300C.

Because other whiskers show no signs of attack at interface cavities, the observed profile is not an artifact of preferential ion etching. Oxygen apparently reaches the internal whiskers by diffusion along interfaces and grain boundaries and, after the onset of cavitation, by gaseous diffusion through cavities. The products of the reaction include silica glass and graphitic carbon, and eventually mullite and carbon monoxide. The rate-limiting step in these reactions is probably the diffusion of oxygen through processing defects or the amorphous material that results from oxidation [11].

Oxidation of SiC whiskers exposed at surfaces produces silica glass which subsequently penetrates the composite along internal boundaries. Evidence of interface glass films is shown in Figure 7, a high resolution image of a whisker side, viewed edge-on. The thickness of the glass film is about 0.5 - 1 nm, and it increases where grain boundaries intersect the interface. Sections taken from tensile regions show substantially more glass phase than those taken from compressive regions, implying that tensile stresses on boundaries facilitate penetration by the low-viscosity glass (and vice versa). Glass tends to accumulate at grain boundary-interface junctions, after which cavities nucleate within the glass and grow, permitting the matrix grains to debond and pull away from the whisker.

SUMMARY

Flexural creep tests were performed on alumina reinforced with 33% SiC whiskers at temperatures of 1200 and 1300C. At both test temperatures, increasing the applied stress caused the measured stress exponent for steady-state creep to change from $n=1$ to $n=5$, suggesting a change in the creep deformation mechanism. TEM observations of crept specimens cooled under load revealed accumulations of silica glass at internal boundary junctions and very low dislocation densities. It was concluded that at low stress levels, the primary creep deformation mechanism was diffusional creep and grain boundary sliding, facilitated by viscous flow of intergranular silica glass. The silica glass originated primarily from the oxidation of SiC whiskers. At higher stresses, cavitation occurred within glass pockets at interfaces and grain boundaries, enabling matrix grains to separate from the whiskers and causing an increase in the creep rate. With

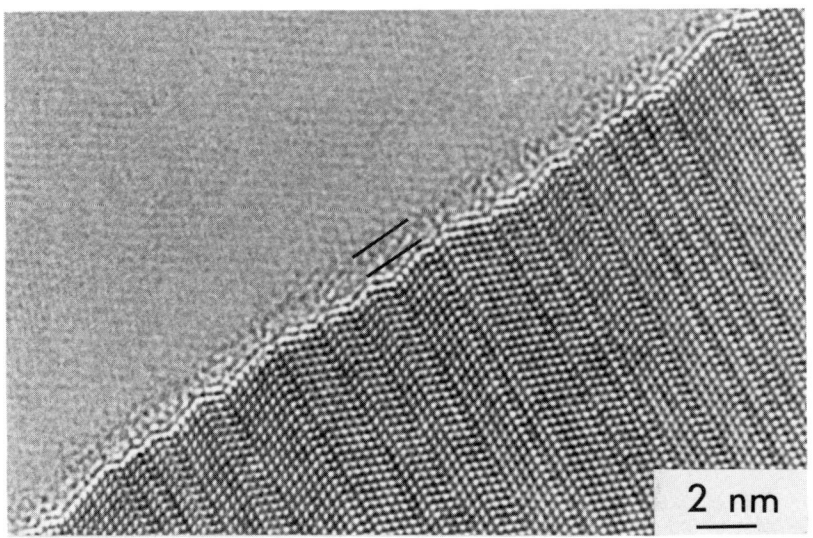

Figure 7. Noncrystalline film at interface between SiC fiber and alumina matrix after creep at 1200C.

increasing creep strain, cavities linked together and formed cracks, resulting in composite failure.

The measured activation energy for steady-state creep was higher for the composite than for unreinforced alumina with a similar grain size, and the creep rates for the composite were much lower.

REFERENCES

1. P.F. Becher and G.C. Wei, J. Am. Ceram. Soc. 67, C-267 (1984).
2. T.N. Tiegs and P.F. Becher, J. Am Ceram. Soc. 70, C-109 (1987).
3. A.H. Chokshi and J.R. Porter, J. Am. Ceram. Soc. 68, C-144 (1985).
4. G.W. Hollenberg, G.R. Terwilliger, and R.S. Gordon, J. Am. Ceram. Soc. 54, 196 (1971).
5. R.M. Cannon and R.L. Coble, in Deformation of Ceramic Materials, edited by R.C. Bradt and R.E. Tressler (Plenum, New York, 1975), pp. 61-100.
6. R.M. Cannon, W.H. Rhodes, and A.H. Heuer, J. Am. Ceram. Soc. 63, (1980) 46.
7. K. Jakus and S.V. Nair, to be published in Ceram. Eng. Sci. Proc. (1988)
8. S.R. Nutt, J. Am. Ceram. Soc. 71, (1988) 149.
9. R.H. Doremus, Glass Structure (Wiley, New York, 1973), p. 105.
10. J.R. Porter and A.H. Chokshi, in Ceramic Microstructures '86, ed. by J. Pask and A. Evans, Plenum.
11. K.L. Luthra, Ceram. Eng. Sci. Proc. 8, (1987) 649.

FRACTURE MECHANISMS IN SiC-WHISKER REINFORCED ALUMINA

CHRISTOPHE H. BOULANGER, YIH-CHERNG CHIANG, AZAR P. MAJIDI AND TSU-WEI CHOU

Center for Composite Materials and Department of Mechanical Engineering, University of Delaware, Newark, DE 19716

ABSTRACT

The fracture mechanisms involved in the toughening of alumina by whisker reinforcement are studied at room temperature. The fracture toughness of a hot pressed SiC-whisker/alumina composite is measured and good agreement is found between the experimental data and a model that takes into account the effects of crack deflection and whisker pullout mechanisms. From the model, it is seen that the role of whisker pullout is negligible compared to that of crack deflection.

INTRODUCTION

The incorporation of SiC-whiskers in an alumina matrix has been shown to significantly increase the fracture toughness of the unreinforced alumina [1,2,3], as well as to enhance the high temperature strength [4], the static fatigue [5], the wear resistance [6] and the creep properties [7]. At room temperature, two toughening mechanisms, crack deflection and whisker pullout, play an important role. Toughening due to crack deflection in particulate or whisker reinforced ceramics has been modelled by Faber and Evans [8,9]. This model assumes that crack deflection is the only toughening mechanism in the whisker reinforced composite and that it always occurs. Moreover, it assumes a random three-dimensional whisker orientation, while the hot-pressed SiC-whisker/alumina composites are virtually transversely isotropic.

This paper examines the fracture toughness and crack propagation mechanisms of hot-pressed SiC-whisker reinforced composites and correlates the experimental data with a fracture toughness model that is based on both crack deflection and fiber pullout.

FRACTURE TOUGHNESS TESTING AND RESULTS

Short bend bars of SiC-whisker/alumina, 3 by 5 by 50 mm, were obtained from a commercial source* with whisker volume fractions of 17, 23 and 29%. The composite with 23% whiskers was a lower grade material and, therefore, is not used in comparisons with the other two. The fracture toughness was measured in a plane normal to the plane of isotropy, for a crack propagating in the direction of the hot pressing axis. The test method selected was the chevron-notched four-point-bend technique. In this technique, the crack initially propagates in a slow and stable manner, then becomes unstable and propagates catastrophically through the rest of the cross section. The fracture toughness was measured from the maximum load which marks the transition between stable and unstable crack propagation. The data reduction was done through the slice synthesis method of Bluhm [10]. After testing, the fractured surfaces were studied by scanning electron microscopy (SEM)[+]. The SEM observations were then used for the assumptions of the model.

Table 1 presents the fracture toughnesses measured at room temperature. In addition to the room temperature studies, tests were also done at elevated temperatures up to 1250°C. The high temperature fracture mechanisms and toughness data have been discussed elsewhere [11].

Table 1: Room temperature fracture toughness data

V_f	Fracture toughness (MPa $m^{1/2}$)
17%	8.21
23%	7.6
29%	8.44

FRACTOGRAPHY

The fracture surfaces at room temperature showed the evidence of crack deflection (Figure 1) and whisker pullout (Figure 2). Crack deflection was clearly the dominant mechanism, while whisker pullout was limited to a few whiskers, and involved pullout lengths of only about twice the whisker diameter. A significant number of whiskers broke in or near the plane of crack propagation and, therefore, did not contribute to the pullout energy.

---

* Advanced Composite Material Corp., Greer, NC
[+] Amray 1200C, Amray, Bedford, MA

It was observed that with increasing temperature, the crack deflection gradually decreased, while the whisker pullout became more significant [11].

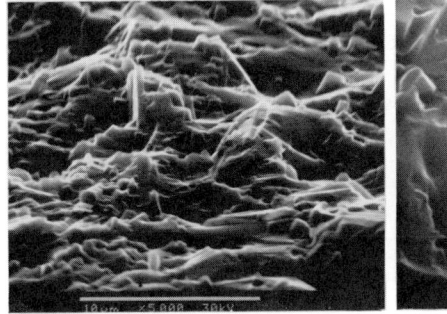

Figure 1: Crack deflection patterns (V_f=17%).

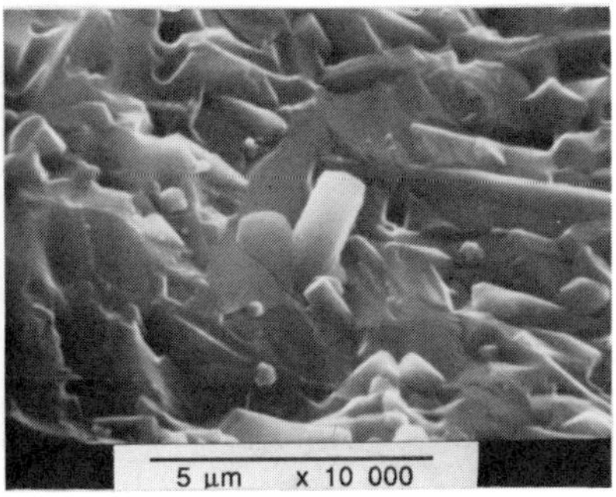

Figure 2: Whisker pullout (V_f=17%).

MODELING AND DISCUSSION

A fracture toughness model is developed [12] that combines the effects of whisker pullout and crack deflection, and that takes into account the anisotropy of the composite. The model is based on the assumption that the increase in the strain energy release rate, G, due to crack deflection is directly proportional to the increase in the fracture surface area:

$$\frac{G_{comp}}{G_{matrix}} = \frac{A_{defl}}{A_{undefl}}$$

Figure 3 gives a schematic of the tilting mechanism for a crack propagating in the x-y plane, in the direction of the x-axis. When the crack reaches a whisker of random orientation described by the angles ϕ and η, it tilts with an angle ϕ in the x-z plane, to propagate in the x'-y plane. The first assumption of the model is that the crack cannot tilt with an angle ϕ greater than a maximum angle ϕ_{crit}.

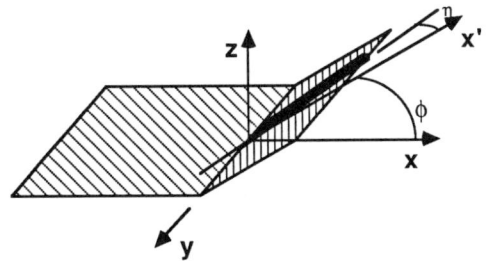

Figure 3: Tilt angle for a crack deflected by whisker of random orientation

The second assumption deals with the whisker orientation and distribution in the composite. The simplest approximation to describe the transverse isotropy due to hot pressing is to assume that all the whiskers are perpendicular to the hot pressing axis, and are randomly oriented in the plane of isotropy. If the schematic on figure 3 is used to describe this situation, for a crack propagating in the direction of the hot pressing axis, it is seen that the angle ϕ will always be 90°, while η is random. Using the first assumption would thus mean that there is no crack deflection. In reality, however, all whiskers do not lie in the plane perpendiular to the hot pressing axis but some may show a slight misalignment from that plane. The second assumption for the model is thus that the angle of the whiskers to the plane of isotropy is randomly distributed between μ and $-\mu$. In this case, crack deflection can occur with an initial tilt angle ϕ that is between (90°-μ) and ϕ_{crit}.

A whisker can be pulled out of the fracture surface if one of its extremities is less than half a critical length away from the fracture plane, and if its angle to the fracture plane is close to 90°. The third assumption defines a critical angle β such that a whisker can be pulled out if its angle to the plane of fracture is between 90°-β and 90°. Figure 4 shows the possible whisker orientations and the cone defined by β in which a whisker can be pulled out.

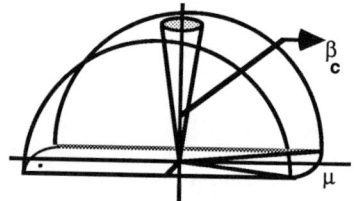

Figure 4: Whisker orientation and pullout angle

The fourth assumption is that, in all cases that are not covered by assumptions 2 and 3, the whiskers will break in the plane of fracture without any toughening effect. The model thus neglects the effect of fiber bridging.

Three angles, β, μ and ϕ_{crit}, are thus needed in order to determine the fracture toughness of the composite from the model. The misalignment angle, μ, can be obtained from the "interlaminar" fracture toughness. Indeed, consider a crack propagating in the plane of isotropy. If all the whiskers are perfectly aligned in the plane, the toughness measured for that crack will be equal to that of the unreinforced matrix material. If there is a slight misalignment, there will be a toughening contribution from crack deflection, but not from whisker pullout. In that case, the only parameter in the model is μ. The relative toughness of the composite can thus be plotted as a function of the misalignment angle μ. In Figure 5, it is shown that the model matches the experimental data from Becher et al. [1] for μ equal to 23°.

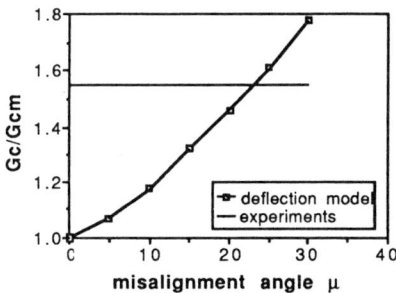

Figure 5: Determination of the misalignment angle μ

Figure 6 shows the good agreement between the model and the experimental results for $\beta = 90° - \phi_{crit} = 15°$ (Only the two high grade materials are considered for the comparison). Using these values, it is possible to compute the toughening contribution from the crack deflection process alone, and for the whisker pullout process alone. These two contributions are plotted on Figure 7, where it can be seen that the toughening effect due to whisker pullout is negligible compared to that due to crack deflection.

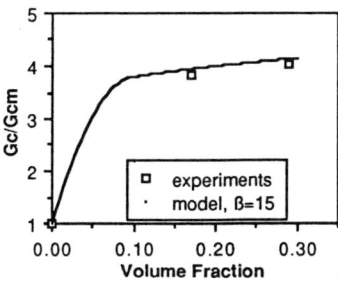

Figure 6: Comparison between the model and the experimental data

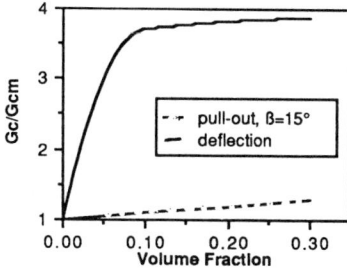

Figure 7: Toughening effects of crack deflection and whisker pullout

CONCLUSION

Toughening of alumina by SiC-whisker reinforcement at room temperature is dominated by the crack deflection mechanism, while whisker pullout plays only a minor role. Therefore, whisker reinforcement is effective in increasing the fracture toughness of alumina, despite a very strong mechanical bonding between whisker and matrix.

ACKNOWLEDGEMENTS

The authors wish to acknowledge support from Air Force Office of Scientific Research (contract no. AFSOR-87-0383, Dr. Alan Rosenstein is the program manager) and the Center for Composite Materials at the University of Delaware for this work.

REFERENCES:

[1] J. Homeny, W. L. Vaughn and M. T. Ferber.
    Am. Ceram. Soc. Bull., vol 67 [2], pp 333-338, (1987)
[2] G. C. Wei and P. F. Becher
    Am. Ceram. Soc. Bull., vol 64 [2], pp 298-304, (1985)
[3] P. F. Becher and G. C. Wei
    Comm. Am. Ceram. Soc., vol 67 [12], pp 267-269, (1984)
[4] P. F. Becher, T. N. Tiegs, J. C. Ogle and W. H. Warwick
    in "Fracture Mechanics of Ceramics", vol 7
    Plenum Press, NY, 1986
[5] C. A. Tracy and M. G. Slavin
    89th annual meeting of the Am. Ceram. Soc., Pittsburg, PA, (1987)
[6] M. T. Sykes, R. O. Scattergood and J. L. Routbort
    Composites, vol 18, pp 121-124, (1987)
[7] A. H. Chokshi and J. P. Porter
    J. Am. Ceram. Soc., vol 70, pp 393-395, (1987)
[8] K. T. Faber and A. G. Evans
    Acta Metall., vol 31, pp 565-576, (1983)
[9] K. T. Faber and A. G. Evans
    Acta Metall., vol 31, pp 577-584, (1983)
[10] J. L. Bluhm
    Eng. Fract. Mech., vol 7, pp 593-604, (1975)
[11] C. H. Boulanger, S. C. Danforth, and A.P. Majidi
    To be published.
[12] Y. C. Chiang and T. W. Chou
    To be published

IMPACT BEHAVIOR OF FIBER REINFORCED GLASS MATRIX COMPOSITES

D.F. HASSON* AND S.G. FISHMAN**
*Department of Mechanical Engineering, U.S. Naval Academy, Annapolis, MD 21402
**Office of Naval Research, Arlington, VA 22217

ABSTRACT

Ceramic matrix composites with continuous fibers in glass matrices were tested with instrumented impact apparatuses. The composite architectures were unidirectional ($0°$) and crossply ($0/90°$). For the $0/90°$ laminates, interlaminar and edge on orientation specimens were tested. An orientation dependence was observed. The CMC material with a weaker fiber/matrix interfacial bond had longer fiber pullout, and hence due to the frictional sliding mechanism higher dynamic work to fracture. In the fracture analysis discussion it is suggested that the use of the LEFM, K, parameter should be qualified in the fracture testing of CMC materials. These qualified toughness K values were found to be in the range of those reported for metal matrix composites.

INTRODUCTION

It has been almost twenty years since Bowen [1] suggested ways to improve the strength and toughness of fiber reinforced ceramics. One year after Bowen's paper, mechanical properties from three point flexure tests of a carbon fiber reinforced silica composite were reported [2]. Currently, processing developments have yielded ceramic matrix composites (CMC's) with improved toughness and tensile strength which has led to a resurgence of interest in the use of ceramics for structural applications [3]. In ceramic matrix composites which exhibit tough behavior and high tensile strength, strong fibers remain intact during matrix fracture, resulting in periodic matrix cracks and a post-cracking load sustaining capability. It has been shown that the bond between fiber and matrix, in tough composites, is weak, and energy is absorbed, during crack propagation by frictional sliding of the fiber through the matrix [4]. In unidirectionally reinforced composites, the matrix cracking stress, the interfacial frictional sliding resistance and the composite ultimate stress are characteristic stresses and are not sensitive to matrix damage, because the fibers have sufficient strength to support the load in the presence of the matrix cracks. Properly processed composites, therefore, exhibit the damage tolerant behavior which is imperative if ceramic composites are to find application in structures subjected to tensile loading [5,6]. Unfortunately, the low fiber-matrix bond strength important in these composites results in inferior transverse properties. As is the case in the reinforced epoxy composites, such considerable anisotropy can be overcome by the use of laminates. Recent investigations of the tensile and compressive response of symmetric ($0/90°$) laminated SiC reinforced lithium aluminosilicate glass ceramics have shown that the stress-strain characteristics of these laminates are predictable, based on strength measurements for the individual laminate layers in combination with knowledge of residual stresses present in the matrix [7,8]. Although the mechanical response of continuous fiber reinforced CMC's is being extensively studied under conditions of static loading, only limited information is available on the response to dynamic loading [9,10]. Also, the effect of notch orientation on various CMC materials under dynamic load has not been studied in detail. The objective of the present study, therefore, is to determine the effect of dynamic load on the toughness of CMC materials with various notch orientations.

EXPERIMENTAL DETAILS

All the CMC materials were fabricated by United Technologies Research Center (UTRC). The CMC's materials were high modulus untreated (HMU) carbon fibers in Corning Code 7740 borosilicate glass, Thornel-300 carbon fibers in borosilicate glass and Nicalon SiC fibers in modified Corning lithium aluminosilicate (LAS III). In addition a HMU carbon fiber/Nb_2O_5 modified borosilicate material was produced specifically to increase the fiber-matrix interfacial bond strength to determine what changes occur in composite interlaminar shear strength and toughness [11]. The architectures were unidirectional ($0°$) and crossply ($0/90°$). The $0/90°$ crossply layers were 200 microns thick. The fiber volume percents were about 42 for the HMU/borosilicate, 50 for the T-300/borosilicate and 44 for the SiC/LAS matrices, while the fiber diameters were 8, 7 and 15 microns, respectively.

All the impact specimens had a span of 40 mm, the breadth, B and thickness W were kept as close to 10mm as allowed by the material thickness. The notch depth, a, to thickness ratio a/W was 0.2 with a notch tip radius of 178 microns for all specimens. The primary specimen orientations, as shown in figure 1, were L-S (interlaminar) and L-T (edge on), Some L-S type specimens were cut at $45°$ to the top ply-$0°$ axis which is the reference axis for the longitudinal, L, orientation.

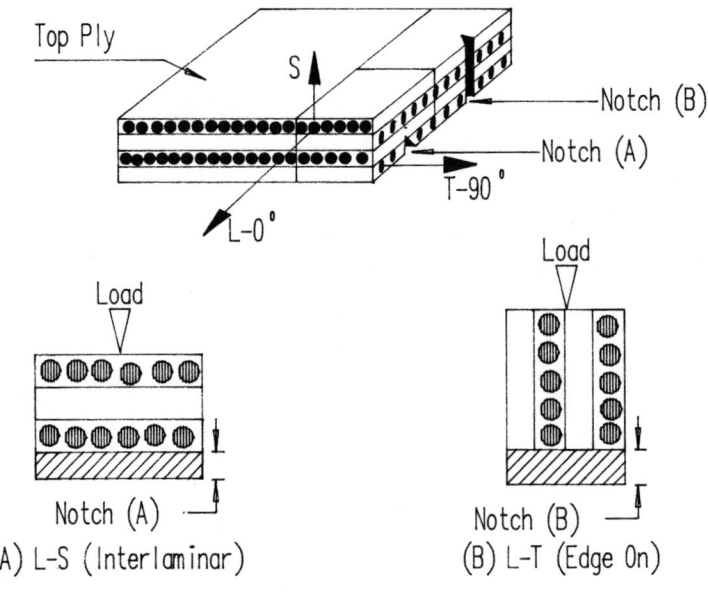

Figure 1. Specimen Orientations

Most impact tests were performed with an instrumented drop tower, while some utilized an instrumented low blow Charpy pendulum apparatus. Test velocities were in the neighborhood of 1 m/s. The test temperature was $20°C$.

RESULTS AND DISCUSSION

As noted by Prewo [9], the span, L, to specimen thickness, W, ratio (L/W) must be in excess of 20 to determine the exact flexure strength. Due to use of the Charpy geometry, this criteria is not satisfied, and for the specimens in the present study variable compression and tension stresses occur in the upper and lower surface regions with shear a maximum at the neutral axis. The data analysis to follow should, therefore, be interpreted in terms of these stresses. The maximum stresses reported, herein, are calculated from the simple beam formula and utilize the maximum recorded load with the notched Charpy geometry. The results of impact tests for unidirectional fiber CMC's are presented in table I.

Table I. Impact Characteristics of Unidirectional Fiber Reinforced CMC's

Material	Orientation	Maximum Stress MPa	Kq MPa(m)$^{1/2}$	Ud J/cm^2
Boro./HMU	L–T	1039	51.0	8.3
Boro./HMU	L–S	417	17.6	4.4
LAS II/SiC	L–T	754	36.8	21.5

The impact load–time behavior of these materials exhibited the usual elastic behavior, followed by a decrease in slope during matrix cracking up to maximum stress where fiber bundle failure causes a rapid drop in stress to a point where a gradual decrease in stress occurs due to frictional sliding fiber pull–out [10]. The L–T maximum stress for borosilicate/HMU compares to that reported for flexure stress by Prewo and Nardone [12]. They also reported the tensile strength to be 574 MPa and the interlaminar shear strength to be 22 MPa. The effect of orientation on the maximum stress values for the borosilicate/HMU material in table I is due to the moment of inertia effect on the bending stress, and hence the shear stress, because for the L–S specimen it is one–third that for the L–T orientation. The L–T maximum stress for LAS II/SiC is lower than that previously reported [13].

Since the quasi–linear elastic fracture mechanics (LEFM) toughness parameter, Kq, is based on maximum load, the fracture toughness is likewise higher for L–T. The terminology, quasi–LEFM toughness parameter is used, because the nonlocalized complex failure involving a multiplicty of cracks at and away from the separation zone in the matrix and the failure of individual fibers and bundles of fibers does not comply with the macroscopic scale of the LEFM analysis. Simply put, there is no singular crack and hence crack length. As noted by Phillips and Davidge [14], Kq is useful for guidance in material development.

It is noted that the Kq value for borosilicate/HMU material in the L–S orientation is comparable to that of 21.4 MPa(m)$^{1/2}$ previously reported by Prewo for borosilicate/HMS material [9]. A comparison of the borosilicate/HMU L–T result with the LAS II/SiC indicates that both materials have favorable strength and toughness levels. These Kq toughness values are in the range of those reported for aluminum metal matrix composites [15].

Another toughness parameter, U_d, the dynamic rupture work is obtained by integrating the load–time signal while taking into account the variation in the instantaneous velocity. The dynamic rupture work values from table I show the same trend as Kq.

Macrofractographic examination of all the unidirectional CMC specimens reported above showed interlaminar shear failure between the fibers and matrix and some fiber tensile failure and pull out especially in the region of the notch. Interlaminar shear failure is illustrated in the left-hand part of the specimen shown in figure 2(a).

The impact tests results for the 0/90° crossply laminates of the various CMC materials are given in tables II and III for the L-T and L-S orientation specimens, respectively. All CMC materials exhibit improved toughness characteristics compared to monolithic ceramics which essentially have no dynamic rupture resistance[10].

The LAS III/SiC material is a little stronger and tougher than the borosilicate/carbon fiber materials. The higher toughness of the LAS III/SiC, especially the dynamic work to fracture, is due to a higher fiber frictional sliding contribution. For specimens which exhibited a primarily planar brittle fracture appearance, as shown in figure 2(b), fiber pullout was about 4 mm for the LAS III/SiC compared to about 3 mm for the borosilicate/HMU carbon material. Also, it should be noted that since fiber pullout is associated with the fiber/matrix interfacial bond, the LAS III/SiC material has a better designed interface. Also, it is noted that the maximum stress for the LAS III/SiC T-S orientation specimens is higher than the reported [16] flexural strength value of 380 MPa.

Table II. Impact Characteristics of Various 0/90° CMC's (L-T Orientation)

Material	Maximum Stress MPa	Kq MPa(m)$^{1/2}$	Ud J/cm^2
Borosilicate/HMU Carbon	298	14.8	3.4
Borosilicate-Nb$_2$O$_5$/HMU Carbon	593	25.0	5.0
Borosilicate/Thornel-300	224	11.3	0.6
LAS III/SiC	617	31.5	6.4

Table III. Impact Characteristic of Various 0/90° CMC's (L-S Orientation)

Material	Maximum Stress MPa	Kq MPa(m)$^{1/2}$	Ud J/cm^2
Borosilicate/HMU Carbon	281 130*	11.8 5.5	6.0 7.4
Borosilicate-Nb$_2$O$_5$/HMU-Carbon	282	12.2	5.8
Borosilicate/Thornel-300	254	11.0	1.8
LAS III/SiC**	474	11.8	9.2

* L-S/45 orientation
** T-S orientation

A comparison of the borosilicate matrix materials reinforced with either HMU or T-300 carbon fibers shows the improvement in processing techniques on impact characteristics because the T-300 carbon fiber material had almost no fiber pullout and hence undesirable interfacial bond characteristics. The Nb_2O_5 doped borosilicate matrix was an attempt to see the effect of increasing the interfacial bond on strength and toughness, but the present results are not conclusive on this modification.

The effect of orientation on the impact characteristics, as determined by comparing tables II and III, is that the interlaminar L-S orientation has a slightly higher dynamic work to fracture. This higher value is attributed to the interlaminar shear mode of fracture as shown in figure 2(a) as compared to the tensile like behavior shown in figure 2(b). The "edge on" orientation, L-T, also reduces the fiber pullout contribution and allows for an almost planar brittle fracture in the matrix from the alternate laminate as shown in figure 2(c).

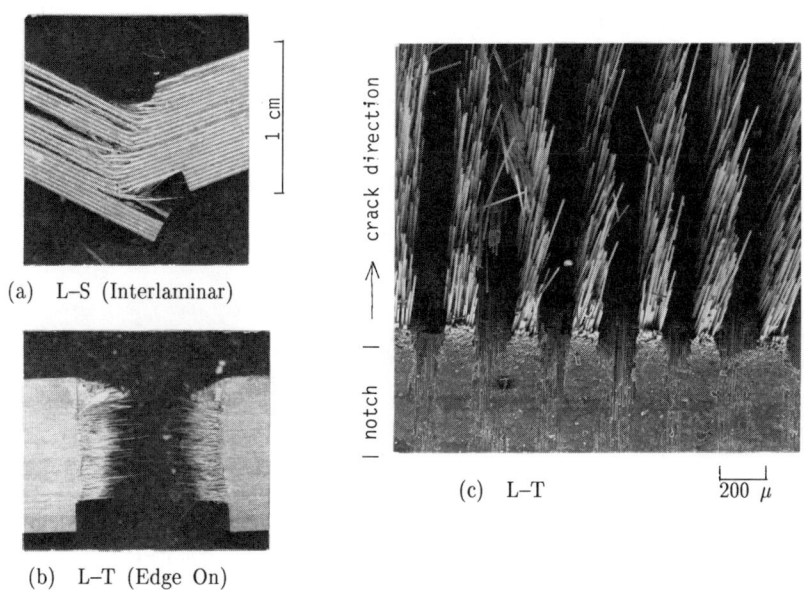

(a) L-S (Interlaminar)

(b) L-T (Edge On)

(c) L-T

Figure 2. Fractographs of Borosilicate/HMU Carbon 0/90° CMC

CONCLUDING REMARKS

The Kq impact fracture toughness characteristics of various CMC materials are in the range of those reported for aluminum metal matrix composites. The LAS III/SiC material has more favorable toughness characteristics than the borosilicate/carbon materials. This is attributed to a better fiber/matrix interfacial bond design which results in longer fiber pullout in the LAS III/SiC material. Fracture toughness properties in laminated CMC materials were found to be orientation dependent. Also, it was noted that monolithic ceramics do not have comparable toughness levels. An overall observation about fracture testing of CMC's is that the use of LEFM to report the fracture toughness parameter, K, should be qualified since the conditions for LEFM are not entirely satisfied in the fracture of CMC materials.

ACKNOWLEDGEMENTS

This work was supported by the Office of Naval Research. The continued laboratory expertise of W. Umlandt is also noted. The graphic preparation is due to the kind efforts of Captain Paul B. Stumbo, United States Air Force. The careful manuscript preparation of Ruth Chittum is also appreciated.

REFERENCES

[1] D.H. Bowen, Fibre Sci. Tech., 1, No. 2, 85, (1968).

[2] I, Crivelli–Visconti and G.A. Cooper, Nature, 221, 754 (1969)

[3] K.M. Prewo, J.J. Brennan and G.K. Layden, Am. Ceram. Soc. Bull. 65(2), 305 (1986).

[4] D.B. Marshall and A.G. Evans, J. Amer. Cer. Soc., 68(5), 225 (1985).

[5] J. Aveston, G.A. Cooper and A. Kelly, in The Properties of Fiber Composities, (IPC Science & Technology Press Ltd., Surrey, UK), 15 (1971).

[6] A.G. Evans, M.D. Thouless, D.P. Johnson – Walls and E.Y. Luh, in Fifth International Conference on Composite Materials ICCM-V, edited by W.C. Harrigan, Jr., J. Strife and A.K. Dhingra (The Metallurgical Society/AIME, Warrendale), 543 (1985).

[7] J. Lankford, Ibid., 587 (1985).

[8] O. Shaizero and A.G. Evans, J. Amer. Cer. Soc., 69, 481 (1986).

[9] K.M. Prewo, Phil. Trans. R. Soc. Lond. A 294, 551 (1980).

[10] D.F. Hasson and S.G. Fishman, in Sixth International Conference on Composite Materials, Volume 2, edited by F.L. Matthews, N.C.R. Buskell, J.M. Hodgkinson and J. Morton (Elsevier Applied Science, London), 2.40 (1987).

[11] W.K. Tredway and K.M. Prewo, UTRC Ann. Rept. R87–917470–1, ONR Contr. N00014–85–C–0332 (1987).

[12] K.M. Prewo and V.C. Nardone, UTRC Ann. Rept. R86–917161–1, ONR Contr. N00014–85–C–0332 (1986).

[13] K.M. Prewo, J. Mater. Sci. 21, 3590 (1986).

[14] D.C. Phillips and R.W. Davidge, Br. Ceram. Trans. J., 85, 123 (1986).

[15] D.F. Hasson, S.M. Hoover and C.R. Crowe, J. Mater. Sci. 20, 4147 (1985).

[16] K.M. Prewo, G.K. Layden, E.J. Minford and J.J. Brennan, UTRC Interim Rept. R85–916629–1, ONR Contr. N00014–81–C–0571 (1985).

PART V

Composite Interfacial Effects

STRUCTURE AND CHEMISTRY OF METAL/CERAMIC INTERFACES

M. RÜHLE and A. G. EVANS
Materials Department, College of Engineering, University of California, Santa Barbara, California 93106

ABSTRACT

In this paper, the present state of knowledge is reviewed concerning the structure and chemistry of metal/ceramic interfaces. Experimental observations are described for several model systems and open problems concerning different aspects of structure and properties of heterophase boundaries are discussed.

INTRODUCTION

The use of ceramics as structural components, as well as in chemical technology and in electronic devices is steadily increasing because of improved mechanical integrity afforded by enhanced toughness and by process control. Ceramic components must typically be connected to other materials, mainly metals. The requirements that the bonded couple must fulfill are dictated by the functions of the ceramic: physical, chemical, electrical, mechanical. However, in all cases, adequate mechanical integrity is a technical prerequisite, as reflected in the fracture resistance of the interfaces. Metal/ceramic bonded couples are presently being used in electron tubes, multilayer substrates and capacitors, metal matrix composites, automotive power sources, etc.[1-9]

Systematic studies of metal/ceramic interfaces started in the early 1960's. Such studies were directed toward the identification of general rules that govern bonding and interface behavior, both theoretically and experimentally, including the thermodynamics of interfacial reactions, crystallographic relationships and the atomistic structure at the interface. The intent of this article is to review the present state of knowledge concerning the physics, chemistry and structure of interfacial regions between metals and ceramics.

THE WORK OF ADHESION

The driving force for formation of a metal/ceramic interface is the yield in energy when intimate contact is established between the metal and ceramic surfaces.[10] For a high rate of interaction, the surfaces have to be brought into excited states. Therefore, temperature and atmosphere are important variables, as well as the properties and structures of the surfaces.

The simplest description of the physical interaction between a metal and a ceramic is the <u>work of adhesion</u>, W_{ad}. Specifically, when clean, defect-free surfaces are brought into contact, energy is released in accordance with the Dupre equation,

$$W_{ad} = \gamma_c + \gamma_m - \gamma_{mc} \qquad (1)$$

where γ_c and γ_m are the free energies of the relaxed surfaces of the ceramic and the metal, respectively, γ_{mc}, represents the energy of the relaxed interface between the metal and the ceramic. The quantity W_{ad} is thus the reversible work released per unit area of interface formed by two free surfaces. Direct measurement of W_{ad} is not possible.[11] Consequently, in practice, W_{ad} is deduced by measuring the contact angle θ established by a <u>solid</u> metal in contact with a ceramic,

$$W_{ad} = \gamma_m (1 + \cos \theta) \qquad (2)$$

Adequate measurement of θ and of γ_m constitutes a non-trivial experimental task. Often γ_m is anisotropic and hence, the crystallography of the surface has to be determined. Furthermore, true equilibrium has to be established by allowing sufficient mass transport and the associated morphological evolution. Measurements on small particles are preferred, although contamination during annealing is always a problem.[12] The most acceptable approach involves the deformation and heat treatment of alloys containing particles of ceramic formed by internal oxidation, etc.[13,14] Plastic straining of the alloy causes particle decohesion. Subsequent annealing then allows mass transport to create an equilibrium void from the initial debond. The angle θ can then be measured on cross-sections through the particles.[11] Different authors[12,14,15] measured and calculated values of γ_{mc} and W_{ad}.

Alloying additions strongly influence the thermodynamic quantitites.[10] Furthermore, certain alloying additions segregate at the interface, by Gibbsian absorption. As an example, the segregation of chromium at various metal/Al_2O_3 interfaces results in a rearrangement of the interface into a more relaxed structure with a lower interfacial energy, resulting in a lower work of adhesion. Such segregant effects are a major issue in metal/ceramic bonded couples.[12,20,21]

BONDING MODELS

A rudimentary understanding of interfaces can be achieved by adopting phenomenological models. Such models are capable of correlating trends in bonding between different material couples and provide insight into some of the broad issues. However, the detailed understanding of trends in interface structure and properties with alloy composition, segregation, etc., requires more sophisticated atomistic models.

Elucidation of the essential issues, especially the prediction of trends in the work of adhesion (and, eventually, in fracture resistance) with such variables as alloy additions and segregation, requires that bonding be examined at all levels. The eventual objective would be the judicious coupling of information obtained from the most rigorous, but compute bound, quantum mechanical supercell approaches with the results of cluster calculations and of simple continuum, thermodynamic formalisms.

Continuum Models

Interaction across the interface first occurs without charge exchange. Such interactions develop between induced dipoles (London), between neutral atoms polarized by a dipole molecule (Debye) and between dipole moments (Keesom). Together, these interactions constitute the Van der Waals attractions (see, e.g., Ref. 22). The London term is generally the most pronounced. For a pair consisting of a metal atom and an oxygen ion, the interaction energy E_p has the form

$$E_p = -(3/2)\alpha_m \alpha_A / R^6 \, [l_m \cdot l_A / (l_m + l_A)] \qquad (3)$$

where R is the distance between the centers of the interacting atoms/ions, α is the polarizability and l the ionization potential, with m referring to the metal and A the anion in the ceramic.

Charge exchange allows ion pairs to form and interact across the interfaces. For examle, the interactions between ions of the metal and of oxygen (or other anions) in the ceramics related to the free energy of metal oxide formation, ΔG^o.[26,27] Furthermore, when the cations of the ceramic are soluble in the metal, dissolution from the interface allows ionic interaction between dissolved cations and the anions in the ceramic.

McDonald and Eberhart[28] examined interactions involving various metals in contact with the (0001) plane of sapphire. For this purpose, they assumed that the (0001) sapphire surface terminates with a layer of close-packed oxygen ion. The metal atoms (to be bonded to Al_2O_3) are then offered two sites; those above the Al ions located below the top layer of oxygen ions and those above empty sites. The first site results in attractive dipole forces, as described in Eqn. (3), which are about constant for all metal/Al_2O_3 couples. The second metal site forms ionic oxygen-metal bonds having a free energy proportional to ΔG^0. By further assuming that all interactions of the dense-packed oxygen plane are occupied with metal atoms, the calculated trends in W_{ad} agree quite well with experimental data for the bonding of Al_2O_3 to simple metals.[31] The very strong bonding of Pt and Pd[19] to alumina is evidently at variance with the simple model. However, Al possesses a very high heat of solution in these metals. Consequently, as noted above, bonds could be formed between the oxygen ions and Al ions dissolved in the metal. Alternatively, a thin segregated Al layer could form between the metal and the Al_2O_3 to enhance the bonding.[31]

The McDonald-Eberhart approach provides helpful generalizations. However, a more fundamental, atomistic understanding of the nature of the bonding is needed to adequately understand critically important alloying and segregation effects, as well as trends in the fracture resistance.

Atomistic Models

An understanding of the fundamental physics of bonding between a metal and a ceramic requires that quantum mechanical models be developed. The simplest approach involves cluster calculations.[30] Such calculations have established that the primary interactions at metal/oxide interfaces involve the metal (d) and oxygen (p) orbitals, to create both bonding and antibonding orbitals. For Cu and Ag in contact with Al_2O_3, both states are about equally occupied, resulting in zero net bonding. However, for Ni and Fe, fewer antibonding states are occupied and net bonding occurs. The calculations also reveal that a transfer of valence charge occurs, resulting in a contribution to the net ionic bonding which increases in strength as the metal becomes more noble. Consequently, metal-to-alumina bonding strengths are predicted to increase in the order: Ag-Cu-Ni-Fe. This order is generally consistent with the measured trends in sliding resistance as well as with the energies of adhesion. However, it is emphasized that the calculations approximate the interface by an $(AlO_6)^{9-}$ cluster and one metal atom. The selection of the charge to be assigned to this cluster is non-trivial and the choice influences the predicted magnitudes of the energies.[32] To further examine this issue, Anderson et al.[33] performed calculations for the Al_2O_3/Pt couple that included more atoms: 31 close-packed Pt atoms and the corresponding numbers of Al and O ions. Then, by applying a quantum-chemical superposition technique, including an electron delocalization molecular orbital method, bonding energies were calculated for different atomic configurations of the Pt/Al_2O_3 interface. These calculations confirmed that the bond was strongest when oxidized Pt atoms opposed close-packed oxygen ion planes. However, further quantitative insights did not emerge. Indeed, the preceding models all have the deficiency that they do not fully account for the heterogeneous nature of the interface and cannot, therefore, be expected to accurately predict energies, segregant effects, etc.

Ab initio calculations seem to be essential for a full understanding of the bonding. Louie et al.[34,35] have performed such calculations on metal-semiconductor interfaces. In these calculations, the metal was described by a jellium, so that insight emerged regarding the bonding mechanisms, but not on the atomistic structure. More recently, supercell calculations have been performed that include an interface area and adjacent regions large enough to incorporate the distorted (relaxed) volumes of both crystals. With this approach, the electron distribution around all atoms has been calculated and the atomic potentials evaluated. In a next step, interatomic forces may be calculated and strains determined. Such calculations have been performed rather successfully for the interface between Ge-GaAs[36] and Si-Ge.[37] The crystals adjacent to those interfaces are isomorphous and very nearly commensurate, such that the misfit between lattice planes is very small. However, misfits between metals and ceramics are typically rather large so that extremely large supercells are required. Therefore, only preliminary calculations have been conducted thus far.[38] Nevertheless, the calculations, performed for MgO/Ag, have allowed determination of the atomic potentials surrounding the different atoms, as well as a separation of bonding into different contributions (ionic, covalent and polarization).

With the advent of a new calculational scheme, (Car and Parrinello[39]) involving a combination of molecular dynamics (see, e.g., Rahman[40]) and density functional theory (Kohn and Sham[41]), it should be possible to conduct computations of <u>relaxed</u> interfaces much more efficiently. The scheme should also allow equilibrium computations of metal/ceramic interfaces at finite temperatures. The conduct of such analysis on model interfaces should greatly facilitate the basic understanding of the bonding phenomenon and allow judicious usage of both cluster calculations and continuum thermodynamic formulations.

STRUCTURE OF INTERFACES

In thermodynamic equilibrium, the atoms and/or ions close to an interface occupy positions that minimize the <u>total energy</u> of the system. However, the proximity to equilibrium depends on the conditions used to form the interface. For example, diffusion-bonded interfaces have atomistic arrangements influenced by the orientation of the two surfaces <u>prior</u> to bonding. Additionally, when the bonding is performed at high temperature, chemical gradients often develop and influence the structure and residual strains form upon cooling. The interface structure thus involves geometric, as well as atomistic considerations, conditioned by relaxation mechanisms inherent in each bonding process. Consequently, interface structures are conveniently described by firstly defining generalized geometric parameters for the unrelaxed interfaces. Then, the relaxation mechanism pertinent to each bonding process may be considered. Finally, various geometric and atomistic bonding models, and associated experimental results, may be evaluated.

Geometrical Parameters

An unrelaxed interface can be described by nine geometrical parameters. The required number was evaluated by a thought experiment, similar to that previously used for grain boundaries.[42] Six parameters describe the relative orientation and translation of the two crystals. The description of the interface orientation with respect to the crystal requires three additional parameters.

Process Relaxations

The bonding method governs the actual geometrical parameters that describe the interface, by virtue of the imposed geometry and the allowable relaxations. During <u>diffusion bonding</u>,[6] intimate surface contact at elevated temperatures, subject to a small pressure, generates the bonded interface. This technique, <u>pre-selects</u> the (macroscopic) rotation, two components of the translation and the interface orientation. However, some of these geometrical parameters are relaxed by <u>local</u> deviations. Specifically, bonding is usually performed at temperatures wherein at least one component may undergo plastic deformation. Consequently, local geometrical relaxation may be accommodated by small angle grain boundary formation adjacent to the interface. Furthermore, mass transport may allow interface facets to develop that relax the constraint on interface planarity.

Interfaces may also be formed by <u>internal oxidation</u>.[43,44] Such interfaces are <u>not unique</u>, but are related to the precipitate <u>morphology</u>, as governed by thermodynamic principles. In particular, since the total energy of the precipitate depends not only on the interfacial energy but also on coherency strains, the interface structure and the shape of the precipitate depend usually on its size.[45] For coherent precipitates, orientation relationships are governed by constraints imposed by interfacial energies, the solubility and lattice parameters, and the most stable morphologies. These relationships should be an integral part of the analysis of the structure of interfaces formed by internal oxidation.

A third way of producing metal/ceramic interfaces is by <u>evaporation</u> of the metal onto a clean ceramic surface. The evaporated species is usually highly mobile[46] at high temperature. Consequently, if the metal wets the ceramic, a thin layer forms in an <u>equilibrium</u> interfacial configuration. For non-wetting configurations, islands are formed. Then, when the substrate is a single crystal and a one-to-one orientation relationship exists between the two components, the islands may grow together and form a single crystal film.[46] Alternatively, when different equivalent orientation relationships exist, the islands develop with slightly different orientations. A range of different behaviors is thus expected for interfaces formed by evaporation.

Geometrical Models

Models described by <u>geometrical</u> parameters may sometimes be insightful as a basis for the description of <u>unrelaxed</u> interfaces. However, it must be appreciated that the direct correlation between such models and interface properties (e.g., energy) is not possible.[47] Foremost among such models is a "lock-in" model developed by Gleiter et al.,[48] deduced from experiments wherein small metallic single crystals were sintered onto a single crystal ceramic substrate. Interfaces between several alkali halides or rock salt structure oxides and various noble metals were formed in this manner. It was observed that the spheres rotated into orientations which tend to minimize the energy of the sphere/plate boundary. Consequently, the orientation relationships determined by X-ray diffraction may be supposed to reflect a low energy configuration of those interfaces. Gleiter et al.[48] accounted for the observed "stable" configurations by means of a model which describes low energy interfaces in terms of densely packed atom rows of one crystal nesting in the grooves between similar rows of the other crystal. The requisite periodic matching between (small) multiples of lattice plane spacings was achieved by imposing small displacements upon certain of the atom rows. The lattice strains from such imposed displacements have thus far been neglected. Nevertheless, in some cases, "lock-in" seemingly describes observations, even when large misfits exist. Conformity with the "lock-in" structure has been established for MgO/Au interface formed by evaporation[49] and for Au on KCl (35 percent misfit);[48] in both cases, the predicted cube-on-cube orientation was observed. The "lock-in" concept thus appears to be a useful first order description in some cases. However, more detailed investigations have revealed important discrepancies. Various observations indicate that small but definite angular off-sets occur between close-packed directions and/or planes. Klomp et al.[50] studied orientation relationships in diffusion bonded sandwiches of polycrystalline Cu and Pt foils between identically oriented (parallel) sapphire crystals. As expected, the metal grains developed a preferred orientation, with (111) of the metal parallel to (0001) of the sapphire. However, the grains were rotated within a wide angular range about the close-packed directions. Discrepancies have also been found for the Nb/Al_2O_3 system.[51,52] Notably, the close-packed planes are not aligned: instead, the (0001) plane of Al_2O_3 and (011) of Nb are tilted by $\sim 3°$ around an axis parallel to $[11\bar{2}0]$ Al_2O_3 and $(101)_{Nb}$. This tilt occurs in specimens grown by epitaxy of Nb on (0001) Al_2O_3[51] and in specimens formed by diffusion bonding,[52] even when initially the

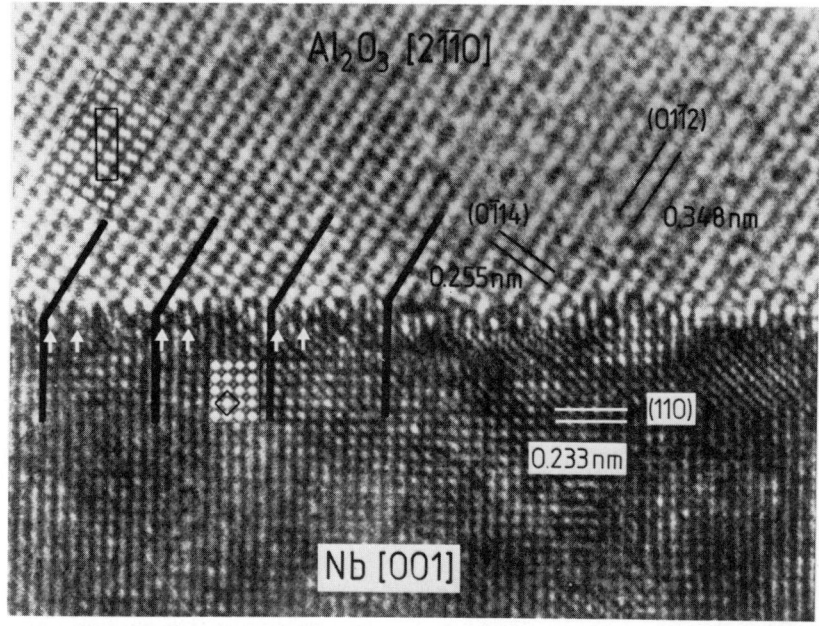

Fig. 1. Lattice image of the regions close to an interface coopho Nb and Al_2O_3. Orientation relationship between the ceramic and the metal: $(0001)Al_2O_3 | |(110)Nb$ and $[2\bar{1}\bar{1}0]Al_2O_3| |[001]Nb$. The foil orientation is parallel to $[001]Nb$. Insets of the HREM images of Nb and Al_2O_3 represent simulated lattice images. Only four lattice planes within the Nb next to the interface are distorted. Facets can be observed. The periodicity of the interface is indicated.

299

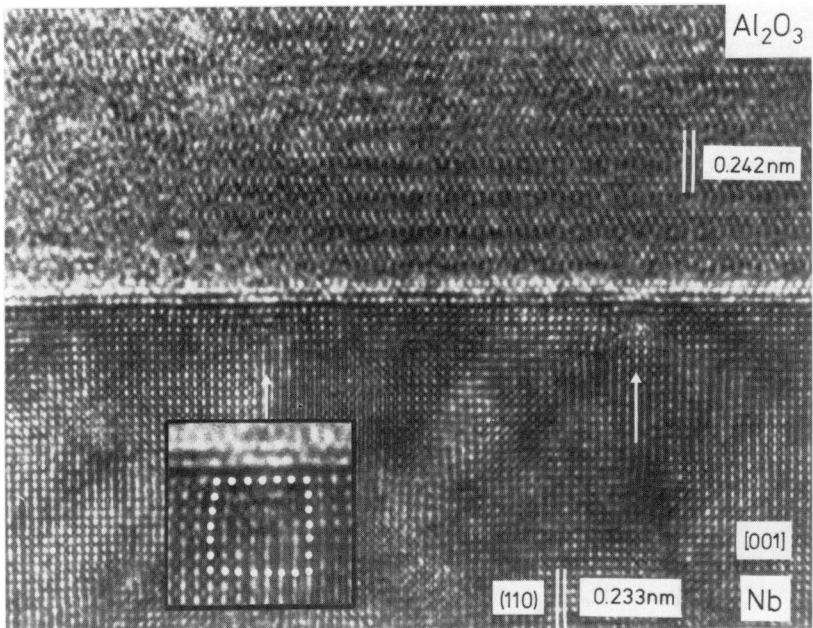

Fig. 2. Lattice image of Nb/Al$_2$O$_3$ interface of an internally oxydized Nb/Al alloy. Interface plane parallel to (0001)Al$_2$O$_3$||(110)Nb and [0$\bar{1}$10]Al$_2$O$_3$ ||[001]Nb. Misfit dislocations in Nb can be observed (see inset). The extra half plane forming the misfit dislocation does not terminate at the interface.

specimens were adjusted to have parallel close-packed planes. It is also noted that even though MgO/Au has the cube-on-cube orientation predicted by "lock-in", misfit dislocations are observed, indicative of interface strains. Such dislocations are not consistent with the "lock-in" concept.

Relaxation effects are presumably responsible for discrepancies with the "lock-in" model. For example, as noted above, subgrain boundaries in diffusion bonded Nb/Al$_2$O$_3$ provide freedom for some rotation of the Nb lattice. Also, the interface develops facets while retaining the 3° tilt of the two lattices. Clearly, the facets alleviate the misfit strain which would develop if close-packed planes were kept in contact everywhere across the interface.

Atomistic Models

The <u>relaxed</u> structure of interfaces is obviously of greater significance than the unrelaxed state. Adequate models of this structure do not yet exist, but await comprehensive experimental insight. Experimental studies can be performed either by diffraction[53,54] or by high resolution (direct imaging) techniques.[55] With the new generation of electron microscopes, direct imaging of <u>atomic columns</u> is possible. Specifically, by aligning along the axis of an interface which possesses only a tilt component, each column of atoms is imaged in one spot, giving a projection of the atomic arrangements in the interface. However, to derive reliable atomic positions, a series of observed images, taken at different focus settings, must be matched quantitatively with computed images, based on an assumed set of atomic coordinates. For matching purposes, image computation and position adjustment is repeated until the best possible fit is reached for the entire through-focus series. This technique is referred to as quantitative high resolution electron microscopy. The interpretation is, of course, most difficult close to the interface where deviations from the perfect lattice are most extreme. Fully quantitative HREM studies have been performed on grain boundaries in germanium[55] and on semiconductor/metal interfaces such as Si/NiSi$_2$.[57]

Quantitative studies of metal/ceramic interfaces have been initiated for the Nb/Al$_2$O$_3$ and Au/MgO systems formed using each of the three processes described above. Interpretation of high resolution images of regions close to a <u>diffusion bonded</u> interface[58] (Fig. 1) requires that image simulations for perfect lattices first be performed. For this purpose, the projected atomic positions and potentials of Nb and Al$_2$O$_3$ in the selected orientation have been determined and used to construct the simulated image for a foil having a thickness (12 nm) which corresponds with that for the actual test specimen. It is evident that lens aberration (C_s = 1.1 mm, 200 kV) conceals some details of the perfect lattice structure. Nevertheless, it is still apparent (Fig. 1) that the perfect lattice is preserved in the Al$_2$O$_3$ up to the interface. Conversely, in the Nb, strong deviations occur for distances up to four lattice planes from the interface. However, a periodicity along the interface can be recognized within the distorted region. Faceting parallel to the interface also occurs and the interface is slightly inclined with respect to the (0001) plane of Al$_2$O$_3$. It is expected that imaging of the same interface with instruments possessing superior resolution would result in a better resolution of the distorted atomistic structure close to the interface. When the results of such studies become available, insightful atomistic models of the interface might be developed.

Interfaces created by <u>internal oxidation</u> have well defined geometrical characteristics. Precipitates of α-Al$_2$O$_3$ formed in Nb/Al alloys are small and penny-shaped (diameter ~ 300 nm) and exhibit a fixed orientation relationship,

$$(0001)_{Al_2O_3} || (110)_{Nb} \text{ and } [0110]_{Al_2O_3} || [001]_{Nb}$$

High resolution TEM investigation (Fig. 2) has revealed, however, that the spacings of the (110)Nb planes (d = 0.23 nm) and of the (0001) Al$_2$O$_3$ planes (d$_2$ = 0.39 nm) are sufficiently different that misfit dislocations would be predicted with spacing

$D \equiv d_1 d_2/(d_2-d_1)$ = 8.8 nm. Weak beam images confirm such dislocations (Fig. 3) with D = 9.1 ± 0.5 nm. To fully analyze the interface structure, as well as the atomistic structure of the misfit dislocations, superior resolution (<0.2 nm) would be needed.

Finally, it is noted that conventional TEM provides complementary information on defects at or close to the interface, such as dislocation spacings, ledge densities, etc. It is emphasized, however, that a more complete analysis of such defects, including Burgers vector determination, is generally not possible since the materials adjacent to the interface possess (usually) different lattice parameters and few "common" diffraction vectors exist.[60]

CHEMISTRY OF INTERFACES

Theoretical Background

In multicomponent two-phase systems, nonplanar interfaces or two-phase product regions can evolve from initially planar interfaces,[61] even at constant temperature and pressure. Under the same conditions, the interface stays planar in binary systems. This difference originates with the thermodynamic degrees of freedom, f. For a binary system f = 0, whereas for ternary or higher order systems, f > 0 (Gibbs phase rule). Consequently, in the latter, interface compositions are in part controlled by the kinetics.

Not all (higher order) interfaces necessarily develop an unstable morphology during a high temperature treatment. Analysis of the phenomenon is needed to assess susceptibility. Similar problems exist for solidification and for the oxidation of alloys.[62] Mathematical treatments predict the time evolution of small interface perturbations. The perturbations may occur either due to initial roughnesses or upon small transport fluctuations caused by changes in temperature or by the presence of defects. If perturbations increase in amplitude with time, initially planar interfaces become morphologically unstable. The critical conditions for instability depend primarily on the mobility of the constituents and the thermodynamic properties of the system.

The formalism previously developed for ternary systems[61] can be adapted to metal/ceramic couples, with the three independent components being the two cations and the anion. A schematic ternary phase diagram and the expected concentration profiles are shown in Fig. 4.[63] In general, the problem is complicated by having several phase fields present, such that intermediate phases form: usually intermetallics with noble metals and spinel (or other oxides) with less noble metals. The actual phases depend on the geometry of the tie lines, as well as on the diffusion paths in the ternary phase field, and cannot be predicted a priori.

The <u>diffusion</u> problem has thus far been examined[63] for the simple case wherein no product phases formed, the interfacial stresses were negligible, mass transport occurred by bulk diffusion and local equilibrium was imposed everywhere. Even then, a general analytical solution was not possible. However, for several metal/ceramic systems, some further simplifications are appropriate. The oxygen and the metal atoms diffuse on different sublattices allowing the interaction term in the diffusion coefficients to be neglected. Negligible solubility of the metal in the oxide (grad μ_{BO} = 0) allows point defect relaxation only in the ceramic. Therefore, for a stoichiometric ceramic, the remaining defect fluxes are small compared with the fluxes in the metal and can be ignored. Subject to the above simplifications, solutions have been obtained for Nb/Al_2O_3.[63,64]

Rather high and probably unrealistic values of solute concentrations are calculated at the interface. In particular, the Al concentration profile in the Nb is steep and reveals an enrichment very near the interface, whereas the O concentration profile is more extended, but with small absolute values. Such results do not agree with experimental observations, as elaborated below, suggesting that several assumptions are invalid. More sophisticated numerical

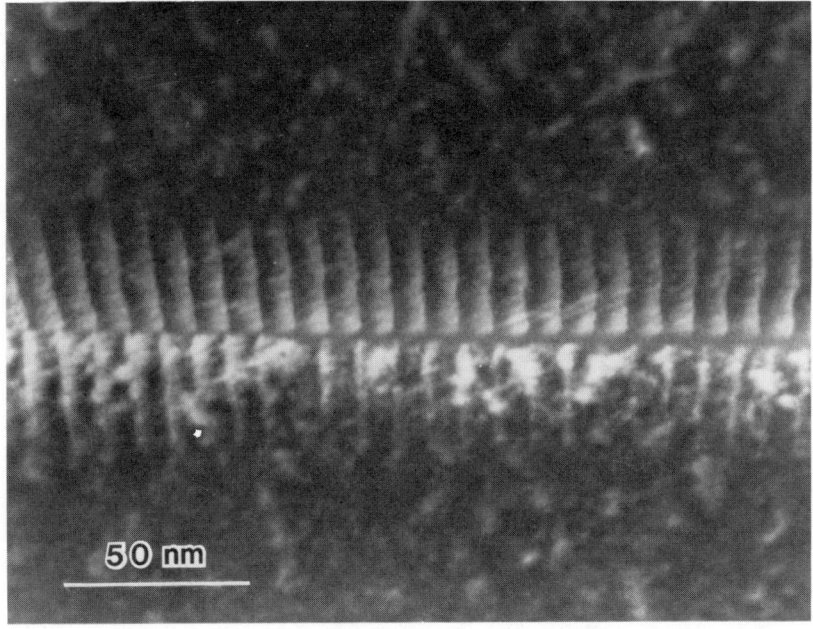

Fig. 3. Weak-beam dark field image of inclined interfaces of Fig. 7 limiting the α-Al$_2$O$_3$ precipitate. Imaging with Nb(1$\bar{1}$0) reflection. Light lines represent the contrast form misfit dislocations.

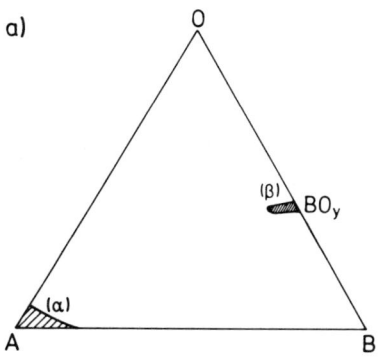

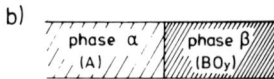

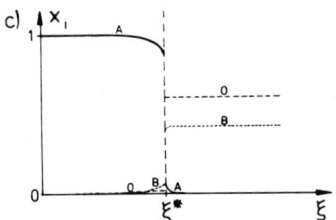

Fig. 4. (a) Schematic ternary phase diagram. Extended phases fields exist near metal A and oxide BO_y; (b) Equilibrium concentration profiles during diffusion bonding between metal phase (A) and oxide BO_y.

models will thus require development as additional experimental results become available.

Interfaces Without Reaction Layers

Detailed SEM and TEM studies performed for Nb/Al$_2$O$_3$ [64-67] have shown that no reaction layer forms (see Fig. 1). Concentration profiles determined on cross-sections of <u>rapidly cooled</u> specimens revealed that, close to the interface, the concentration of Al is below the limit of detectability. However, with increasing distance from the interface, the concentration of Al, c_{Al} increases to a saturation value, c_{Al} = 0.65 wt%, at a distance of ~2.5 μm (Fig. 5). The c_{Al} remains at that level up to a distance of d = ~16 μm after bonding for two hours and then decays exponentially. The magnitude of d depends on the bonding time. The corresponding oxygen content is below the limit of detectability. These measurements suggest that at the bonding temperature, c_{Al} at the interface possesses a value governed by the solubility limit. This limiting concentration would then extend into the Nb to a characteristic distance that depends on the bonding temperature and time. Upon cooling, the solubility of Al in Nb decreases, causing some of the dissolved Al (as well as O) to diffuse back to the interface and condense as Al$_2$O$_3$.

"<u>Slow</u>" cooling after bonding resulted in completely different observations. Instead, small precipitates of θ-Al$_2$O$_3$ form in the Nb at distances between 8 μm and 14 μm from the interface.[65,66] Furthermore, close to the zone wherein precipitation occurs, c_{Al} is very small. The precipitation presumably occurs during slow cooling, because the time requirements for precipitate nucleation are satisfied.

Bonding between Pt and Al$_2$O$_3$ subject to an inert atmosphere also occurs without chemical reaction. Specifically, no Al can be detected by Auger spectroscopy on the Pt side of an interfacial fracture surface.[18,19] However, for bonds formed subject to a H$_2$ atmosphere containing ~100 ppm H$_2$O, Al is detected in the Pt, indicative of Al$_2$O$_3$ being dissolved by Pt. More detailed studies concerning local chemical compositions are clearly required for a better understanding of the bonding processes involved.

Interfaces With Reaction Layer

For systems that form interphases, it is important to be able to predict those product phases created during diffusion bonding (given the possible phases present in the phase diagram). However, even if all the thermodynamic data are known, so that the different phase fields and the connecting tie lines can be calculated, the preferred product phase still cannot be unambiguously determined. The problem involves kinetic considerations. Specifically, the diffusion paths in phase space are controlled by different diffusion coefficients and, consequently, interface compositions depend also on the diffusivity ratios. Sometimes, small changes in the initial conditions can influence the reaction path dramatically, as exemplified by the Ni-Al-O system.[68] Under high vacuum conditions (activity of oxygen $a_0 < 10^{-12}$) the diffusion path in the extended Ni phase field follows that side of the miscibility gap rich in Al and low in O, (path I in Fig. 6), caused by the more rapid diffusion of O than Al in Ni. This interface composition is directly connected by a tie line to the Al$_2$O$_3$ phase field, such that no product phase forms. However, whenever Ni contains sufficient O (about 500 ppm solubility), the Ni(O)/Al$_2$O$_3$ diffusion couple yields a spinel product layer. Under these conditions, spinel forms, because the new diffusion path in the Ni phase field requires that the tie line connects the metal and spinel field (path II in Fig. 6). The associated thermodynamic and atomistic consideration pertinent to spinel layer formation have been addressed.[67,68] However, available observations do not unequivocally identify the operative mechanism. It is also noted that the interface between spinel and nickel seems to be unstable: morphological instabilities becoming more apparent with increasing spinel layer thickness.

Bonding of Cu to Al_2O_3 seems to require a thin layer of O on the surface of Cu prior to bonding and Cu/Al_2O_3 or Cu/Al_2O_4 form.[69] The spinel thickness can be reduced by annealing under extremely low oxygen activities leading first to a "nonwetting" layer of Cu_2O and then to a direct Cu/Al_2O_3 bond. The mechanical stability of spinel-free Cu/Al_2O_3 specimens has not yet been investigated.

Bonding of Ti to Al_2O_3 results in the formation of the intermetallic phases TiAl or Ti_3Al, which probably also include oxygen. The thickness of the reactive layer increases with increasing bonding time and morphologically unstable interfaces develop. A detailed study is again required for the identification of the different stable phases.

Similar studies have been performed for other ceramic partners such as simple cubic oxides (MgO, NiO,....), sesquioxides (Cr_2O_3, Mn_2O_3...), Si_3N_4 and SiC. The situation is much more complicated for the latter materials, since impurities or sintering additives quite frequently diffuse to the interface and form a glassy film. The bonding is then governed by the interfaces between the glass film and both the ceramic and the metal.

SINTERING OF INTERFACIAL FLAWS

Scratches and other defects frequently exist on surfaces which have to be bonded. Long bonding times and high bonding temperatures are required when such flaws are present, because residual defects at the interface restrict the mechanical properties of the metal/ceramic interface. Investigation of the sintering of interface flaws is thus of great importance. The closure of interfacial flaws at homogeneous bonds involves several mechanisms[70]: surface diffusion, volume diffusion, diffusion along the interface, power law creep and plastic yielding. Models for homogeneous interfaces based on these mechanisms qualitatively describe the experimental observations. However, a quantitative comparison has been inhibited by the scarcity of reliable experimental data. For a heterogeneous bond and especially for a metal/ceramic interface, additional processes have to be considered[64]; namely, the diffusion of metal atoms into the ceramic, dissolution of the ceramic at the interface and diffusion of the species of the ceramic into the metal, chemical reactions between the metal and the ceramic to form a product phase and recondensation of a dissolved ceramic at the interface. The inherent complexity demands a vital role of experiments in ascertaining the most important variables for typical metal/ceramic couples.

For evaluation of the governing transport mechanisms, artificial interfacial defects of different dimensionality and sizes may be introduced into the metallic surfaces[66,67] prior to bonding: Linear flaws by photolithography and other shapes by indentation. After bonding, cross sections of interfacial fracture surfaces studied by SEM allow determination of the change in flaw size. The depth variations may be measured with a profilometer.

Inspection of results obtained for Nb/Al_2O_3 revealed that the surface of Nb hardly changes, whereas condensation products of Al_2O_3 could be detected on the sapphire. The amount of condensed Al_2O_3 depends on the cooling. From chemistry studies it has already been established that Nb dissolved Al_2O_3 and, upon cooling, Al_2O_3 recondenses at the interface. This mechanism also seems to be involved in the "filling" of the flaws. However, the experimental, as well as the theortical, studies are not yet sufficient to derive a quantitative description of the process. Studies of this type on other systems are also needed to obtain a comprehensive view of the important issues in diffusion bonding.

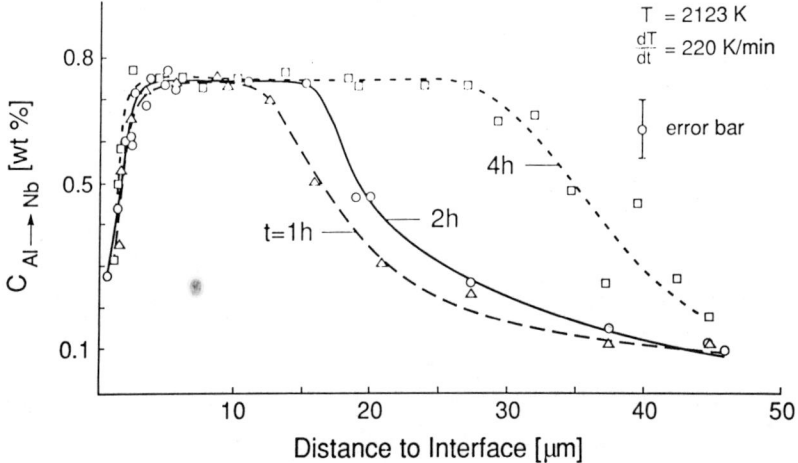

Fig. 5. Measured concentration profiles of Al in Nb as a function of the distance of bonding time and of the interface. Diffusion bonding conditions: 2073K, 2h, dynamic vacuum (10^{-4} torr), "fast" cooling 215K/min. (a) distances 0 to 150 µm; (b) distances 0 to 20 µm (higher spatial resolution).

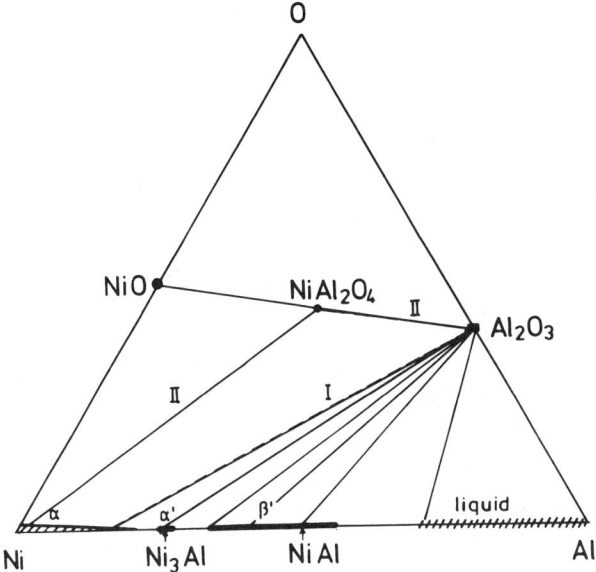

Fig. 6. Ni-Al-O phase diagrams (schematically for T = 1600K). Two reaction paths are possible when Ni is bonded to Al_2O_3: (I) Low oxygen activity: no reaction product forms, (II) High oxygen activity: spinel forms.

CONCLUSION

Although considerable progress has been made in the understanding of structure and chemistry of metal/ceramic interfaces, still many open questions exist. Detailed and careful experimental studies will reveal more insight in the mechanisms of bondings and these experimental studies may challenge further theoretical studies.

REFERENCES

[1] E. A. Giess, K. N. Tu and D. R. Uhlmann (eds), Electronic Packaging Materials Science, MRS Symp. Proc. Vol. 40 (1985).

[2] K. A. Jackson, R. C. Pohanka, D. R. Uhlmann, D. R. Ulrich (eds), Electronic Packaging Materials Science II, MRS Symp. Proc. Vol. 72 (1986).

[3] J. M. Gibson and L. R. Dawson (eds), Layered Structures, Epitaxy and Interfaces, MRS Symp. Proc. Vol. 37 (1985).

[4] A. K. Dhingra and S. G. Fishman (eds), Interfaces in Metal-Matrix Composites, The Metallurgical Society, Inc. (1986).

[5] M. Rühle, R. W. Balluffi, H. Fischmeister and S. L. Sass (eds), Proc. Intern. Conf. on the Structure and Properties of Internal Interfaces, J. Phys. Colloq. 46-C4, (1985).

[6] Y. Ishida (ed), Interface Structure, Properties and Diffusion Bonding, Elsevier Science Publisher (1987), in press.

[7] J. A. Pask and A. G. Evans (eds), Ceramic Microstructure '86: Role of Interfaces, Plenum Press, New York and London (1987).

[8] E. A. Almond, C. A. Brookes and R. Warren (eds), Science of Hard Materials, Inst. Phys. Conf. Series 75, Adam Hilger Ltd., Bristol and Boston (1986).

[9] G. S. Upadhyaya (ed), Sintered Metal-Ceramic Composites, Elsevies Amsterdam (1984).

[10] K. L. Mittal (ed), Adhesion Measurement of Thin Films, Thick Films, and Bulk Coatings, STP640, ASTM Philadelphia (1978).

[11] B. V. Derjaguim, Recent Advances in Adhesion, Gordon and Breach, New York (1971) p. 513.

[12] E. D. Hondros, "Physicochemical Measurements in Metal Research" in (R. A. Rapp ed): Techniques of Metals Research Series, Interscience Publishers, N. Y., NY, Vol. IV 2 (1970) 293-348.

[13] J. W. Hancock, I. L. Dillamore, R. E. Smallman, 6th Plansee Seminar (1968) p. 467.

[14] K. E. Easterling, H. F. Fischmeister, and E. Navara, Powder Metall. 16 (1973) 128.

[15] M. Nicholas, J. Mat Sci; 3 (1968) 571.

[16] I. G. Palmer and G. C. Smith, Proc. 2nd Bolton Landing Conference on Oxide Dispersion Strengthening, Gordon and Breach N. Y. (1968) 351.

[17] M. McLean and E. D. Hondros, J. Mat. Sci 6 (1971) 19.

[18] J. T. Klomp in: Ceramic Microstructure '86: Role of Interfaces, Plenum Press (1987), in press (J. A. Pask and A. G. Evans eds).

[19] J. T. Klomp in: Interface Structure, Properties and Diffusion Bonding (Y. Ishida, ed) Elsevier Science Publisher (1987), in press.

[20] H. F. Fischmeister, E. Navara, K. E. Easterling, Metal. Science, J. 6 (1972) 211.

[21] J. L. Smialek, R. Browning, 168th Meeting Electrochem. Soc., Las Vegas Oct. 1985 (NASA Techn. Memor 87/68).

[22] L. Pauling, General Chemistry 3rd edt., Freeman, San Francisco (1970) 118.

[23] E. D. Doyle, J. G. Horne and D. Tabor, Proc. Roy. Soc. London A366 (1979) 173.

[24] A. M. Stoneham, Appl. Surf. Sci. 14 (1982/83) 249.

[25] A. M. Stoneham and P. W. Tasker, Harwell-Report T1166 (1985) to be published.

[26] W. K. Kingery, H. K. Bowen and Dr. R. Uhlman, Introduction to Ceramics, Wiley New York (1976) 210.

[27] J. T. Klomp in [1] p. 381.

[28] J. E. McDonald and J. G. Eberhart, Trans AIME 233 (1965) 512.

[29] S. V. Pepper, J. Appl. Phys. 47 (1976) 801.

[30] K. H. Johnson and S. V. Pepper, J. Appl. Phys. 53 (1982) 6634.

[31] H. F. Fischmeister, in Ceramic Microstructure '86 Role of Interfaces (J. A. Pask and A. G. Evans, eds) Plenum Press (192``87) in press.

[32] A. B. Anderson, S. P. Mehandru and J. L. Smialek, J. Electrochem. Soc. 132 (1985) 1695.

[33] A. B. Anderson, C. Ravimohan and S. P. Mehandru, J. Electrochem. Soc. 134 (1987) in press.

[34] S. G. Louie and M. L. Cohen, Phys. Rev. B13 (1976) 2461.

[35] S. G. Louie, J. G. Chelikowsky and M. L. Cohen, Phys. Rev. B15 (1977) 2154.

[36] K. Kunc and R. M. Martin, Phys. Rev. B24 (1981) 3445.

[37] C. G. Van de Walle, R. M. Martin, J. Vac. Sci. Techn. B3 (1985) 1256.

[38] P. Bloechl and O. K. Anderson, to be published.

[39] R. Car and M. Parrinello, Phys. Rev. Lett. 55 (1985) 2471.

[40] A. Rahman, in: Correlation Functions and Quasiparticle Interactions in Condensed Matter (J. W. Halley, ed), NATO Adv. Study Ser. 35 Plenum; New York (1977).

[41] W. Kohn and L. J. Sham, Phys. Rev. 140 (1965) A1133.

[42] G. Kalonji, J. de Physique 46 (1985) C4-249.

[43] C. Wagner, J. Electrochem. Soc. 103 (1956) 571.

[44] J. L. Meijering, in: Advances in Materials Research (H. Herman, ed.) Wiley Interscience 5 (1971) 1.

[45] A. G. Khachaturyan, Theory of Structural Transformations in Solids, Wiley New York (1983).

[46] J. Matthews, Epitaxial Growth, Vol. I, Academic Press, New York (1975).

[47] A. P. Sutton and R. W. Balluffi, Acta metall. 35 (1987).

[48] H. Gleiter and H. S. Fecht, Acta Metall. 33 (1985) 577.

[49] R. H. Hoël, Surf. Sci. 169 (1986) 317.

[50] C. A. M. Mulder and J. T. Klomp, J. de Physique 46 (1985) C4-111.

[51] D. B. McWhan, Mat. Res. Soc. Symp. 37 (1985) 493.

[52] M. Florjancic, W. Mader, M. Rühle and M. Turwitt, J. de Physique 46 (1985) C4-129.

[53] J. Budai, P. D. Bristowe and S. L. Sass, Acta Metall. 31 (1983) 699.

[54] J. Vitek and M. Rühle, Acta Metall. 34 (1986) 2095.

[55] A. Bourret, J. de Physique 46 (1985) C4-27.

[56] R. W. Balluffi, M. Rühle and A. P. Sutton, Mat. Science and Engineering 89 (1987) 1.

[57] J. M. Gibson, R. T. Tung, J. M. Phillips and R. Hull, J. de Physique 46 (1985) C4-369.

[58] W. Mader and M. Rühle, Acta Metall., submitted.

[59] W. Mader, Mat. Res. Soc. Symp. Proc. 82 (1987) 403.

[60] R. C. Pond, J. of Microscopy 135 (1984) 213.

[61] M. Backhaus-Ricoult and H. Schmalzried, Ber. Bunsenges. Phys. Chem. 89 (1985) 1323.

[62] W. W. Mullins and R. F. Sekerka, J. Appl. Phys. 34 (1963) 323; 35 (1964) 444.

[63] M. Backhaus-Ricoult, Ber. Bunsenges. Phys. Chem. 90 (1987) 684.

[64] K. Burger and M. Rühle, Adv. in Ceramics, in press.

[65] K. Burger, W. Mader and M Rühle, Ultramicroscopy 12 (1987) 1.

[66] M. Rühle, K. Burger and W. Mader, J. Micros. Spectrosc. Elecron. 11 (1986) 163.

[67] M. Rühle, M. Backhaus-Ricoult, K. Burger and W. Mader, in [7], p. 295.

[68] J. A. Wasynczuk and M. Rühle, in [7], p. 341.

[69] M. Wittmer, Mat. Res. Soc. Symp. Proc. 40 (1985) 393.

[70] B. Derby and E. R. Wallach, Mat. Sci. 15 (1982) 49; 18 (1984) 427.

INTERFACIAL CHEMISTRY-STRUCTURE AND FRACTURE OF CERAMIC COMPOSITES

L. H. SCHOENLEIN[*], R. H. JONES[*], C. H. HENAGER[*], C. H. SCHILLING[*]
AND F. GAC[**]
[*] Pacific Northwest Laboratory[(a)], Richland, WA 99352
[**]Los Alamos National Laboratory, Los Alamos, NM 87545

ABSTRACT

The interfacial chemistry and phases of SiC-reinforced Si_3N_4 composites have been evaluated by transmission electron microscopy (TEM) with associated x-ray energy dispersive spectroscopy (EDS) microanalysis, and Auger electron spectroscopy (AES). Hot-pressed Si_3N_4 (HPSN) composites reinforced with Nicalon™ SiC fibers or Tateho SiC whiskers and reaction-bonded Si_3N_4 (RBSN) composites reinforced with uncoated or coated VLS SiC whiskers have been evaluated. In the Nicalon™ fiber-reinforced HPSN, an interfacial phase composed of a layer of amorphous carbon and an adjacent layer of graphitic carbon was observed and is believed to assist fiber pullout during fracture of the composite. However, the fracture strength and toughness of these composites were considerably less than those of unreinforced HPSN. HPSN composites reinforced with Tateho SiC whiskers contained an interfacial phase believed to be similar to the intergranular phase found in the HPSN matrix. In RBSN composites fabricated with an Fe_2O_3 sintering aid, the VLS SiC whiskers were severely faceted by a reactive iron silicide phase despite C, BN, or SiO_2 coatings on the whiskers. When no sintering aid was used, the uncoated whiskers were not degraded and appeared to be strongly bonded to the RBSN matrix. The composites reinforced with SiO_2-coated whiskers possessed the highest fracture strength and toughness, and the composites reinforced with the BN-coated whiskers possessed the lowest fracture strength and toughness.

INTRODUCTION

The brittle nature of ceramic materials is responsible for their susceptibility to catastrophic failure. The reinforcement of ceramic bodies with high-strength, high-modulus ceramic whiskers (or fibers) is currently under intensive investigation as a means to toughen these materials. Current theories suggest that microcracking (1), crack deflection (2), and crack bridging (3,4) are the predominant toughening mechanisms, although the relative contributions of each mechanism to toughening have not been clearly delineated. However, there is general agreement that the interfaces between the whiskers and the matrix are of prime importance in enhancing toughness. Weakly bonded interfaces significantly enhance the toughness of composites by facilitating crack bridging and whisker pullout during crack propagation, while strongly bonded interfaces facilitate whisker fracture which yields little or no increase in toughness.

It is often necessary to coat the whiskers with an inert substance to prevent interfacial reactions between the whiskers and the matrix during fabrication of the composites, as these reactions are likely to cause strong interfacial bonding. The type of coating material and its thickness may be varied to yield interfacial phases which, in addition to preventing interfacial reactions, optimize the toughness of the composite. It is anticipated that composites will be used at elevated temperatures in severe environments, so interfacial phases which also have good oxidation and corrosion resistance will be required to minimize slow crack growth.

(a) Operated for the U.S. Department of Energy by Battelle Memorial Institute under Contract DE-AC06-76RLO 1830.

The optimum whisker coating or interfacial phase is not presently known for Si_3N_4 composites reinforced with SiC whiskers. The initial goal of this work is to evaluate the interfacial phases and structures in composites reinforced with uncoated and coated whiskers and relate them to the whisker coatings, and to the fabrication parameters and mechanical properties of the composites. A future goal of this work is to gain sufficient understanding of these variables so as to be able to optimize the mechanical properties of these materials in ambient and severe environments.

EXPERIMENTAL

HPSN composites reinforced with 10 volume % Nicalon™ SiC fibers or Tateho SiC whiskers were fabricated at Pacific Northwest Laboratory by hot-pressing at 1750°C in a nitrogen atmosphere. Five weight % MgO was added to both systems as a sintering aid. RBSN composites reinforced with uncoated, C-coated, BN-coated, or oxidized VLS SiC whiskers were fabricated by F. Gac at Los Alamos National Laboratory by hot pressing silicon powder containing 5 or 10 volume % whiskers at 1300°C in a nitrogen atmosphere. Four weight % Fe_2O_3 was added to the specimens to aid the reaction sintering process. An additional sample containing 5 volume % uncoated whiskers was fabricated without a sintering aid.

Specimens for TEM analysis were sectioned from the composites with a low-speed diamond saw and mechanically polished to a thickness of ~50 μm. Circular discs 3 mm in diameter were core drilled from the polished sections, mounted on Mo washers, and ion milled with 6 kV Ar ions at an angle of incidence of 15° until perforation was achieved. A light carbon film was evaporated on the fiber-reinforced HPSN thin sections to prevent charging in the electron microscope. Carbon coating was not required on the whisker reinforced composite thin sections; the conductivity of the whiskers was adequate to prevent charging in the electron microscope. TEM analysis was performed with a Philips EM400T electron microscope equipped with a Kevex Ultra-Thin Window x-ray detector and a Tracor-Northern multichannel analyzer. Auger Electron Spectroscopy (AES) was also performed on the fiber-reinforced HPSN composites fractured in-situ under high vacuum.

RESULTS AND DISCUSSION

Hot-Pressed Si_3N_4 (HPSN) Composites

In the Nicalon™ fiber-reinforced HPSN, the matrix had an average grain size of 50 μm and contained 70 % $\beta-Si_3N_4$ and 30 % $\alpha-Si_3N_4$. EDS analysis of matrix grain boundaries showed that the MgO sintering aid resided in the intergranular glassy phase typically found in sintered Si_3N_4.

The Nicalon™ fibers used to reinforce the HPSN are 15 - 20 μm in diameter and have an average length of 1 mm. In the as-received condition, the fibers have the approximate composition of $SiC_{1.4}O_{1.3}$ and contain very small crystallites of β-SiC with an approximate diameter of 1.7 nm (5). The microstructure of the fibers in the composite is considerably different (6); the grain size is increased to ~50 nm diameter and many of the grains contain small plates of α-SiC. In addition, a large amount of amorphous intergranular material containing Si, Mg, and O is present. Since the fibers did not initially contain Mg, it must have diffused into the fibers from the matrix during fabrication of the composite. In some areas of the fibers, Mg enrichment was sufficient to crystallize the amorphous intergranular material to forsterite, Mg_2SiO_4.

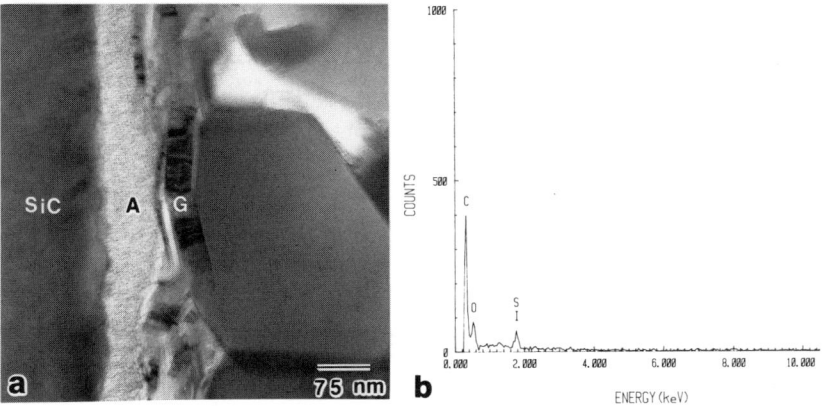

Figure 1. (a) Interfacial microstructure of Nicalon™ fiber-reinforced HPSN and (b) associated EDS spectrum. Amorphous (A) and graphitic (G) carbon layers are present at the fiber-matrix interface.

Two interfacial phases are present between the fibers and the matrix, as shown in Figure 1a. An amorphous carbon layer was always observed adjacent to the fiber and a crystalline carbon layer was always observed adjacent to the matrix. Both phases were continuous along the fiber-matrix interface and each layer ranged in thickness from 30 - 70 nm. Microdiffraction patterns revealed that the crystalline carbon layers were turbostratically stacked, which is typical of graphitic carbon structures. EDS microanalysis revealed that both phases contained a small amount of Si and O (Figure 1b), but no Mg was detected in either phase, although it must have diffused through them since it was found in the fibers. It is believed that the carbon layers formed in situ during fabrication due to the exsolution of excess carbon from the fiber. The graphitic structure of the crystalline carbon layer probably formed from the amorphous carbon by surface diffusion during fabrication of the composite.

The microstructure shown in Figure 1a is consistent with the microchemistry observed by AES analysis of the fracture surface of these composites (7). A carbon-rich layer was detected on the as-received fibers and in the composites on both the fiber and the matrix surfaces of the fiber-matrix interface. This carbon-rich layer is obviously the interfacial phase assemblage shown in Figure 1a, although it is not possible to distinguish between the graphitic and amorphous layers with AES. The presence of carbon-rich layers on both the fiber and the matrix surfaces following fiber pullout suggests that either fracture occurred between the amorphous and graphitic carbon layers or the fracture path alternated between the fiber-carbon and matrix-carbon interfaces.

HPSN composites reinforced with 10 volume % Tateho SiC whiskers were fabricated in the same manner as the fiber-reinforced specimens. The as-received Tateho whiskers were 0.1 - 0.5 μm in diameter and 10 - 40 μm long and were composed of 98 % SiC, of which 95 % was β-SiC and 5% was α-SiC. The remainder of the whiskers was amorphous SiO_2, which was located on the surfaces (8) and in the hollow centers of the whiskers (9). Similar to the fibers, the microstructure of the whiskers in the composite was considerably different from their initial microstructure (10); the β-SiC was almost completely transformed to α-SiC and the amorphous SiO_2 phase in the hollow of the whiskers was enriched with Mg. In some whiskers Mg enrichment was sufficient to crystallize the amorphous phase to forsterite, Mg_2SiO_4. Since the as-

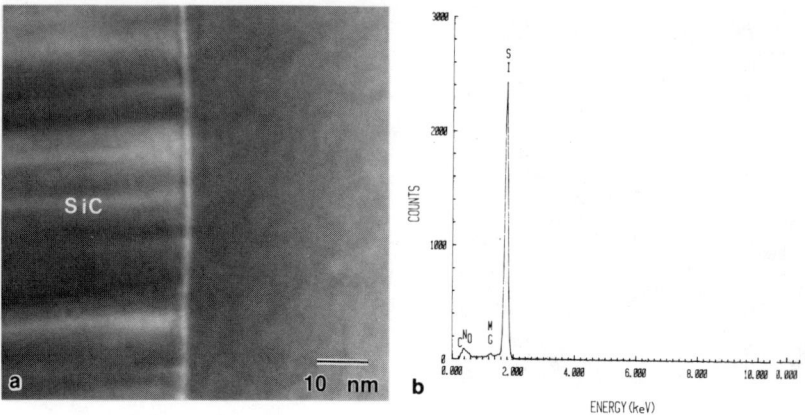

Figure 2. (a) Interfacial microstructure of Tateho SiC whisker-reinforced HPSN and (b) EDS spectrum of region containing interfacial phase. A thin layer of amorphous material is present at the whisker-matrix interface.

received whiskers did not contain any Mg it must have diffused from the matrix into the whiskers during fabrication of the composite.

The interface between a Tateho whisker and the HPSN matrix is shown in approximately edge-on orientation in Figure 2a. The alternating contrast in the SiC whisker is due to different polytypes of α-SiC. An interfacial phase ~1 nm wide is observed at the whisker-matrix interface. Microdiffraction patterns from the area near the interface did not reveal any reflections other than those attributable to SiC or Si_3N_4 which suggests that the intergranular phase is amorphous. EDS microanalysis of the interfacial region (Figure 2b) performed with a 10-nm probe reveals that Mg and O are present in addition to the expected Si, C and N, suggesting that the interfacial phase may have a composition similar to the glassy intergranular phase found in the HPSN matrix. There appear to be no interfacial reactions between the whiskers and the matrix in these composites. This contrasts with similarly fabricated HPSN composites containing VLS SiC whiskers, in which faceting of {111} SiC whisker planes occurs and is believed to be due to the dissolution of SiC during hot pressing at 1850°C (11). The absence of interfacial reactions with the Tateho whiskers may be due to the SiO_2 coating initially present on their surfaces.

Reaction-Bonded Si_3N_4 (RBSN) Composites

The matrix microstructure of RBSN composites reaction sintered with Fe_2O_3 sintering aid contained mostly submicron-size grains of $\alpha-Si_3N_4$ and a small fraction of $\beta-Si_3N_4$ grains. In composites sintered without the Fe_2O_3 sintering aid the matrix appeared similar, but micron-size areas of unreacted Si were present due to incomplete nitridation. The VLS whiskers (12) used to reinforce the RBSN composites were approximately 4 - 8 μm in diameter and 1 mm in length and were predominantly single-crystal β-SiC. SEM examination of the as-received whiskers showed that the whisker surfaces were smooth, although microfacets 1 - 3 nm in height may be present on the surfaces (11). Large facets, as will be discussed, were never observed on the whisker surfaces. In all RBSN composites, the whiskers remained as single crystal β-SiC after fabrication, but structural defects were observed in the whiskers which

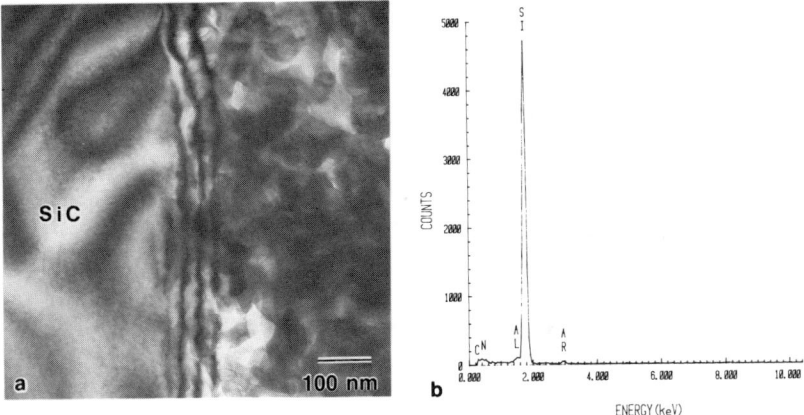

Figure 3. (a) Interfacial microstructure of uncoated VLS SiC whisker-reinforced RBSN fabricated without Fe_2O_3 sintering aid and (b) associated EDS spectrum. The whisker surface does not appear to be degraded.

may have been introduced during fabrication.

The interface between the matrix and whisker in a RBSN composite reinforced with uncoated SiC whiskers reaction sintered without the Fe_2O_3 sintering aid is shown in Figure 3a. The surface of the whisker is smooth and does not appear to be degraded by the reaction sintering process. The whisker also appears to be directly bonded to the RBSN matrix; TEM images of the interface in an edge-on orientation formed with diffusely scattered intensity did not reveal the presence of an amorphous interfacial phase. However, EDS microanalysis (Figure 3b) showed that a trace of Al is present at the interface and may be associated with an interfacial phase. The Ar detected is an artifact due to the ion thinning process. However, the matrix is very heterogeneous due to incomplete nitridation and it is likely that some whisker-matrix interfaces may contain unreacted Si or SiO_2.

The remaining RBSN composites reinforced with uncoated, C-coated or BN-coated SiC whiskers were reaction sintered with 5 weight % additions of Fe_2O_3 sintering aid. Although the sintering aid had the beneficial effect of allowing nitridation of the Si powder to completion, it also reacted with the coated and uncoated SiC whiskers, causing the formation of an amorphous interfacial reaction product and faceting of the whisker surfaces. The whisker-matrix interface of a RBSN composite reinforced with uncoated SiC whiskers is shown in Figure 4a. An amorphous interfacial phase, consisting of Si and a small amount of Fe (Figure 4b), that contains an FeSi particle is present between the RBSN matrix and the SiC whisker. The planes bounding the interfacial phase on the whisker side of the interface are {111} SiC, indicating that they have dissolved preferentially. The extent of dissolution may be quite severe; in other areas of the whisker, facets up to 0.5 μm in height were observed. Apparently the Fe_2O_3 additions, which react with Si to form a liquid iron silicide phase during nitridation (13), react similarly with the SiC whiskers resulting in their dissolution and the formation of facets on the whisker surfaces. The solid solubility of Fe in Si is very small at room temperature and the FeSi particle probably precipitated from the liquid iron silicide on cooling from the fabrication temperature.

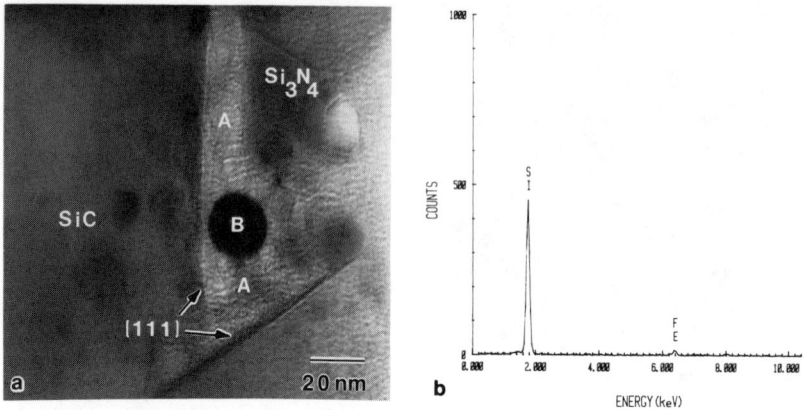

Figure 4. (a) Interfacial microstructure of uncoated VLS SiC whisker-reinforced RBSN fabricated with Fe_2O_3 sintering aid and (b) EDS spectrum of the amorphous interfacial phase. The amorphous interfacial phase (A) contains an FeSi particle (B).

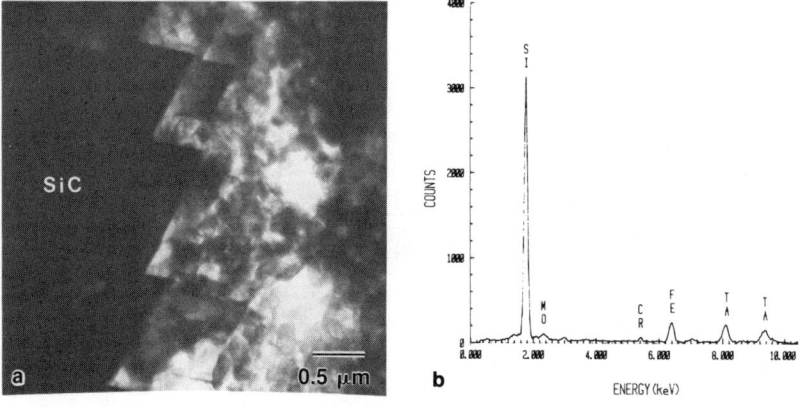

Figure 5. (a) Interfacial microstructure of BN-coated whisker-reinforced RBSN fabricated with Fe_2O_3 sintering aid and (b) EDS spectrum of amorphous interfacial phase. Large facets on {111} SiC planes are present on the whisker surfaces.

The interface between a BN-coated SiC whisker and the RBSN matrix is shown in Figure 5a. Large facets ~0.5 μm in height on {111} SiC planes of the whiskers were also observed in these composites. An amorphous material containing mostly Si with Fe and Cr impurities (Figure 5b) is present at the whisker-matrix interface, suggesting that similar decomposition of the whiskers by liquid iron silicide during nitridation has occurred. The Cr impurity originates from a whisker-growth catalyst used in fabricating the VLS whiskers and the Ta and Mo impurities are artifacts of the ion thinning process. Similar

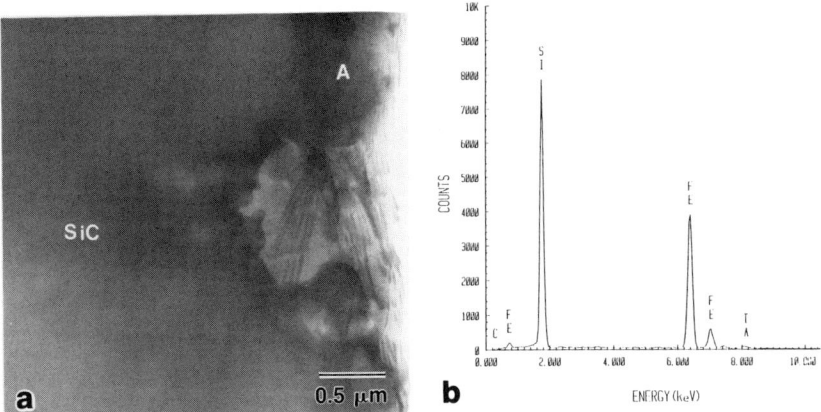

Figure 6. (a) Whisker surface in SiO_2-coated whisker-reinforced RBSN fabricated with Fe_2O_3 sintering aid and (b) EDS spectrum of material at bottom of pit. The SiO_2 coating (A) and whisker were pitted in the composite.

decomposition and faceting of C-coated whiskers in RBSN composites was also observed.

The SiO_2-coated whiskers in the remaining RBSN composite were produced by oxidizing the whiskers in air at 1000°C for 1 hr. Although faceting of the whiskers in these composites did not occur, the whisker surfaces were pitted (Figure 6a). EDS microanalysis of the region near the bottom of the pit (Figure 6b) shows that a considerable amount of Fe is present, suggesting that the pits were formed by dissolution of the SiC by liquid iron silicide during nitridation. The absence of whisker faceting may be related to the SiO_2 coating on the surfaces of the whiskers. The whisker surfaces, in both pitted and unpitted areas, appeared to be bonded to the matrix by the SiO_2 coating.

Fracture Properties

The flexure strength and fracture toughness values for HPSN composites reinforced with Nicalon™ fibers or Tateho SiC whiskers and SiC whisker-reinforced RBSN are shown in Table I. In general, the strength and toughness of both types of composites are lower than those of their respective unreinforced matrices. The toughness of Nicalon™ fiber-reinforced HPSN is lower than that of unreinforced HPSN for two reasons. First, the unreinforced HPSN has a β/α Si_3N_4 ratio of 4, whereas the HPSN composite possessed a lower β/α ratio of 2.3 (7), so the toughness of the composite matrix would be expected to be lower than that of the unreinforced material. Secondly, the fracture properties of the fibers have been degraded since the grain size of the SiC crystals has increased and amorphous and crystalline intergranular phases have developed in the fibers during fabrication. The fracture toughness of the Tateho SiC whisker reinforced HPSN composite was greater than that of unreinforced HPSN. Apparently the interfacial bonding by the amorphous phase was weak enough to permit whisker debonding during crack propagation sufficient to toughen the composite.

Table I. Fracture Properties

Material	σ_f (MPa)	K_{IC} (MPa/m$^{1/2}$)
HPSN	700	5.8
HPSN/10 vol.% Nicalon™ Fibers	250	3.1
HPSN/10 vol.% Tateho Whiskers	307	7.8
RBSN	221	2.0
RBSN/5 vol.% SiO_2-Coated Whiskers	138	3.5
RBSN/10 vol.% C-Coated Whiskers	138	2.0
RBSN/10 vol.% As-Beneficiated Whiskers	124	2.0
RBSN/5 vol.% As-Beneficiated Whiskers; No Fe_2O_3	104	1.5
RBSN/10 vol.% BN-Coated Whiskers	83	0.5

The RBSN composites reinforced with SiO_2-coated whiskers exhibited the highest strength and toughness, and are followed in order of decreasing strength and toughness by RBSN reinforced with C-coated whiskers, RBSN reinforced with uncoated whiskers, RBSN reinforced with uncoated whiskers reaction sintered without the Fe_2O_3 sintering aid, and RBSN reinforced with BN-coated whiskers. Since the composite containing the SiO_2-coated whiskers had the least severe whisker degradation (i.e., pitting) and possessed the highest strength and fracture toughness, it may be concluded that the whisker facets are the most significant strength- and toughness-limiting defect. The whiskers were not faceted in the RBSN composite reaction sintered without the Fe_2O_3 sintering aid, but this composite contained large areas of unreacted Si, which likely acted as the most significant strength- and toughness-limiting defect.

CONCLUSIONS

Correlations have been made between the fracture properties and interfacial phases and structures of HPSN composites reinforced with Nicalon™ SiC fibers or Tateho SiC whiskers and RBSN composites reinforced with VLS SiC whiskers. Results indicate that processing parameters may greatly influence the microstructures of the fibers, whiskers, matrices and their respective interfaces. In the fiber-reinforced HPSN composite, an amorphous-graphitic carbon interfacial structure which developed during hot pressing allowed a small amount fiber pullout to occur during fracture. However, the fracture properties of the composite were inferior to those of unreinforced HPSN due to degradation of the fibers' fracture properties caused by SiC grain growth and the development of an intergranular phase in the fibers during hot pressing. In addition, the MgO sintering aid diffused from the matrix into the fiber during hot pressing which further degraded the properties of the fibers and the matrix. In the Tateho SiC whisker-reinforced HPSN composites, an SiO_2 layer initially present on the whisker surfaces may have prevented facetting of the whisker surfaces during hot pressing and also resulted in the formation of a thin amorphous interfacial phase between the whisker and the matrix. The interfacial bonding provided by this amorphous phase was weak enough to permit whisker debonding during fracture sufficient to toughen the composite. The SiO_2 amorphous core of the whiskers was enriched by the MgO sintering aid from the matrix. In some whiskers enrichment was sufficient to crystallize the amorphous SiO_2 to Mg_2SiO_4. The resulting depletion of MgO in the matrix probably reduced the amount of $\beta-Si_3N_4$ and decreased the fracture properties of the matrix.

The Fe_2O_3 sintering aid added to enhance nitridation of the RBSN composites was found to cause dissolution of VLS SiC whisker surfaces despite protective C, BN, or SiO_2 coatings. This dissolution caused faceting and pitting of the whisker surfaces and the formation of an amorphous reaction product at the whisker-matrix interface. Consequently, the fracture properties of these composites were, for the most part, inferior to those of unreinforced RBSN.

ACKNOWLEDGMENTS

The authors would like to thank Dr. M. Ruhle for useful discussions. This work was supported by the U.S. Department of Energy, Fossil Energy Materials Program, under contract DE-AC06-76RLO 1830.

REFERENCES

1. A.G. Evans and K.T. Faber, J. Amer. Ceram. Soc. 67 (4), 255-260 (1984).

2. K.T. Faber and A.G. Evans, Acta Metall. 31 (4), 565-576 (1983).

3. D.B. Marshall, B.N. Cox and A.G. Evans, Acta Metall. 33 (11), 2013-2021 (1985).

4. A.G. Evans and R.M. McMeeking, Acta Metall. 34 (12), 2435-2441 (1986).

5. G. Simon and A.R. Bunsell, J. Mater. Sci. 19, 3658 (1984).

6. R.H. Jones, C.H. Henager, Jr., C.H. Schilling, L.H. Schoenlein, W.J. Weber and F. Gac, to be published in proceedings of the 12th Annual Conference on Composites and Advanced Ceramics, Cocoa Beach, Fla., The American Ceramic Society (1988).

7. J.L. Bates, C.W. Griffin and W.J. Weber, "Improved Ceramic Composites Through Controlled Fiber-Matrix Interface", in proceedings of the Fossil Energy Materials Conference, ORNL/FMP-87/4 (1987).

8. G. Krug and S.C. Danforth, Ceram. Eng. Sci. Proc. 8 (7-8), 712-716 (1987).

9. M. Ruhle, private communication.

10. R.H. Jones, L.H. Schoenlein, C.H. Henager, Jr. and C.H. Schilling, "Improved Ceramic Composites Through Controlled Fiber-Matrix Interfaces", to be published in the Annual Progress Report for the DOE AR § TD Fossil Energy Materials Program for the period ending April 30, 1988.

11. S.R. Nutt and D.S. Phillips, in *Interfaces in Metal-Matrix Composites*, edited by A. Dingra and S. Fischman (ASM/AIME Proc., New Orleans, LA 1986) pp. 111-120.

12. J.V. Milewski, F.D. Gac, J.J. Petrovic and S.R. Skaggs, J. Mater. Sci. 20 1160-1166 (1985).

13. D.R. Messier and P. Wong, J. Amer. Ceram. Soc. 56 (9), 480-485 (1983).

THERMAL OXIDATION OF Al_2O_3–SiC WHISKER COMPOSITES: MECHANISMS AND KINETICS

F. LIN, T. MARIEB, A. MORRONE, and S. NUTT
Brown University, Div. of Engineering, Box D, Providence, R.I. 02912

Abstract

It has been shown that SiC whisker-reinforced Al_2O_3 is susceptible to thermal oxidation because of a series of reactions in which SiC is oxidized and then reacts with Al_2O_3 to form a whisker-free layer of mullite and silica glass. In this paper, the kinetics of the reaction are analyzed as a function of the composition and whisker content of the composite. The mechanism of the reaction is described in terms of the evolution of phases within the reaction layer on the basis of transmission electron microscope observations and concentration profiles determined by electron microprobe analyses. The reaction is initiated by oxidation of SiC, forming silica glass and graphitic carbon. The silica glass penetrates the Al_2O_3 matrix along grain boundaries, and reacts to form mullite. As the reaction proceeds, Al_2O_3 is consumed, leaving residual intragranular Al_2O_3 islands, and a population of transgranular microcracks in the mullite. It has also been shown that Al_2O_3-SiC composites are susceptible to crack nucleation and growth under cyclic compressive loads. The presence of the mullite reaction layer on the surface of the composite reduces its resistance to crack initiation under these conditions. The threshold for crack initiation in cyclic compression is measured as a function of reaction layer thickness.

Introduction

Whisker reinforcement has been shown to improve the mechanical behavior of ceramics, hence increasing their potential for structural applications in severe environments. In particular, SiC whisker-reinforced Al_2O_3 has a significantly higher room temperature fracture toughness [1], as well as better resistance to creep [2] and thermal shock [3] than unreinforced alumina. However, when SiC whisker-reinforced Al_2O_3 is exposed to air at elevated temperatures the whiskers react with oxygen resulting in the formation of a porous whisker-free surface layer [2,4]. The presence of this scale is detrimental to the service life of the composite in high-temperature structural applications. Consequently, in addition to the extensive research that has been directed successfully toward improving the mechanical performance of ceramics, the thermochemical integrity of reinforced ceramics must be addressed. This investigation was undertaken to study the the reactions taking place in Al_2O_3–SiC whisker composites exposed to air at high temperatures.

Few reports are available on the thermochemical integrity of structural ceramic composites with oxidizable reinforcements. Borom et al. [4] studied high-temperature oxidation reactions in different combinations of ceramic matrices and reinforcements. They found that the reactions produced a surface layer resulting from the oxidation of the reinforcement and the subsequent combination of this oxide with the matrix. The thickening rate of that layer can be either linear or parabolic with time, depending on the final reaction products. For alumina matrix composites with 20 v/o SiC whiskers (SiC_W) and 20 v/o SiC particles, Borom et al. found the thickening rate to be parabolic, with an activation energy of 502 kJ/mol. They measured an activation energy for the oxidation of pure SiC of 420 kJ/mol, comparable to that of the composite, although the rate of reaction for pure SiC was 10 to 20 times lower. This comparison points to the importance of reinforcement-matrix interactions during the reaction. Porter and Chokshi [2] also measured parabolic thickening rates during creep experiments on $Al_2O_3-18\text{v/o SiC}_W$. They reported an activation energy of 350 kJ/mol. The difference in activation energies may result from the presence of impurities, and the fact that in creep experiments the reaction takes place under stress. The investigation presented here studies the kinetics and mechanisms of the thermal oxidation reaction scale formed in SiC whisker–reinforced Al_2O_3. The kinetics are determined as a function of composition and the volume fraction of SiC_W. The evolution of phases within the scale thickness is determined from TEM observations and electron microprobe analysis. Furthermore, the effect of the scale thickness on the material resistance to crack initiation under cyclic compressive loads is discussed.

Materials and Procedures

The materials used in this investigation were SiC whisker–reinforced Al_2O_3 obtained from three sources. Samples with 15 v/o and 20 v/o SiC_W were provided by Dr. J. Porter* (RISC composites), and materials containing 1 v/o, 4 v/o, 10 v/o, 30 v/o, and 50 v/o SiC_W were provided by Dr. P. Becher† (ORNL composites). The processing and characterization of these materials are described in [2] and [5], respectively. Finally, 33 v/o SiC_W –Al_2O_3 was purchased from the Greenleaf Corp. The latter, commercially available composites, were found to be quite pure, with few pockets of amorphous phase confined to interface–grain boundary junctions. Energy dispersive x-ray analysis in the TEM showed that the amorphous phase was silica–rich, with traces of Ca, K, Fe, Mo, and Y.

The specimens were cut to dimensions at least 6mm × 3mm × 2mm, then polished, and finally heated in an air furnace at a rate of 18°C per minute to temperatures between 1300°C and 1600°C. The temperature was maintained for predetermined times, at the end of which the specimens were furace cooled. During the heat treatment, an apparent reaction layer formed on the surface of the samples. The depth of that scale was measured in cross-section by light microscopy. Thin foils were prepared from different depths of the scale, as well as cross-sectional thin foils, for analytical electron microscopy observations.

For the crack initiation experiments, single–edge notched bars were machined from plates of the 33 v/o SiC_W material. The notches had a 60° profile and a root radius of 100μm. After heat treating the samples as described above, the scale was polished off from one of the surfaces normal to the notch, thus exposing unreacted material. This method allowed observations of the crack growth on reacted and unreacted material under the same experimental conditions. Cyclic compressive loads were applied along the axis of the bars, normal to the plane of the notch. The cyclic load was sinusoidal, with a frequency of 10 Hz, and the ratio of the minimum to the maximum load was 10. The load amplitude was incremented by 3% every 5000 cycles until the nucleation of a fatigue crack was observed. Then the load was kept constant until the crack propagated well into the unreacted core of the specimen.

Results

1. Kinetics

The scale formed on the surface of the $SiC_W-Al_2O_3$ samples during heat treating in air was easily observed by light microscopy of scale cross–sections. As shown in Fig. 1, a dark boundary layer separates the porous whisker–free scale from the unreacted core. The scale thickness was measured from the free surface to the center of the dark boundary

---

* Rockwell International Science Center, Thousand Oaks, Ca. 91360
† Oak Ridge National Laboratory, Oak Ridge, Tn 37831

layer. The thickness as a function of time obtained at different temperatures for the 33 V/o material and for the RISC composites is shown in Fig. 2.

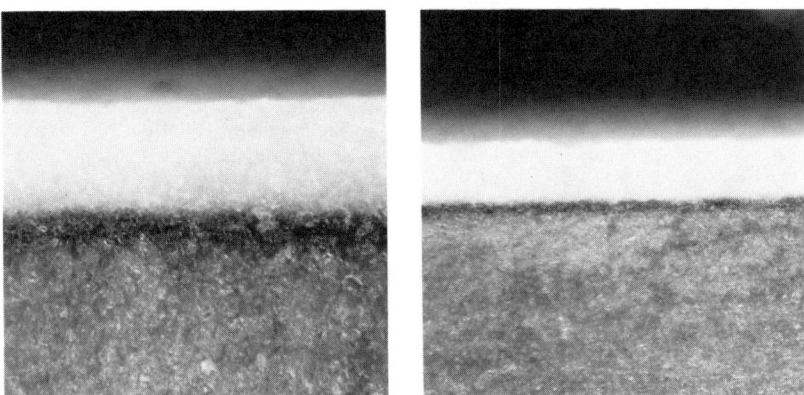

Fig. 1: Optical micrograph of scale developed after 60h. at 1500°C in 10 V/o (left) and 30 V/o (right) SiC$_W$ composites (200×)

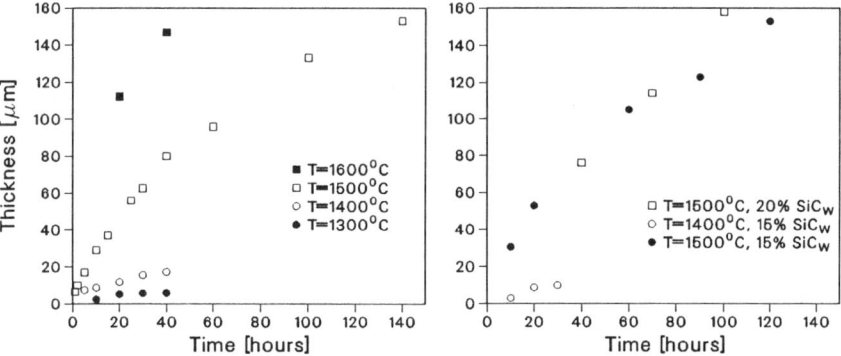

Fig. 2: Scale thickness as a function of time for 33 V/o (left), and RISC composites (right).

Figure 2 suggests that the reaction rate kinetics is parabolic. In fact, plots of the scale thickness squared as a function of time are linear, as shown, for example, in Fig. 3 for 33 V/o material at 1500°C . Hence, the thickness x can be expressed as a function of time by:

$$X^2 = K_P\, t \qquad (1)$$

where the constant of proportionality K_P is the reaction rate. This representation implies that the reaction rate is controlled by a diffusion step. Therefore, K_P is related to the diffussivity, D, of the oxidizing species in the reacted media by

$$K_P = \frac{2DN_0}{n}, \qquad (2)$$

where N_0 is the concentration of oxidant on the free surface of the sample, and n is the density of oxidant incorporated to the oxidation product. Since diffusion is thermally activated, the reaction rate can be expressed by an Arrhenius equation:

$$K_P = A \exp(-E/RT), \qquad (3)$$

where A is a constant that includes a frequency factor, and R is the universal gas constant. On the basis of equation (3), the activation energy for the reaction is readily calculated from an Arrhenius plot of K_P as a function of the inverse of temperature, as shown in Fig. 4. The activation energy values obtained with that procedure in the temperature range 1300–1600°C are 530 kJ/mol for the 33 v/o material, and 919 kJ/mol for the RISC composites with 15 v/o SiC_W.

The ORNL composites were heat treated at 1500°C to study the effect of SiC_W content on the reaction kinetics. The results of that set of experiments are shown in Fig. 5, along with the data for 33 v/o for comparison. Not included in the figure is data from the materials with 1 v/o and 4 v/o SiC_W, since the samples reacted completely after 30 hours at 1500°C. It is apparent from Fig. 5 that the reaction rate is strongly dependent on two factors. One factor is the SiC_W content of the composite; faster reaction rates are attained with decreasing SiC_W content. The other factor is the processing, as shown by the difference in reaction rates of materials from different sources but with nearly the same SiC_W content, such as the ORNL 30 v/o composite and the 33 v/o composite from Greenleaf.

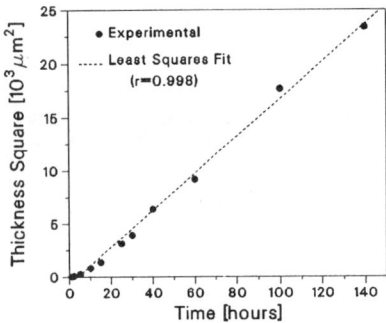 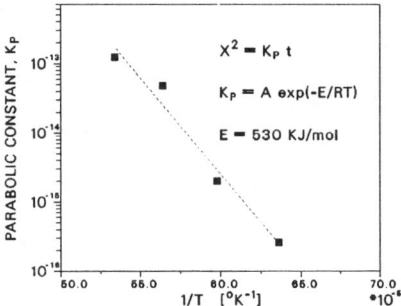

Fig. 3: Scale thickness squared vs. time for 33 v/o SiC_W at 1500°C.

Fig. 4: Arrhenius plot for 33 v/o composite.

2. Mechanisms

The structure of the scale fomed on the surface of the composites was analysed with an X-ray diffractometer. The analysis showed that the crystalline phases present were mostly alumina and mullite ($3(Al_2O_3)2(SiO_2)$). Other crystal phases appeared to be present as well, but these could not be identified by X-ray diffraction because the corresponding peaks were too weak and the instrument resolution was insufficient. However, the data clearly indicated that the relative amount of mullite found in the scale increased with the volume fraction of SiC$_W$ previously present in the composite. The 1% SiC$_W$ composite reacted completely after 30 hours at 1500°C, and the crystalline product of the reaction was rich in alumina, but contained no mullite. On the other hand, the scale on the 50% SiC$_W$ resulting from the same treatment was mostly mullite, and contained no alumina.

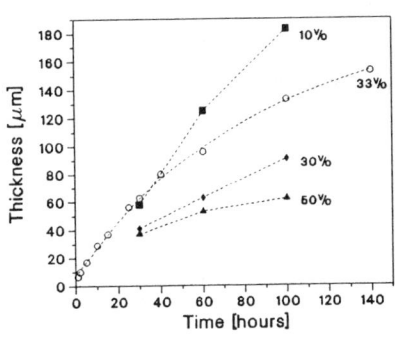

Fig. 5: Effect of SiC$_W$ content on reaction rate of ORNL composites.

Fig. 6: Transgranular cracks in mullite grains.

Analytical electron microscopy (AEM) observations of the 33% material oxidized for 10hrs showed the different features of the scale as a function of depth. Near the outer surface, the scale appeared to be mostly mullite with glass pockets, as shown in Fig. 6. The mullite grains in this region present a population of transgranular cracks. These cracks are open, and hence easily identified as such, in grains that contain a large number of them. In deeper regions, where the cracks are closed and only a few are present they are more difficult to identify, possibly these are other planar defects associated with shear. Deeper regions within the scale revealed the presence of residual intragranular alumina in the mullite grains, as shown in Fig. 7. The alumina was often connected to the boundary of the surrounding grain by a line of defects identified as dislocations that in some cases formed low angle grain boundaries. Pockets of glassy phase that penetrated the grain boundaries were more common in this region than in the previous one. In both cases, EDS showed that the glass was silica with small amounts of Ca. The early stages of the reaction take place at the reaction front represented by the black region shown in Fig. 1. Taken from that region, the TEM micrograph of Fig. 8 shows a whisker whose edge has become very irregular as a result of the oxidation. Surrounding the whisker are the initial products of the oxidation: silica glass and graphitic carbon. The silica glass was seen to penetrate the grain boundaries of neighboring Al_2O_3. In some areas within the black layer, silica had reacted with alumina, and small mullite grains were found to coexist with SiO_2 glass, Al_2O_3, and SiC$_W$, as shown in Fig. 8. The graphitic carbon was identified in AEM by electron diffraction, and this finding was confirmed by high resolution electron microscopy (HREM) and electron microprobe analysis. The HREM micrograph of Fig. 9 shows the typical graphitic structure around a whisker edge, with a line spacing of 3.4 angstoms corresponding to the basal planes of graphitized carbon. Also shown in the figure is the SiO_2 glass, Al_2O_3, and a new crystalline phase surrounded by graphite that could not be identified but is presumed to be a precursor of mullite.

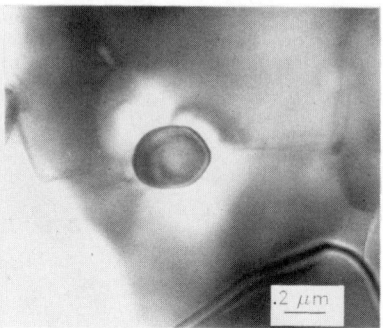

Fig. 7: Intragranular residual Al_2O_3 in a mullite grain.

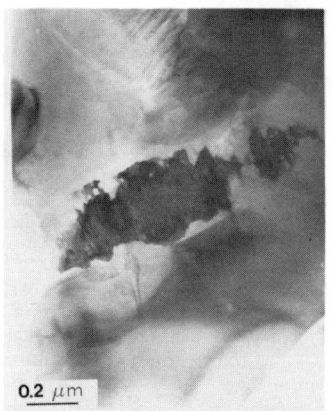 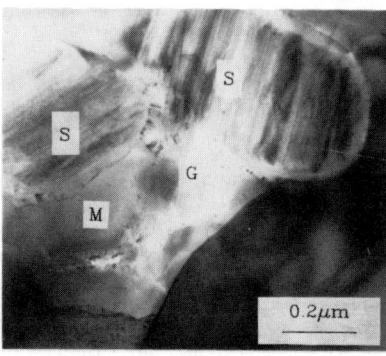

Fig. 8: The oxidation starts at the surface of the whiskers (left). Several phases coexist at the reaction front, (G) silica and graphite, (A) mullite , (S) SiC_W, and a large Al_2O_3 grain.

Concentration profiles of C, Si, O, and Al across the scale were measured by electron microprobe analysis. Two different spot sizes were used, a $10\mu m$ spot with readings taken at $5\mu m$ intervals, and a $1\mu m$ spot with readings taken at every $1\mu m$. The results are shown in Fig. 10 for a specimen treated for 40 hours at $1500°C$. The distance '0' in the plots indicates the onset of the the reaction at the begining of the black layer, with the latter being about $10\mu m$ wide. While the smaller spot gave superior spatial resolution, it also resulted in large oscillations due to the discrete nature and small distances involved in the process. On the other hand, the larger spot size gave a smooth curve but with lower spatial resolution. However, both sets of data show that the carbon concentration increases within the black layer and then decreases rapidly to zero farther in the scale. Also, there is a clear build–up of oxygen throughout the scale, which confirms the nature of the species necessary for the reaction to occur.

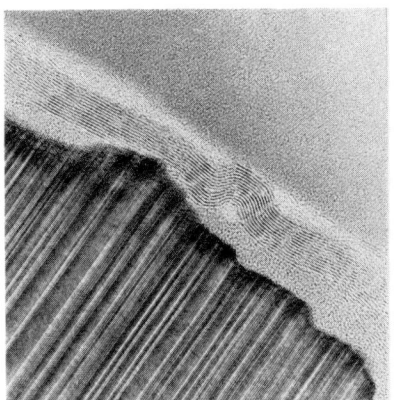 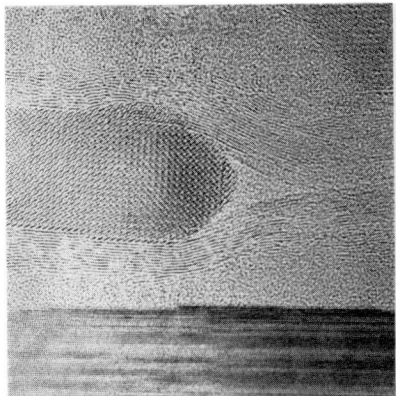

Fig. 9: HREM micrographs of whisker edges surrounded by glass, graphitic carbon, and possibly a crystalline precursor of mullite (right).

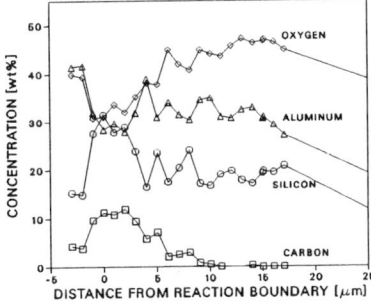 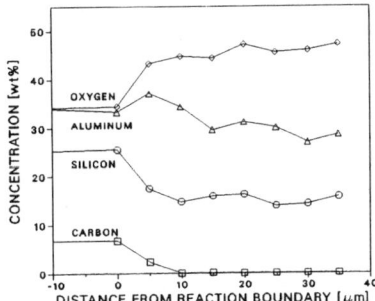

Fig. 10: Concentration profiles obtained with $1\mu m$ (left), and $10\mu m$ (right) electron microprobe spot size (33 V/o SiC_W, 40h. at 1500°C).

3. Crack Initiation

The applied compressive load amplitude necessary to initiate a fatigue crack, as described earlier, was used to calculate the applied stress intensity factor amplitude shown in Fig. 11 as a function of the scale thickness. For the range of thickness shown, there is a decrease of 18% in the loads necessary to initiate a crack, thus the presence of the scale increases the susceptibility of the material to crack initiation. Once a crack starts, it is not confined to the scale. Upon continued loading, the cracks that initiated in the scale grew longer and wider, and penetrated the unreacted material. Thus, the surface reaction can be expected to reduce the material lifetime when the service requirements involve both high temperature and variable amplitude compreessive loads.

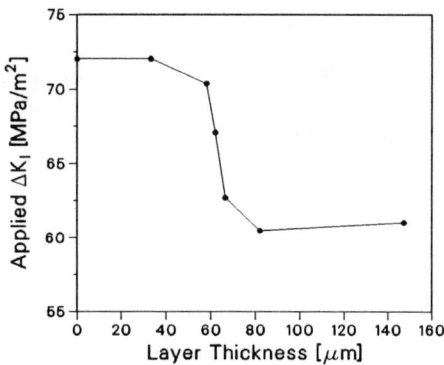

Fig. 11: Decrease in applied load amplitude for crack initiation vs. scale thickness.

Discussion

The final product of the reaction taking place in SiC_W–Al_2O_3 composites exposed to air at high temperatures depends on the volume fraction of the reinforcement. Mass-balance considerations predict that in composites with up to 24.9 v/o SiC_W the product will be mullite and alumina, while in composites with higher volume fraction of whiskers the product will be mullite and silica. Results of X-ray diffraction scans of reacted composites with SiC_W reinforcement in the range 1-50 v/o are in agreement with this prediction. Also in agreement with that prediction is the AEM observation of residual intergranular silica glass in the mullite–rich regions of the scale in 33 v/o composite. However, the reaction does not occur homogeneouly throughout the material, but rather starts at the surface and produces a scale that thickens with time. Our investigation of the mechanisms of this process shows that the reaction proceeds in the following way: I) The SiC whiskers oxidize forming silica glass and graphitic carbon. II) The silica glass penetrates the Al_2O_3 grain boundaries and reacts to form mullite, III) Carbon oxidizes and escapes to the surface as a gaseous product. The latter is supported by the concentration profiles which show a sharp decrease for carbon beyond the vicinity of the black boundary layer.

The kinetics associated with the process of scale thickening was found to be parabolic, hence a diffusion step is rate controlling. Oxygen must be incorporated for the reaction to take place, and this is confirmed by the electron microprobe observations that oxygen concentration is larger throughout the scale than in the unreacted material. Oxygen diffusion through the oxide layer is then the most likely rate controlling step. However, more experimental and analytical information is needed to determine which of the different phases present in the scale controls oxygen diffusion for each particular volume fraction of SiC_W reinforcement.

The activation energy measured for the oxidation of 33 v/o composite of 530 kJ/mol agrees with the value reported by Borom et al [4] of 502 kJ/mol for 20 v/o SiC_W. They also reported a parabolic rate constant of $3.9 \times 10^{-15} m^2/sec$, very close to the $2.0 \times 10^{-15} m^2/sec$ presently measured. That small difference between parabolic rate constants is consistent with the difference in SiC_W reinforcement, in the light of the results shown in Fig. 5. The activation volume of 919 kJ/mol measured for 15 v/o RISC composite is extremely high compared with the above values or with Porter's 350kJ/mol [2] for 18 v/o SiC_W. The reason for such a high value is believed to be the lack of experimental data to obtain a minimum accuracy with the Arrhenius plot. In any case, there are no predicted values to compare to the activation energies measured in order to identify a controlling diffusion mechanism. On the basis of Fig. 5, the reaction appears to be slower when the product is richer in silica, i.e. for the composites with higher SiC_W content, which points to silica as the phase controlling oxygen diffusion. Since the formation of silica has residual graphitic carbon as a by-product, the data in that figure agrees with the suggestion of Luthra [6] and Porter [2] that a concurrent rate controlling mechanisms may be the escape of CO gas. In this case, the residual carbon in the reaction boundary layer will absorb incoming oxygen and prevent it from oxidazing more SiC whiskers. These speculations point to the need for more analytical and experimental studies to understand the oxidation raction in composites.

Summary

When SiC_W-reinforced Al_2O_3 is exposed to air at high temperatures, the whiskers oxidize forming silica glass and graphitic carbon. The silica glass penetrates the Al_2O_3 grain boundaries and reacts to form mullite. The result is a porous whisker-free scale which composition depends on the initial volume fraction of whiskers in the composite. The rate of the scale thickening is parabolic and it also depends largely on the initial SiC whisker content. The scale is not homogeneous, it is rich in silica, carbon, and alumina in the vicinity of the raection front, but very rich in mullite toward the surface, with residual intragranular Al_2O_3 and microcracks in the mullite. The presence of the mullite reaction layer on the composite reduces its resistance to crack initiation under cyclic compressive loads.

Acknowledgements

This research was supported by the Brown University MRG under contract No. DMR-8714665. The authers are greatful to Dr. P. Becher (ORNL) and Dr. J. Porter (RISC) for providing composite samples, and to A. Davis (CAMECA Instruments, Stamford, Ct) for the use of the Camebax SX 50 Electron Microprobe.

References

1. A.A. Morrone, S.R. Nutt, and S. Suresh, accepted for publication J. Mater. Sci.

August, 1987 (unpublished).
2. J.R. Porter and A.H. Chokshi, accepted for publication J. Am. Ceram. Soc., 1987 (unpublished).
3. T.N. Tiegs and P.F. Becher, Comm. Am. Ceram. Soc., **70**, C-109 (1987).
4. M.P. Borom, M.K. Brun, and L.E. Szala, Ceram. Eng. Sci. Proc., **8**, 654 (1987).
5. P.F. Becher and G.C. wei, J. Am. Ceram. Soc., **67**, C-267 (1984).
6. K.L. Luthra, Ceram. Eng. Sci. Proc., **8**, 649 (1987).

THE INFLUENCE OF HEAT TREATMENT UPON FIBER PULL-OUT IN A CERAMIC COMPOSITE

M.D. THOULESS*, O. SBAIZERO*, E. BISCHOFF* and E.Y. LUH**
* Materials Department, University of California, Santa Barbara, CA 93106
**Department of Materials Science and Mineral Engineering, University of California, Berkeley, CA 94720.

Abstract

The toughness of ceramic-matrix composites is strongly influenced by fiber pull-out. The extent of the pull-out depends upon the properties of the fiber and the fiber/matrix interface. Samples of a SiC/LAS composite were subjected to different heat treatments in order to systematically vary these properties. The predicted distribution of the fiber pull-out lengths was calculated by combining a shear lag analysis with Weibull statistics for the fiber strengths. Comparison of the analysis with experiments and microstructural observations contribute to an understanding of the role of the fiber/matrix interface upon the mechanical properties.

Introduction

Weak interfacial bonding between the fibers and matrix appears to be a prerequisite for high toughnesses in fiber-reinforced ceramic-matrix composites.[1-3] The weak bonding ensures that fiber failure occurs away from the crack plane so that even broken fibers can contribute to the toughness by sliding against the matrix and resisting crack opening. In one particular system consisting of SiC fibers in a glass-ceramic matrix (a SiC/LAS composite*) the interfacial properties can be varied by heat treating in air at 800 °C. This system was used in the present study to investigate the influence of the interface on mechanical properties. Previous investigations have revealed that even limited exposure to air at elevated temperatures decreases the utility of such composites. The failure mode changes from one in which there is a large strain-to-failure (multiple matrix cracking) to brittle fracture governed by a single dominant flaw. Transmission electron microscopy (TEM) was used to examine the change in the interfacial microstructure with heat treatment.[4] The fibers in the as-received composites are often at least partially debonded from the matrix. This debonding occurs because of the thermal expansion mismatch between the fiber and matrix. Furthermore, the surfaces of the fiber are covered by a layer of carbon approximately 150 nm thick (Fig. 1).[4,5] These two aspects are responsible for the low interfacial shear stress of $\tau \approx 2$ MPa which is measured in the as-received composite.[6] After even relatively short exposure times to air at 800 °C, the carbon layer disappears and is replaced by silica (Fig. 2).[4] No substantial changes occur on further heat treatment; the SiO_2 layer thickens and the interfacial gap decreases. These observed changes can be correlated to the extent of the pull-out and associated mechanical properties of the composite.

---

* United Technologies Research Center, East Hartford, CT.

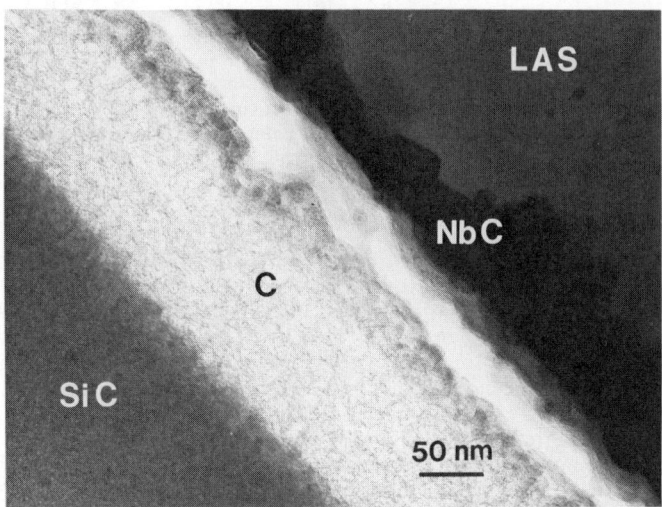

Figure 1 The fiber/matrix interface in the as-received composite. The surfaces of the fiber are covered by a layer of carbon. The debonding between the fiber and matrix results from a thermal expansion mismatch.

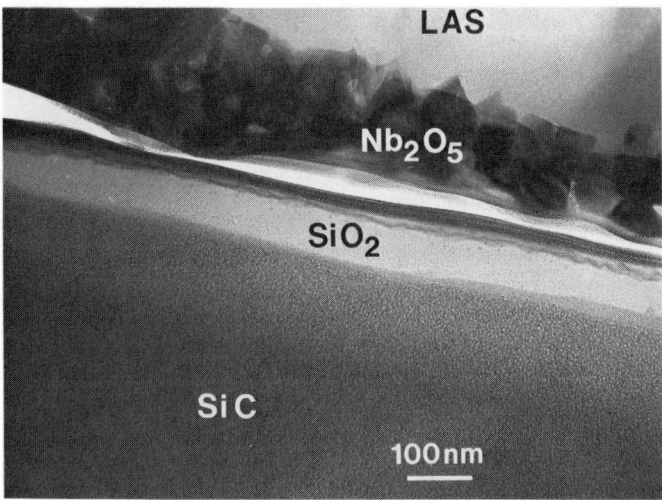

Figure 2 The fiber/matrix interface after heat treatment at 800 °C in air for 4 hours. The carbon layer has been replaced by SiO_2.

THEORY

Following the analysis of Marshall and Evans,[1] it is assumed that the crack opening is resisted by the sliding of fibers against the matrix which occurs at a constant interfacial shear stress, τ. There is therefore a linear distribution of stress in the fiber, decaying from a maximum in the crack plane. An assumption of a unique fiber strength then predicts that fiber failure occurs at the point of maximum stress, *i.e.* in the middle of the crack plane, and no pull-out would be observed. However, it is recognized that there will be a random distribution of flaws in the fibers so that the strengths may be described by Weibull statistics. The probability, $\delta\phi$, that an element of length δz will fail when the stress is less than or equal to σ is given by

$$\delta\phi = \frac{2\pi R}{A_o}\left(\frac{\sigma}{S_o}\right)^m \delta z \tag{1}$$

where R is the fiber radius, m and S_o are shape and scale parameters, and A_o, which has units of area is introduced only for dimensional considerations. A_o is usually equated to unity, but it must be remembered that the choice will affect the value of S_o in such a way that $A_o S_o^m$ remains constant. The analysis pertinent for describing various characteristics of the fiber failure process has been presented elsewhere.[3,7] Only a few salient results will be given in this paper. The mean pull-out length of the fibers can be derived as

$$\bar{h} = \frac{A_o}{4\pi R}\left(\frac{S_o}{\Sigma}\right)^m \Gamma\left\{\frac{m+1}{m+2}\right\} \tag{2a}$$

and the cumulative pull-out length distribution is in the form of an integral:[7]

$$\xi\{h\} = 1 - \int_{\zeta^{m+1}}^{\infty}\left(1 - \zeta\,\beta^{-1/(m+1)}\right)^m \exp\{-\beta\}\,d\beta \tag{2b}$$

where $\zeta = h/\lambda\Sigma$ and Σ, which is a measure of the mean tractions that the fibers exert upon the crack surfaces, is given by

$$\Sigma = \left[\frac{A_o S_o^m \tau(m+1)}{2\pi R^2}\right]^{1/(m+1)}$$

Equation 2b can be integrated numerically and the results are illustrated in Fig. 3. The distribution becomes narrower as m increases; the limit of $m = \infty$ corresponds to the case where the fibers have a unique strength S_o, and there is consequently no pull-out.

EXPERIMENTS

To determine the statistical parameters m and S_0, single fibers with a gauge length of 10 mm were tested in uniaxial tension. Tests were conducted on Nicalon* fibers that had been heat treated in air at 800 °C for 4, 8 and 16 hours. The strength distributions were then used to determine the statistical parameters m and S_0 (Table I). In conjunction with these tests, the fracture toughness, K_f, of the fibers was determined by examining the fracture surfaces and measuring the critical flaw sizes and mirror dimensions.[7] It is apparent from Table I that m, S_0 and K_f all decrease systematically with heat treatment.

Finally, beams made of the composite were heat treated in air at 800 °C for 4, 8, 16 and 100 hours and then tested in four-point bending. The fracture surfaces were examined in a scanning electron microscope (SEM) (Fig. 4) and the pull-out lengths were measured. Initially, these were done by looking at the specimens from one edge and directly measuring all the fibers. The resultant distribution is shown in Fig. 5a. However, it was later appreciated that this approach resulted in the fibers with short lengths being ignored. The technique was therefore varied so that the lengths were measured using one fiber as a reference, and then tilting the specimen to measure the actual length of the reference fiber. A typical distribution is shown in Fig. 5b.

DISCUSSION

The results were analyzed by assuming that the values of m and S_0 determined from the single-fiber tests were appropriate for the fibers in the matrix. The mean value of the pull-out lengths (Eqn. 2a) could then be used to determine the appropriate value of τ and the predicted cumulative distribution compared with Eqn. 2b. This approach relies on the assumption that the characteristics of the fibers and the effects of heat treatment do not change when the fibers are surrounded by the matrix. Unfortunately, the difficulty of extracting undamaged fibers from the composite precluded making meaningful measurements of the statistical parameters from fibers incorporated in the composite.

Preliminary analysis of the results suggested that the pull-out distribution may have contributions from two modes of fiber failure. Some fibers may have been subjected to sliding as described by the analysis; others may have failed at the crack tip, and the crack may have passed through them without substantial interaction. An extreme example of this was observed in a sample that was heat treated for 100 hours and the fracture surface was essentially featureless. Such behavior could arise from localized interfacial bonding by the silica layer near the crack plane. To compensate for this phenomenon, the mean pull-out length was computed for fibers with lengths greater than 5 μm and compared to the theoretical predictions. The resultant estimate of τ was then used to predict the cumulative distribution which is compared with experiment in Fig. 6. The appropriate values for τ and f_f, the fraction of fibers that were assumed to fail without pull-out and therefore do not contribute to the toughening process, are shown in Table I. It should be noted that the excessive value of τ for the sample heat treated for 16 hours could easily arise from small errors in m; the results are extremely sensitive to m when it is less than 2.

---

* United Technologies Research Center, East Hartford, CT.

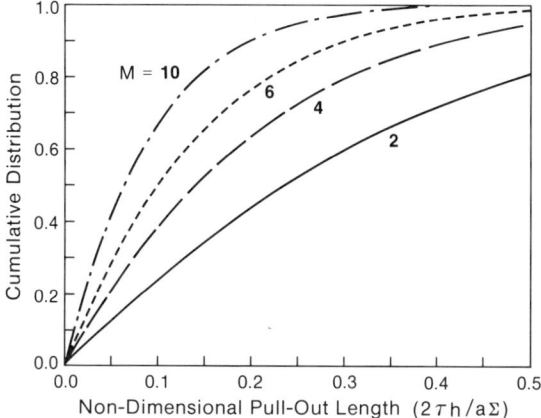

Figure 3   Variation in the predicted cumulative pull-out distribution with the statistical parameter m.

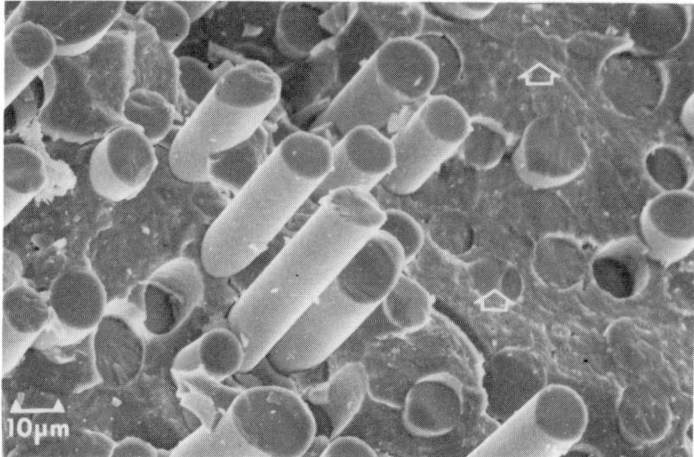

Figure 4   SEM micrograph of a fracture surface from a composite heat treated at 800 °C in air for 16 hours. The arrows indicate fibers that failed without pull-out in the crack plane.

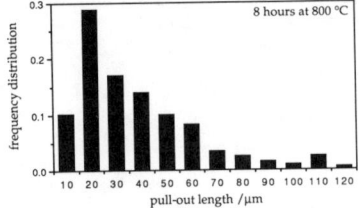

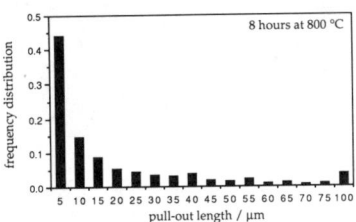

Figure 5 Experimentally observed pull-out distributions.

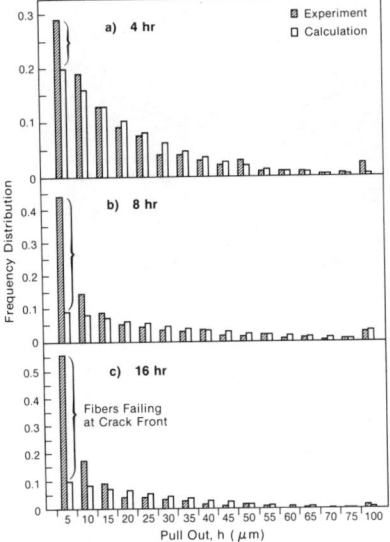

Figure 6 A comparison between the measured and calculated distributions of pull-out lengths. The percentage of fibers failing in the crack plane was obtained by extrapolation.

TABLE 1

Composite properties obtained for single fiber and pull-out tests

Heat-treatment time (hrs. at 800°C)	m	S_0(MPa) ($A_0=1m^2$)	K_f (MPa$\sqrt{m}$)	τ (MPa)	f_f
4	4.2	80	1.03	300	0.09
8	3.0	17	0.79	300	0.35
16	1.7	0.33	0.76	2000	0.46

mean radius of fibers, $R = 7\mu m$

CONCLUSIONS

The effect of heat treatment is to replace the layer of carbon that surrounds the fibers in the as-received composite by silica. The interfacial shear resistance appears to increase by approximately two orders of magnitude upon this change. This is consistent with an increase in the coefficient of friction to the order of unity from the very low levels associated with the carbon layer.[8] Some bonding may occur between the silica layer and the matrix after extended heat treatments, resulting in an increased fraction of fibers that fail at the crack tip and therefore do not contribute to the toughening of the composite.

ACKNOWLEDGEMENTS

The work was supported under ONR contract N⁰· N00014-85-K-0883.

REFERENCES

1. D.B. Marshall and A.G. Evans, *J. Am. Ceram. Soc.*, **68**, (5), 225 (1985).
2. J.J. Brennan and K.M. Prewo, *J. Mater. Sci.*, **17**, (8) 2371 (1982).
3. M.D. Thouless and A.G. Evans, *Acta Metall.*, (in press).
4. E. Bischoff, M. Rühle, O. Sbaizero and A.G. Evans (to be published).
5. J.J. Brennan, *Ceramic Microstructures '86: Role of Interfaces.*, edited by A.G. Evans and J.A. Pask, Plenum, New York.
6. D.B. Marshall, *J. Am. Ceram. Soc.*, **67**, (12), C259 (1984).
7. M.D. Thouless, O. Sbaizero and A.G. Evans (to be published).
8. D.B. Marshall, *Ceramic Microstructures '86: Role of Interfaces.*, edited by A.G. Evans and J.A. Pask, Plenum, New York.

ULTRASONIC PROPAGATION AT CYLINDRICAL METAL-CERAMIC INTERFACES IN COMPOSITES

H. N. G. WADLEY, J. A. SIMMONS, AND E. DRESCHER-KRASICKA
National Bureau of Standards, Gaithersburg, MD

ABSTRACT

The nature of ultrasonic propagation at cylindrical interfaces is being explored as a potential method for inferring the modulus of the interface in metal matrix composites. In many metal matrix composite systems leaky interface waves have been found to propagate along the interface. The velocity of these waves depends sensitively upon the local moduli and provides a potential basis for measuring these quantities. The opportunity exists to exploit the leaky character of these waves for microscopic imaging at the interface.

INTRODUCTION

The interface between matrix and reinforcement is of central importance in composites where it is the dominating factor controlling load transfer and crack resistance. In metal matrix composites opportunities exist for extensive reaction between matrix and reinforcement during processing. For example in Al matrix composites reinforced with SiC, it is not unusual to form a layer of Al_4C_3 at the interface. The extent of these reactions (and thus the ultimate performance of the composite) depends upon composition (in the above example a high matrix silicon concentration suppresses the reaction), temperature and cooling rate. Presently, there are few quantitative models for relating the structure of the interface to micromechanisms of interfacial fracture-in part due to the absence of nondestructive techniques for adequately characterizing the thickness and elastic properties of interphases in composites. Such techniques are also needed as a basis for sensors for controlling the interface during processing and for quality assurance purposes.

The use of ultrasound for determining the elastic properties of bulk materials is well known. By measuring ultrasonic velocity, it is possible to determine elastic constants. The work reported here is exploring the extension of this concept to interfaces where a third phase is formed, and, in particular, focuses on the case of a cylindrical interface geometry representative of that found in fibrous composites.

THEORETICAL BASIS OF RESEARCH

The theoretical framework for interface waves in a composite with a single interface has been developed within the context of the theory of ultrasonic scattering with one dimension of variability of material constants (in the direction normal to the interface) [1].

We shall illustrate this for the fiber-reinforced (cylindrical) geometry. The constitutive relations for such a composite take the form:

$$\sigma_{ij}(\underline{x},\omega) = C_{ijk\ell}(\underline{x},\omega) u_{k,\ell}(\underline{x},\omega) \qquad (1)$$

where $\underline{\sigma}$ is the stress, $\underline{u}$ is the displacement, ω is the angular frequency and $C_{ijk\ell}$ are the elastic constants. These relations comprise the general dynamic expression for a linear viscoelastic solid. In the planar geometry the elastic constants can represent any anisotropic material; however in the cylindrical geometry, rotational symmetry restricts the elastic constants to transverse isotropy (hexagonal symmetry) with the plane of isotropy perpendicular to the cylinder axis.

The elastic constants do not vary in the directions parallel to the interface (assumed to be the θ and z directions) but only in the direction normal to the interface (r direction). Outside the interface zone, we choose both matrix and reinforcement to be homogeneous and simply elastic, so that $C_{ijk\ell}(\underline{x},\omega)$ are constants independent of $\underline{x}$ and ω (but differing in the two regions). The material inside the interface zone, however, may be of a general linear viscoelastic character. Conservation relations yield three invariants of motion (frequency, one slowness with dimensions of inverse velocity, and an integer ,n, dual to the θ variable), which follow from the geometry of the model and the equations of motion. They allow us to express the result of any ultrasonic scattering or interface wave experiment in terms of the superposition of monochromatic eigenwaves characteristic of the composite structure.

In the cylindrical material these eigenwaves take the form

$$\underline{u}(r,\theta,z) = \text{Real } (\underline{q}(r)e^{i\omega(t-\frac{z}{v})}e^{in\theta}),$$

$$= \text{Real } (\underline{q}(r)e^{i\omega t}e^{-i\omega z(\alpha-i\beta)}e^{in\theta}) \qquad (2)$$

where ω, v and n are the invariants of motion, and the complex slowness $\alpha - i\beta = 1/v$. This representation may be substituted into the elastic equations of motion.

$$C_{ijk\ell}\sigma_{k\ell,j} = \rho \frac{\partial^2 u_i}{\partial t^2} \qquad (3)$$

to determine $\underline{u}(r)$ and the interface phase velocity, v, for a particular ω and n. Both in the homogeneous matrix and homogeneous fiber an eigenwave can be represented by means of potential functions. One then only need use the equations of motion to interconnect the two potential function representations through the interface zones. This has been discussed elsewhere [1].

In the special case discussed here for which experiments have been performed, we consider only so called radial-axial waves with no displacements in the theta direction (n = 0), and we replace the interface zone with an interface surface. In that case, the equations of motion at the interface reduce to continuity of normal tractions. This condition together with the required continuity of displacements across the interface leads to a 4 x 4 matrix equation for v as a function of ω [2]:

$$\underline{\underline{S}}\underline{\xi} = 0 \qquad (4)$$

whose solution can be found from the determinant equation $|\underline{\underline{S}}| = 0$.

The frequency dependance of the matrix $\underline{\underline{S}}$ of equation 4 is solely through the parameter fr_o where $f = \omega/2\pi$ and r_o is the radius of the interface. $\underline{\underline{S}}$ also depends on the four parameters $\pm (1 - v^2/b_m^2)^{1/2}$ and $\pm (1 - v^2/a_m^2)^{1/2}$ (m = 1,2), where a_m is the longitudinal velocity and b_m the shear velocity of the appropriate medium. The choice of which square root branch to take leads, in principle, to 16 possible branches for v. Actually, symmetries reduce the number to only four distinct branches corresponding to $\pm (1 - v^2/b^2)^{1/2}$ and $\pm (1 - v^2/a^2)^{1/2}$ for the matrix, since the solid rod can have no divergent solutions. We can label these branches A for - - , B for - +, C for + - and D for + + choices of square root signs. It can also be shown that when v is a solution to equation 4, so also are -v, v*, -v* (where * denotes complex conjugate). We shall only be interested in possibly attenuating waves travelling in the positive z direction, and thus, choose that v which has both non-negative real and imaginary parts (c.f. Equation 2).

Examination of equation 2 shows that particle motion is harmonic with an elliptical trajectory. $\underline{q}(r)$ has two non-zero complex components, q_r and q_z, the θ displacement being zero for radial-axial waves. $|q_r|$ and $|q_z|$ are the magnitudes of the maximum excursions in the r and z directions, respectively. The negative phase angle of the q's determine the positive phase ωt at which this maximum occurs.

Unlike ordinary modes whose velocities are real and whose elliptical trajectory axes are always oriented parallel and perpendicular to the z axis, the ellipse orientation in leaky modes changes as a function of r (but not z), and is inclined at angles θ and $\theta \pm \pi/2$ where:

$$\theta = -\frac{1}{2} \tan^{-1} \left[\frac{\text{Imag}\left(q_z^2 + q_r^2\right)}{\text{Real}\left(q_z^2 + q_r^2\right)} \right] \qquad (5)$$

The direction of precession about an elliptical orbit can also change with r. This change of direction is signaled by the minor orbit of the ellipse degenerating to zero, i.e. inclined rectilinear motion. Such bifurcations in precession are also a common feature of ordinary rod modes and Stoneley waves.

Finally we note that $\underline{q}(r.f)$ can be expressed as:

$$\underline{q}(r,f) = \underline{q}\left(fr_o, \frac{r}{r_o}\right), \qquad (6)$$

and has an asymptotic exponential behavior as $r \to \infty$ dictated by the Bessel functions occurring in $\underline{q}(r)$. These waves whose displacements decay exponentially away from the interface are then seen to be branch D modes. Leaky modes arise from the other three branches. They are eigenfunctions of infinite energy, but if truncated outside of an energy flow line, as discussed below, such modes become "almost" solutions to the wave equation with finite energy.

The elastic power flow in and out of an arbitrary region V is described by the Poynting vector:

$$\frac{d}{dt}\int_V E \, d\underline{v} = \frac{1}{2}\int_V (u_{ij}C_{ijk\ell}u_k + \rho\delta_{ij}\dot{u}_i\dot{u}_j)d\underline{v} = -\int_{\partial V}(\sigma_{ij}\dot{u}_i)n_j dS$$
$$= \int_{\partial V} P_j n_j dS \qquad (7)$$

where E is the energy density.

The energy velocity as defined by

$$\underline{g} = \underline{P}/E \qquad (8)$$

has been used in electromagnetic theory as well as being introduced for plane waves in anisotropic elasticity, where it is seen to contain the concept of group velocity [3,4]. For monochromatic waves, such as used here, both $\underline{P}$ and E are time averaged over one period.

The equivalence of energy velocity and group velocity for plane waves has caused the more fundamental energy velocity field to be neglected in favor of the more easily calculated group velocity. However, in inhomogeneous materials and in non-planar geometries the two concepts do not coincide for low frequency waves and the energy velocity can vary from point to point. Thus, the energy velocity is of more fundamental importance and is the field quantity evaluated here in the study of leaky waves.

If one draws an arrow with components g_j throughout a medium, these arrows not only point in the direction of maximum energy flow, but no energy flows perpendicular to the arrows. Thus if one draws a small circle, or other closed surface, at time t_o, one can use the energy velocity field together with the outer perimeter of the circle to develop a tubular region inside of which energy is conserved and whose velocity is governed by the energy velocity within the tube.

If the above description of leaky waves and energy leakage is valid, then it should be possible to detect that leaked energy. This experiment has been carried out for Al/stainless steel and Al/SiC (systems where leaky waves of types A and C have been detected). A model experiment is shown in Fig. 1.

As an ultrasonic wave travels down the bare fiber and strikes the point at which the fiber enters the matrix, the wave begins to leak. The energy from this leakage will follow an energy flow curve (actually a conical-like surface). Except for possible curvature near the fiber, this surface lies at an angle to the fiber (leakage angle) which is characteristic of the particular mode and the parameter fr_o. Because of the fr_o scaling, experiments carried out at low frequencies and large r_o can be expected to scale to high frequencies and small r_o such as occur in composites.

Detailed numerical methods have been developed for analyzing radial-axial leaky modes in composite systems. An example of some numerical results in the Al/Fe and Al/SiC systems are summarized in Fig. 2. A direct linkage has been established between these leaky waves and the rod modes of the stainless-steel fiber, which acts as a leaky wave guide.

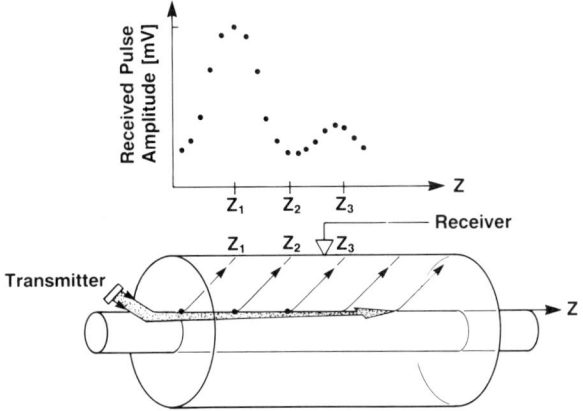

Fig. 1   Schematic diagram showing detection of leaky waves in a specimen composed of a stainless steel rod of radius 3.2mm embedded by shrink fitting into an aluminum cylinder with outer radius 28.2mm. A surface mode was produced using a wave guide on a bare section of the rod protruding from the matrix. The amplitude of the leakage was measured along the outer surface of the aluminum cylinder.

EXPERIMENTS

In order to exploit leaky waves for characterizing interfaces they should ideally be generated from without the matrix. Figure 3 shows four possible schemes for generating and detecting leaky waves. The approach described in Fig. 3a has been used to obtain an average velocity along the entire interface [5]. We report here on the approach described in Fig. 3b, which permits more quantitative detailed information about local state of the interface, such as could perhaps best be obtained by the approach shown in Fig. 3d.

The waveguide used in experiments on the Al/Fe system, as schematized in Fig. 1 produced essentially surface type rod modes. These modes principally converted into the branch C type leaky modes shown in Figs. 2a and 2b. For the experiments, a surface acoustic wave was transmitted along a steel rod using a concave conical waveguide which only contacted the 3.2 mm radius rod at the outer surface. Two frequencies, 2.5 MHz and 5 MHz, were used. The receiving transducer was placed at different points of the matrix and the acoustic response was measured. The resultant signal, which shows a series of peaks, was decomposed to permit comparison with theory. Figures 4 and 5 show the position of the measured maximum and the theoretically calculated displacements for two of the modes in the family. All four modes whose leakage angle permitted measurement were present.

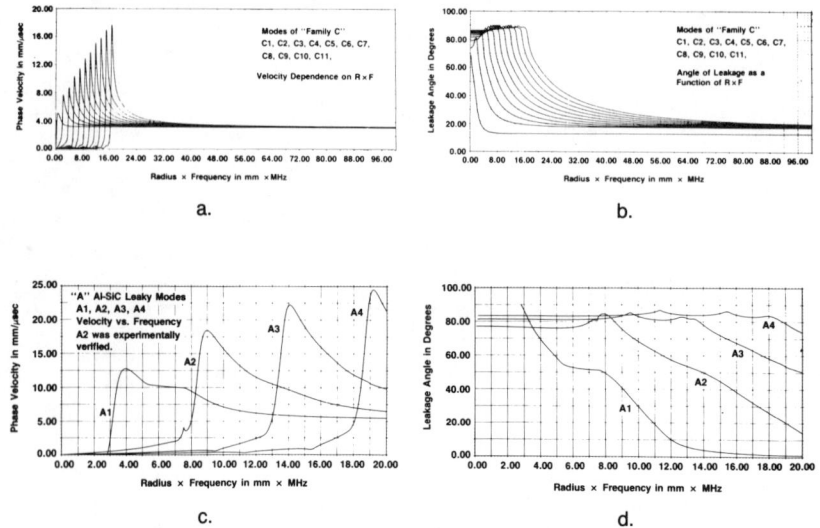

Fig. 2 Figures 2a and 2b show the first eleven members of the C family of modes for the Al/Fe system, and Figs. 2c and 2d show one related group of modes from the A family for Al/SiC. Figures 2a and 2c show the phase velocities and Figs. 2b and 2d show the leakage angles in degrees. There are infinitely many modes in both families, but only those are shown whose phase velocity peaks below $fr_o = 30$ mm/μs.

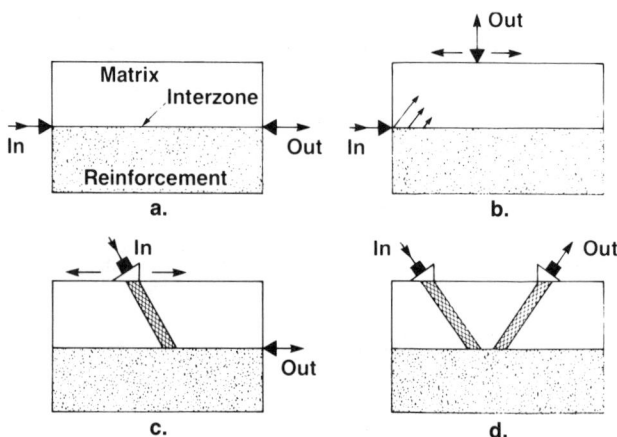

Fig. 3 Methods of excitation and detection of leaky modes.

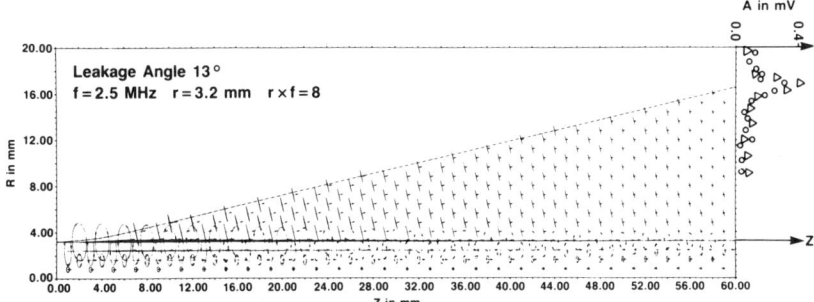

Fig. 4   Amplitude data from the experiment described in Fig. 1 superimposed upon the calculated displacements for the mode C shown in Fig. 2a. The amplitudes are expressed in mV and the frequency was 2.5 MHz. The harmonic elliptical displacement trajectories are shown through cross-section of the upper half of the specimen. The arrows extending from the center of each ellipse give the relative magnitude of the Poynting vector (energy flow per unit area) at that position in the specimen. The curve bounding the top of the displacement plot traces out the energy flow starting at the point where the rod enters the matrix. Beyond this curve the energy of the induced mode rapidly dies out leaving a lacuna.

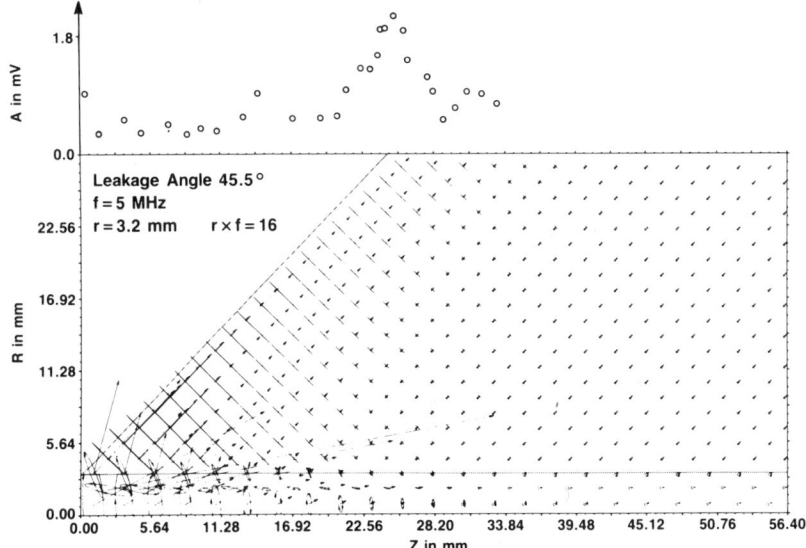

Fig. 5   As with Fig. 4, but for mode C7 carried out at 5 MHz.

Studies were also conducted on the Al-SiC system using a different transduction system to induce other types of leaky modes. The transducer used in this case was ring PZT configuration mounted at the end of a 3.2mm radius SiC rod. A simple scanning system with a small radius conical type transducer for detection is shown in Fig. 6. The movable conical transducer was placed directly against the outer aluminum surface of the Al/SiC sample; a special couplant was used to maintain uniform reproducible results.

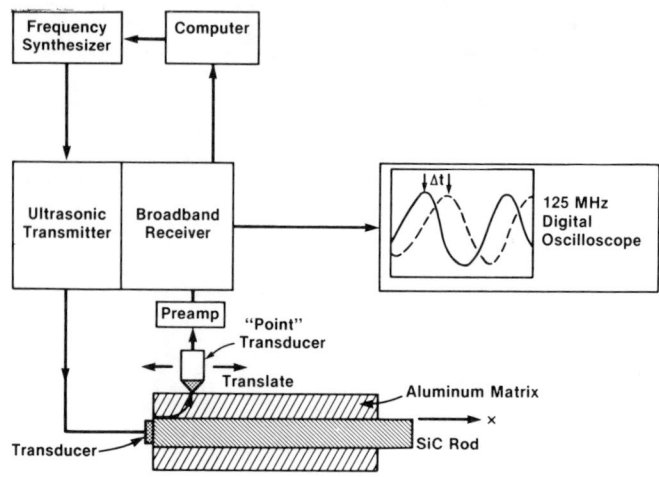

Fig. 6   Schematic diagram of experiment used to detect mode A2 of Fig. 2c. The specimen used was similar to that described in Fig. 1 except that a SiC rod was used and the outer radius of the aluminum matrix was 14.1 mm. The phase velocities were measured using the pulse overlap method.

Two leaky A branch modes were detected and their phase velocities measured by pulse overlap techniques. Figure 7 shows the theoretical and measured dispersion curves for the A2 mode shown in Figs. 2c and 2d.
The dotted lines in this figure shows the attenuation of these modes in dB per MHz per mm traveled down the rod. (The per MHz in the attenuation reflects the fact the attenuation increases linearly with frequency at a fixed radius.) It is worth noting that the maximum measured phase velocity of the A2 mode of 18.65 Km/s is about twice that of the longitudinal velocity in SiC and about three times that of the longitudinal velocity in aluminum. This approach, which exploits the coherent energy leakage of leaky waves, allows one to separate the A2 mode from other leaky modes and scattered waves. The leakage angles of active modes were well enough separated, and the attenuation of the A2 mode was sufficiently small, that it was possible to determine the local velocities of this mode over a section of the interface by measurements carried out over an outside section of the matrix. We were then able to map out the variation in phase velocity along the section of the interface. That

variation in velocity is plotted in Fig. 8. In this model sample, these
changes in velocity can be expected to be related to changes in the elastic
moduli near the interface.

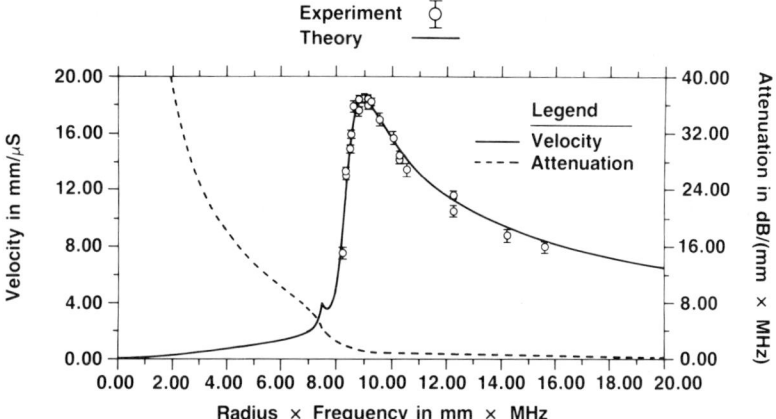

Fig. 7   Comparison of theory and experiment for mode A2 in Fig.
2c. The experiment apparatus used was that shown in Fig.
6, but in this case, the frequency was variable and the
velocity measurements were taken at the position of peak
amplitude. The maximum velocity measured was 18.65±0.04
Km/s. The theoretical velocity at this point was 18.4
Km/s.

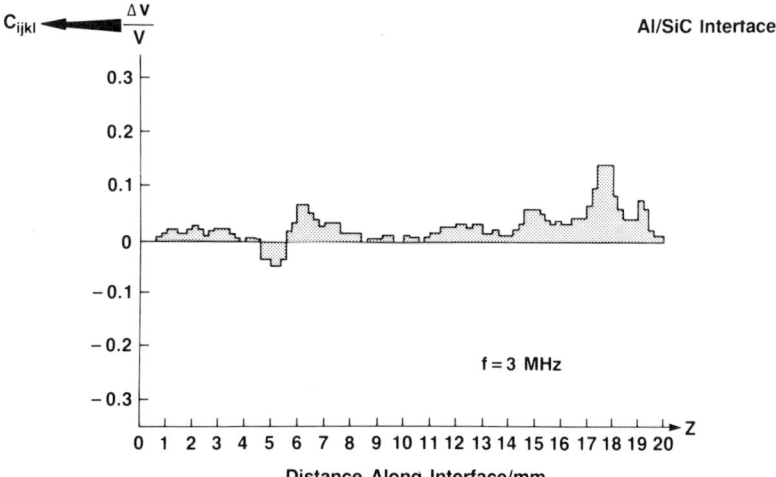

Fig. 8   Plot showing $\Delta v/v$ over a portion of the interface of the
specimen used in Fig. 6.

CONCLUSION

A study of ultrasonic wave propagation at fiber-matrix interfaces without an interface zone has theoretically predicted and experimentally confirmed the presence of leaky modes in both the Al/Fe and Al/SiC systems where the leakage and dispersion curves of SiC different modes were measured. Except for a curved portion near the interface, the energy peaks at a fixed angle to the interface where the exponential divergence of the wave amplitude away from the interface is just balanced by the attenuation of the wave amplitude along the interface.

Two different experimental techniques have been employed to detect energy leakage, and an experimental procedure capable of adaptation to acoustic microscopy was identified. Such a procedure should be capable of detecting a variety of leaky waves even in the presence of an interface zone. The formulation for the effect of the interface zone has been presented elsewhere [1].

The measurements reported here were not carried out on materials with a finite interface zone. However, even though the interface zone has small radial thickness compared to the wave length of ultrasonic waves used, it has a substantial length along the interface. Its elastic properties can then be expected to influence both dispersion and leakage properties of elastic interface modes. The great variety of these waves permit separation of the longitudional and shear properties of the interface zone.

ACKNOWLEDGEMENTS

This work has been supported by the Office of Naval Research, Program Manager, Dr. S. Fishman.

REFERENCES

1. H. N. G. Wadley, et al., Composite Materials Interface Characterization, NBSIR 87-3630, 1988.
2. D.A. Lee and D.M. Corbly, IEEE Trans. on Sonics and Ultrasonics, SU-24 1977, pp. 206-212.
3. J.L. Synge, Proc. Roy. Irish Acad., 58A 1956, pp. 13-21.
4. B.A. Auld, Acoustic Fields and Waves on Solids, Vol. 1, Chap. 7, (Wiley, New York, 1973).
5. E. Drescher-Krasicka, John A. Simmons, and Haydn N.G. Wadley, Review of Progress in Quantitative Nondestructive Evaluation, Vol. 6B 1987, pp. 1129-1136 edited by D.O. Thompson and D.E. Chimenti (Plenum Press).

HREM CHARACTERISATION OF THE INTERFACE IN A SiC FIBERS/Ti MATRIX COMPOSITE

M. LANCIN*, J.S. BOUR*, J.THIBAULT-DESSEAUX**

* Physique des Matériaux, CNRS, 1, Pl. A. Briand, F- 92195 MEUDON
** Physique du Solides, DRF, CENG, 38041 GRENOBLE, FRANCE

ABSTRACT

The microstructure of the interfacial zone in a SCS6 filaments/Ti-6Al-4V matrix composite was determined using TEM and HREM. The interfacial zone (e ~ 5 µm) consists of microstructurally distinct layers. The fiber/matrix interaction resulted in two layers respectively crystallized in grains of about 20 nm and 100 nm. The reaction zones are mainly made of brittle TiC. They are separated from the SiC reinforcement by five layers which form the coating of the SCS6 filament. These five layers may be described as SiC-C composites with either an equiaxe fine grain structure or a distorted lamellar structure. The structure of these layers can explain the toughning of the alloy by the SCS6 filaments despite of the brittle F/M reaction products formed during the processing.

1 INTRODUCTION

Titanium alloy (Ti-6Al-4V) reinforced with SiC filaments (SCS6) appears to be a very interesting metal/matrix composite to make structural components for aerospace engines [1]. The AVCO SCS6 filament is said to consist of SiC coated by four layers of pyrolytic carbon alternatively rich and poor in silicium (cf. the typical electron probe analysis provided by AVCO). The C coating reacts with the matrix during the processing and is partly consumed [1]. Therefore, the mechanical properties of the composite depend on the microstructure and on the chemical composition of the interfacial zone. This zone is composed of (i) the coating layers which remain unaltered after the processing and of (ii) the reaction products formed during the processing.

The major reaction product is TiC [2,3]. Using convergent beam electron diffraction and TEM/EDX, Rhodes and Spurling also found a Ti_5Si_3 layer next to the matrix and a narrow TiC + Ti_5Si_3 layer (70 nm) next to the filament.

In order to confirm the reaction products composition and to determine the microstructure of the whole interfacial zone, a study was performed using transmission electron microscopy (TEM and HREM). The results are described in this paper.

EELS and EDX analyses were also done to distinguish TiC from β-SiC which exhibit identical crystalline structure (including the d spacings). Some results are used in this paper but the detailed analyses will be published elsewhere.

2 EXPERIMENTAL

The composite consisting alternatively of a Ti-6Al-4V foil (100 µm width) and of an unidirectional SCS6 filaments layer, was hot pressed 30 mm at T = 1198K under vacuum.

Specimens were cut perpendicularly to the filament cores, mechanically polished and ion thinned. Observations were performed at 100 kV (TEM) or at 200 and 400 kV (HREM).

3 RESULTS

In 10 different samples, 15 interfacial zones were studied. A typical interfacial zone consists of microstructurally distinct layers as shown in figures 1 and 2. Layers L1 to L7 have been numbered from SiC to Ti. These layers are concentric despite the uneven L5/L6 interface (Fig. 3). Their sequence remains constant even so the interface L4/L6 is observed where L5 is entirely consumed (fig. 3). On the contrary, the layer A does not always exist and besides its location varies; generally A is located between L4 and L5 but it is also observed in L4 and in L2.

The different layers exhibit the following typical features:

- **Layers L1 and L3 :**
Their thicknesses range from .1 to 1 μm. They are composed of equiaxial β-SiC crystallites (≲ 10 nm) embedded in a carbon matrix (fig. 5a). The matrix is a mixture of amorphous carbon (diffuse ring around the beam) and of distorted turbostratic layers which give a broad (0002) ring reinforced in a direction perpendicular to the filament surface (fig. 4b). L1 is richer in β-SiC than L3 (fig. 1b). In L1, the amorphous carbon is predominant next to the interface SiC/L1 as shown on the diffraction pattern (fig. 4a).

The L1/L2 and L3/L4 boundaries are sharp whereas the L2/L3 ones is very uneven.

- **Layers L2 and L4 :**
Their thicknesses vary from .5 to 1.2 μm. They are formed of a turbostratic carbon matrix with large β -SiC inclusions (~ 100 × 10 nm). In the C matrix, the ordered regions correspond to a few interreticular spacings (fig.5b). Despite the misorientation observed between the adjacent ordered blocks, the (0002) planes tend to align parallel to the filament surface (Fig.4c).

- **Layer A :**
This narrow layer (.01 to .2 μm) consists of amorphous C containing crystalline inclusions whose composition (β-SiC or TiC) has to be clarified (fig.6a).

- **Layer 5 :**
Inclusions (≲ 10 nm) are spread in a carbon matrix whose structure varies throughout L5. The C matrix may exhibit the turbostratic structure of L2 and L4 but in general it is more misorientated (fig. 6c and 4d). Amorphous zones were sometimes observed either in L5 or near the L5/L6 interface.The crystalline structure of the inclusions corresponds either to β-SiC or to TiC : some SiC inclusions were identified by EELS.

- **Layer 6 :**
Grains of~ 20 nm in diameter make up this narrow layer 50 to 70 nm thick. Sometimes, crystallites a few nm large, form a border which runs in L4 or in L5 along L6 (fig. 6b). Both L4 and L5 exhibit and amorphous structure next to L6 or to the strip.

In L6 and in the edge, TiC grains were identified by EELS and diffraction. On some TiC diffraction patterns extra rings

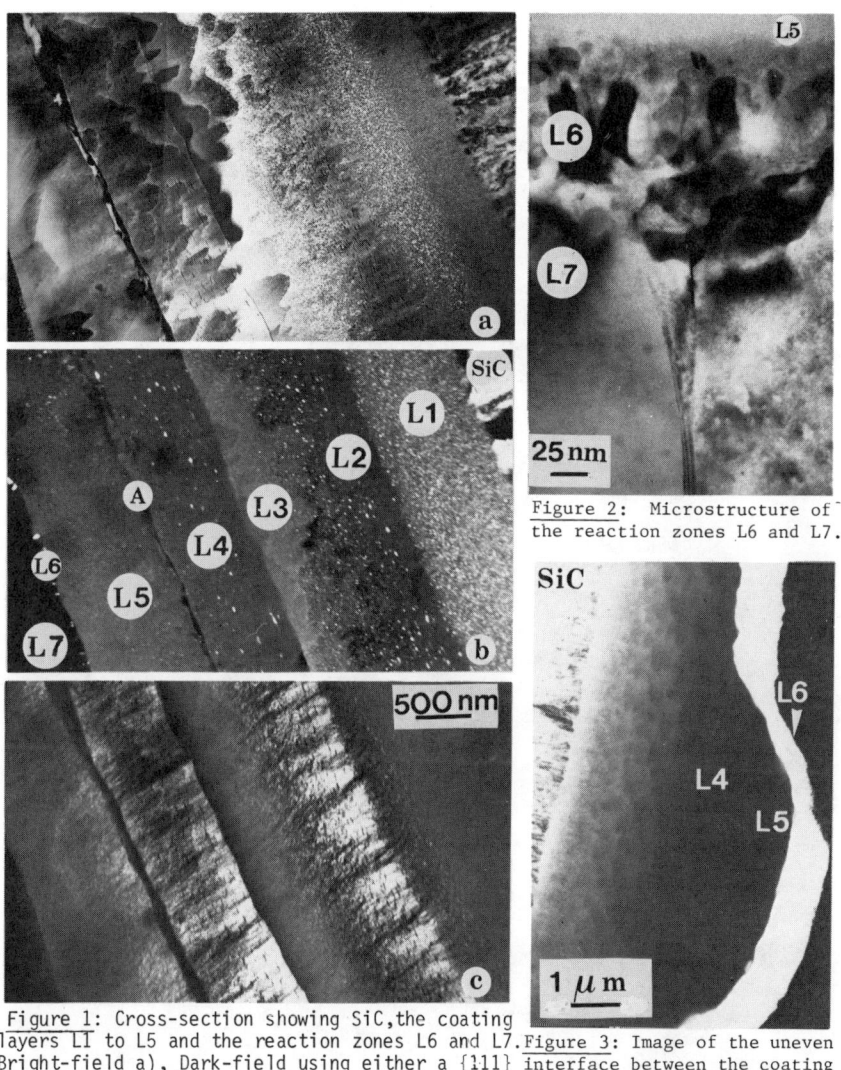

Figure 1: Cross-section showing SiC, the coating layers L1 to L5 and the reaction zones L6 and L7. Bright-field a), Dark-field using either a {111} β-SiC reflexion or a {0002} C reflexion.

Figure 2: Microstructure of the reaction zones L6 and L7.

Figure 3: Image of the uneven interface between the coating layers and the reaction zones.

Figure 4: Typical diffraction patterns of a) L1 next to SiC, b) L1 and L3, c) L2 and L4, and d) L5.

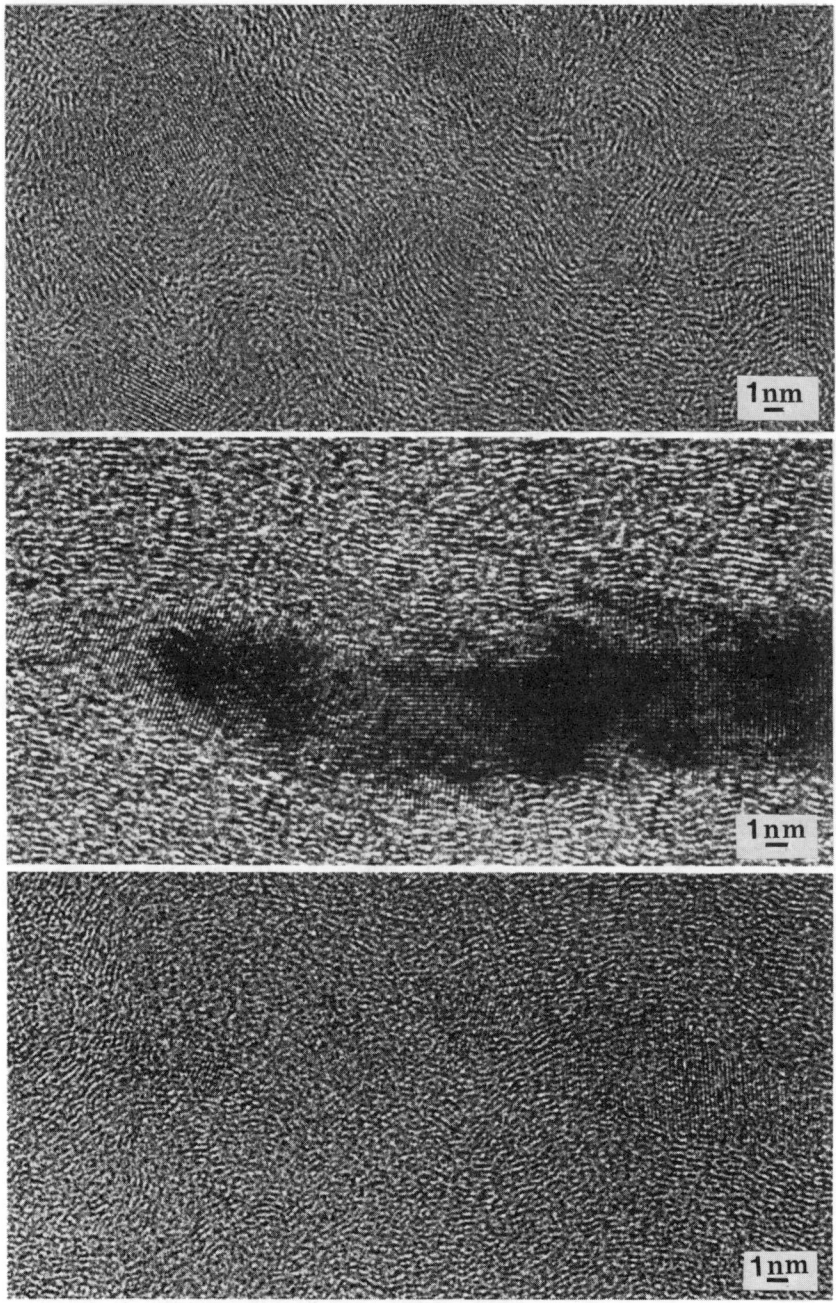

Figure 5: Typical microstructures of the coating layers.
a) L1 and L3 , b) L2 and L4, c) L5.

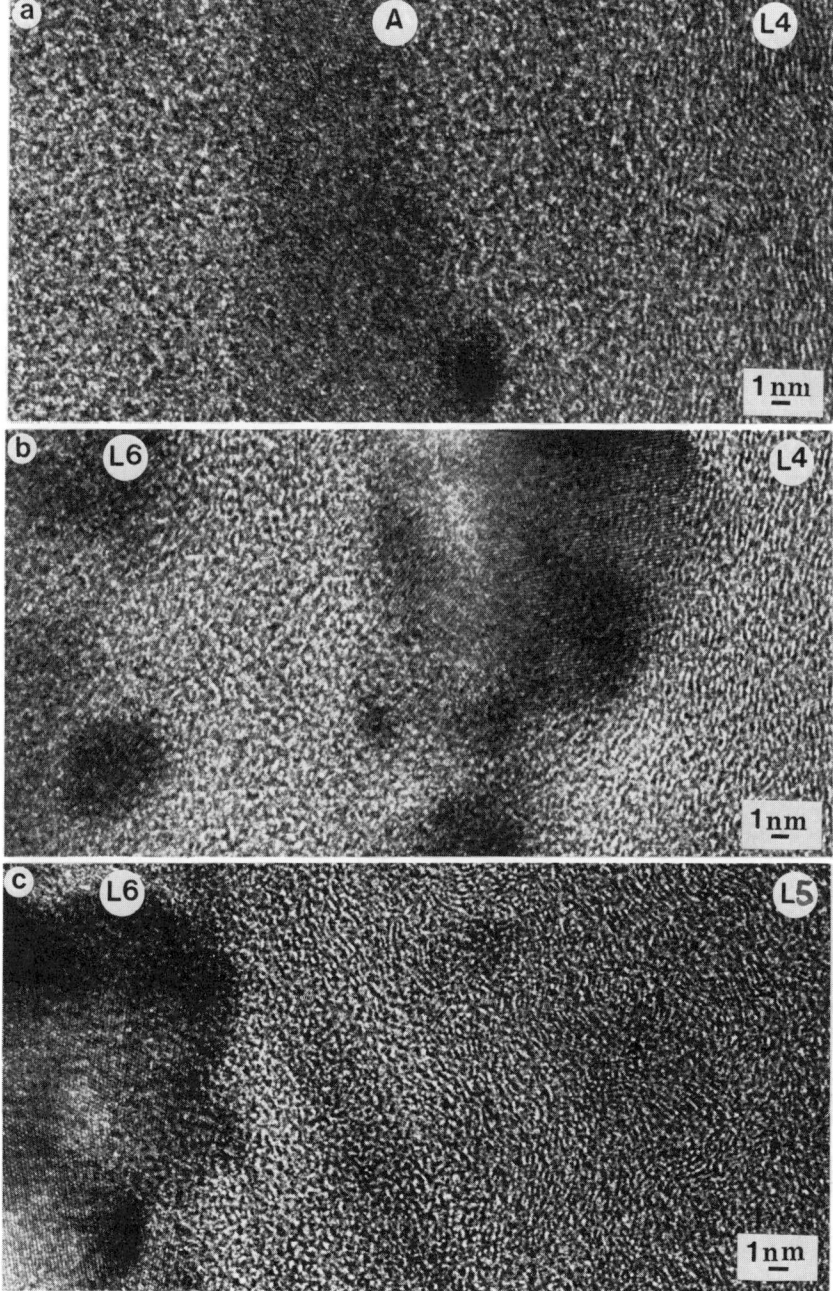

Figure 6: Microstructure of the layer A a) and of the interfaces L4/L6 b) and L5/L6 c).

were observed which could be attributed to a silicide Ti_xSi_y.

- Layer 7 :
L7, whose width varies from 400 nm to 1 µm, is composed of grains 200-400 nm large. Despite its homogeneous morphology, L7 consists of two distinct phases : Ti_xSi_y grains form a continuous pavement (e ~ 200 nm) next to the matrix ; TiC grains are located between this silicide and L6.

4 DISCUSSION AND CONCLUSION :
L6 and L7 are composed of TiC and Ti_xSi_y, which are the reaction products formed during the processing. The width, the grain size and the composition of these reaction zones confirm Rhodes and Spurling results [3]. New analyses have to be done to determine the composition of Ti_xSi_y and to verify if titanium silicide other than Ti_5Si_3 can precipitate (eg. Ti_3SiC_2 [4]).

During the processing, the reaction zones develop at the expend of the matrix and of the coatings layers, the 5th and sometimes the 4th ones. The interdiffusion results in an amorphisation of the layers next to the reaction zone.

The C rich coating of the SCS6 filament is made of five layers (L1 to L5) and not of four ones as generally admitted. In a AVCO SCS2 filament, a TEM study [5] has shown that the C coating (e ~ 2 µm) consists of four layers which exhibit the diffraction patterns respectively of L1, L2, L4 and L5. No A layer was observed and no L3 ones either. L3 may have not been detected by TEM because of its small extend and of the uneven L2/L3 interface.

The alternate microstructure of the layers L1 to L5 is the result of rapid gas composition changes in the CVD chamber. The origin of the A layer is not yet clarified.

The SCS6 coating appears to be very efficient to enhance the mechanical properties of the titanium matrix [1] [6]. Its efficiency originates in the two types of microstructure exhibited by the layers. The distorted lamellar structure (eg.L2) provides the good tensile properties and shear resistance of the composite owing to the weak chemical bonding between the blocks. This lamellar structure may result in a too easy pull out of the filaments but the SiC inclusions may slow down the process. The particulate SiC-C composites (eg.L1) are very efficient to promote crack deflection and crack blunting and to prevent the crack propagation from the matrix to the filament.

REFERENCES
1. P.R. SMITH, F.H. FROES, J. Met. 36(3) 19 (1984)
2. R. PAILLER, P. MARTINEAU, M. LAHAYE, Y. LE PETITCORPS, R. NASLAIN, Materials Science Monograph, 28B, 1112 (1985)
3. C.G. RHODES, R.A. SPURLING, Recent advances in composites, ASTM STP 864 (1985)
4. P. MARTINEAU, R. PAILLER, M. LAHAYE, R. NASLAIN, J. Mat. Sc. 19, 2749 (1984)
5. S.R. NUTT, F.E. WAWNER, J. Mat. Science 20, 1953 (1985)
6. P.R. SMITH, F.H. FROES, J.T. CAMMETT, Failure mode in composites, Ed. J.A. Cornve, F.W. Crossman, AIME (1977)

PHASE STABILITY AND INTERFACE REACTIONS IN THE Al-SiC SYSTEM

DOH-JAE LEE*, MARK D. VAUDIN**, CAROL A. HANDWERKER**
and URSULA R. KATTNER**
*Chonnam National University, Kwangju, Korea
**National Bureau of Standards, Gaithersburg, MD 20899.

ABSTRACT

The Al-SiC system has been used as a model system in an examination of phase stability in the presence of a liquid phase and microstructure development in metal-matrix composites. The Al-Si-C phase diagram has been calculated for temperatures between 500°C and 1500°C. The phases formed between Al(liquid) and SiC at 920°C have been determined experimentally, using analytical electron microscopy, in both fiber and particulate composites and compared with what is predicted from the equilibrium phase diagram. The morphologies and the spatial distributions of phases have also been examined in addition to the phase analysis. The only phases found were Al, Al_4C_3, SiC, and Si. Although Al_4SiC_4 is calculated to be stable at 920°C, it was not found. The SiC grain structure was found to influence strongly the morphology of the Al_4C_3-SiC and Al-SiC interfaces.

INTRODUCTION

Metal matrix composites are prime examples of complex engineering systems where materials are not in thermodynamic equilibrium during initial fabrication, during production of components or in use.[1] Interdiffusion and reaction processes produce interface roughness, which seriously affects the fiber-matrix bond. Because of the practical consequences of these interface reactions, the principles governing these processes need to be understood and their importance in controlling composite properties evaluated.

The Al-SiC system has been used as a model system in an examination of phase stability in the presence of a liquid phase and microstructure development in metal-matrix composites. A proper analysis of non-equilibrium effects caused by diffusion and reactions at interfaces requires that we know the phase composition of the system at equilibrium. Our study of Al-SiC consists of three major parts. First, we have calculated the Al-Si-C phase diagram at temperatures between 500°C and 1500°C. Second, the phases formed between liquid Al and SiC at 920°C have been determined experimentally and compared with what is predicted from the equilibrium phase diagram. Third, the morphologies and the spatial distributions of phases (e.g., Al adjacent to SiC in some regions but separated by intermediate phases in nearby regions) have been determined to complement the phase analysis.

When SiC is in contact with liquid Al but intermediate phases, such as Al_4C_3 and Al_4SiC_4, separate these two phases in an isothermal section of the equilibrium phase diagram, the intermediate phases may form either a continuous layer or isolated precipitates on the original solid. For the case of a continuous layer of Al_4C_3, growth of the Al_4C_3 phase is dominated by solid state diffusion processes through the Al_4C_3 layer. In the second case, both isolated precipitates of Al_4C_3 and SiC remain in contact with liquid Al. Further growth of the isolated Al_4C_3 then occurs by the dissolution of SiC into the liquid Al with extensive undercutting of the SiC. In a composite system, the degradation rate of the fiber and the interface roughness will depend on which of these two scenarios operates.

PHASE STABILITY IN THE Al-Si-C SYSTEM

For the calculation of the Al-Si-C system, the analytical description given by Dörner [2] was used, as will be summarized in detail elsewhere in another paper. For the two binaries Al-C and Si-C, the liquid phase is described as a regular solution and all solid phases are assumed to be stoichiometric. For the Al-Si system the liquid phase is described by three polynomial terms for the concentration dependence of the enthalpy and entropy and the two solid phases are described as Henrian solutions. For an analytical description of the solution phases in the ternary system no ternary interaction terms are used. The two ternary compounds Al_4SiC_4 (space group $P6_3mc$) and $Al_4Si_2C_5$ ($R\bar{3}m$) are described as stoichiometric. The calculations were carried out using the Thermo-Calc Databank System. [3]

According to Dörner's calculation of the high temperature phase diagram in the Al-Si-C system, the ternary compound $Al_4Si_2C_5$ forms peritectically at 1921.7°C and decomposes in a degenerate invariant equilibrium at 1832.8°C. The compound Al_4SiC_4 forms peritectically at 1921.3°C and the temperature of formation observed experimentally by Dörner is 1930°C. Oden and McCune [4] found by thermal analysis that Al_4SiC_4 forms in the quasibinary peritectic reaction (liquid + C → Al_4SiC_4) at 2080°C. They could not confirm the existence of $Al_4Si_2C_5$ but found a new ternary compound of the stoichiometry Al_8SiC_7 ($P3$) which forms peritectically at 2085°C.

Isothermal sections of the Al-Si-C ternary at 500°C and 900°C and the Al-SiC isopleth are shown in Figure 1. It should be remembered that calculated phase diagrams, including the ones presented here, are not the "definitive" phase diagrams for a system, but are best estimates consistent with existing data. The phases predicted to exist between liquid Al and SiC at 900°C are an Al(Si) liquid, Al_4C_3 ($R\bar{3}m$), Al_4SiC_4, and SiC ($F\bar{4}3m$, $P6_3mc$, $R\bar{3}m$ or $R3m$). In several previous studies [5-10], phase formation in the Al-SiC system has been investigated under a variety of conditions, for example, for annealing times ranging from "flash" anneals on the order of one minute [6] to the longest annealing time of three hours.[7] In some cases, elemental C and Si were added near the SiC- Al interface.[5,7,10] As predicted by the 500° and 900°C phase diagrams, Figures 1a and 1b, additions of Si to the Al-SiC interface suppress the formation of Al_4C_3. To examine the phase relationships in this system in materials annealed for longer times and in the presence of a reactive liquid phase, two types of composite were formed, one with a small volume fraction of SiC fibers in contact with Al, and the other with SiC particulate in pure Al at a ratio of Al:SiC of 1:1, for which the stable phases at 900°C are Al(Si) liquid, Al_4SiC_4 and SiC.

EXPERIMENTAL PROCEDURES

The SiC fibers used in this study were uncoated commercial material 141μm in diameter (AVCO Corp.) manufactured by chemical vapor deposition of β-SiC onto a carbon filament substrate. These SiC fibers consist of faulted columnar subgrains 80-100 nm in diameter with a growth pattern extending radially from the substrate. The microstructure of the as-fabricated fibers has been examined in detail previously

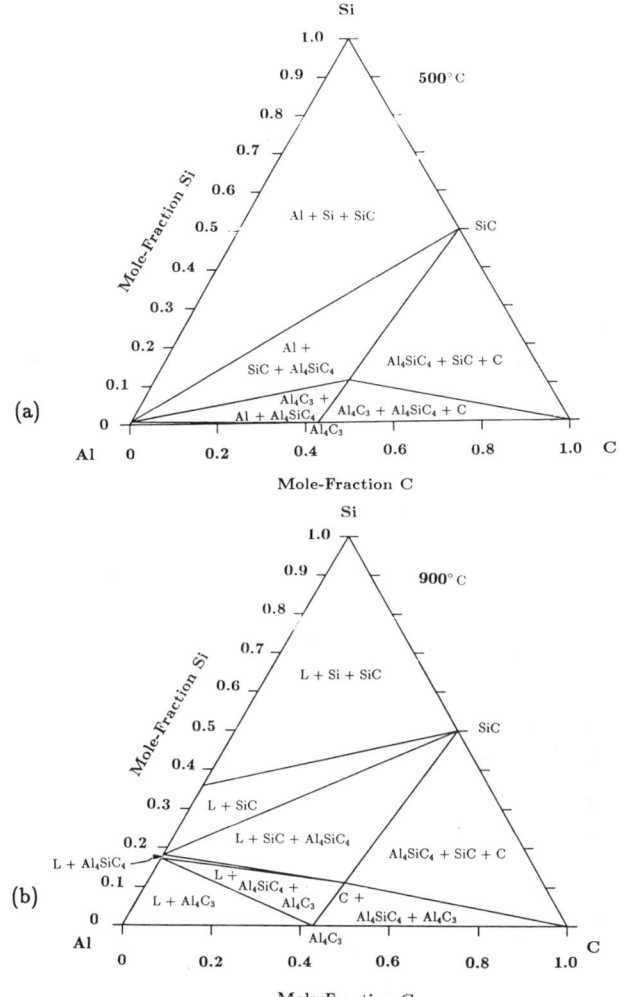

Figure 1. Isothermal sections of the Al-Si-C system at 500°C (a) and 900°C (b), and the Al-SiC isopleth (c) for the temperature range 400°-2600° and and expanded view (d) for the temperature range 550°-600°C and mole-fraction C = 0-0.2. (The four three-phase triangles including the phase Al_4SiC_4 appear to meet at a five-phase invariant, thereby violating the phase rule. This apparent junction of four three-phase triangles at a point results from the assumption that Al_4SiC_4 is stoichiometric. This assumption of absolute stoichiometry for Al_4SiC_4 makes Al_4SiC_4 a line compound, collapses the two-phase regions between the three-phase triangles to lines and creates the apparent five-phase invariant. In reality, the carbides in this system exist over a range of compositions, however small, as would two-phase regions separating three-phase triangles and the five-phase invariant point does not occur.

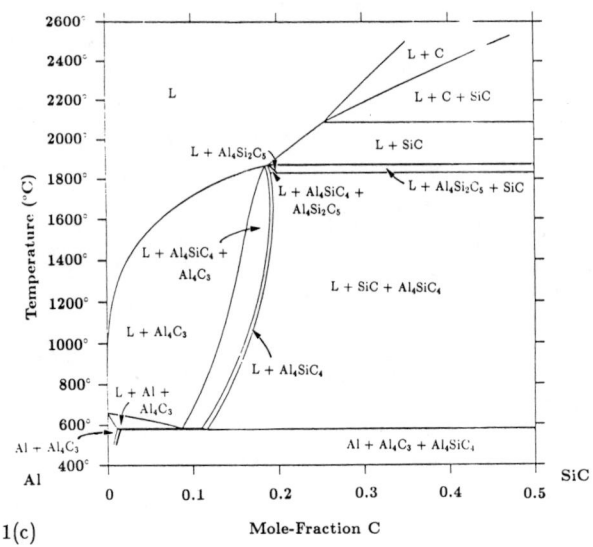

1(c)

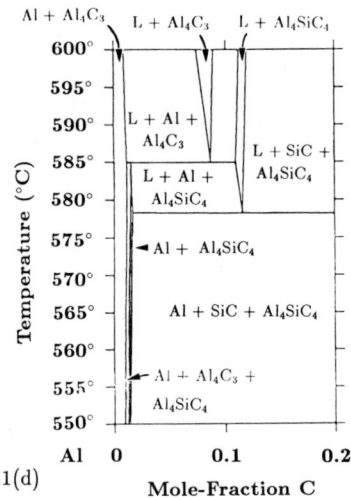

1(d)

by Nutt and Wawner.[5,11] An array of SiC fibers was placed between two aluminum disks cut from Al foil of 99.98% purity, placed in a capped alumina crucible to minimize contamination of the Al and sealed in a Vycor^R tube under vacuum. The specimen was heat treated at 920°C for times up to 6 days to produce a significant reaction zone.

The particulate specimen was prepared by mixing Al powder and SiC powders in a 1:1 Al:Si molar ratio and adding a small quantity of ethanol to prevent de-mixing during handling. The SiC powder consists of dense polycrystalline agglomerates with an average agglomerate size of 100μm and an intra-agglomerate grain size of approximately 1-5μm and is a mixture of α-(4H and 2H) and β-polytypes. The compact was hot pressed in a graphite die for 110 minutes in an argon atmosphere at 20MPa and 620°C. Small pieces of the densified compact were encapsulated and heat-treated as above for 8 days.

Samples were prepared for study in the transmission electron microscope (TEM) by mounting the specimens in epoxy and sectioning perpendicular to the plane of the disk, mechanical polishing, including dimpling, and ion milling with careful monitoring to ensure that the region of interest became electron-transparent. Phase analysis was performed on each grain in the area of interest using a number of electron microscopy techniques. The component elements in each grain were determined by energy dispersive X-ray analysis (EDX) and, in some cases, by electron energy loss spectroscopy. Phase identification was achieved by analyzing selected area diffraction (SAD) patterns from each grain. It was not always possible to tilt a given grain to a high symmetry zone axis. In such cases, since all the non-cubic phases that were expected to be present had at least trigonal symmetry, SAD patterns from zone axes of the $\langle hhil \rangle$ type were recorded. These spot patterns were easily recognized as they are distorted hexagonal nets with orthogonal mirror lines; the actual symmetry of the zone axis was determined by convergent beam diffraction in a number of cases, and this information was useful in the phase analysis. Values for a_o and c_o were calculated from the lengths of the diffraction vectors and the angles between them, making use of the symmetry of the diffraction patterns. Using the results from the various analytical techniques, it was possible to identify all the grains in the region.

RESULTS AND DISCUSSION

Interface regions in both the fiber and particulate specimens were studied in detail in the TEM. Figure 2a is a TEM micrograph of the interface region between a large SiC particle and the surrounding aluminum metal in the particulate specimen. Over thirty grains in the interface region were analyzed by SAD and EDX, and the distribution of phases is indicated on the micrograph. All the grains that were analyzed by EDX as containing aluminum as their major component were determined to be either Al or Al$_4$C$_3$ by SAD. Similarly, all the grains that were analyzed as silicon by EDX were identified as either Si or SiC by SAD. In one region of Figure 2b there is a large grain of Al surrounding elemental silicon which has precipitated out during solidification of the Al(Si) melt during cooling to room temperature from 920°C. Large Si dendrites were identified by optical microscopy and SEM. No aluminum silicon carbides were found. The same results on phase formation were obtained in the fiber composite samples. The calculated free energy of formation of Al$_4$SiC$_4$ from the elements Al(liq), C(solid), and Si(liq) at 900°C is -214 kJ/mole; however, once Al$_4$C$_3$ is formed, the calculated

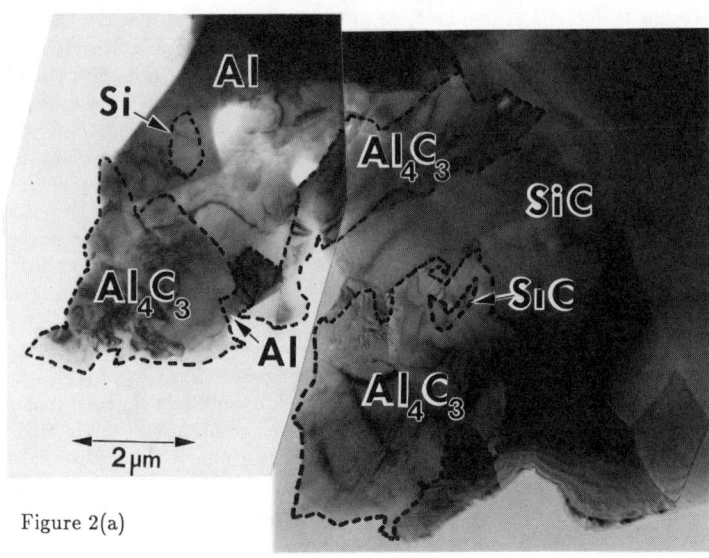

Figure 2(a)

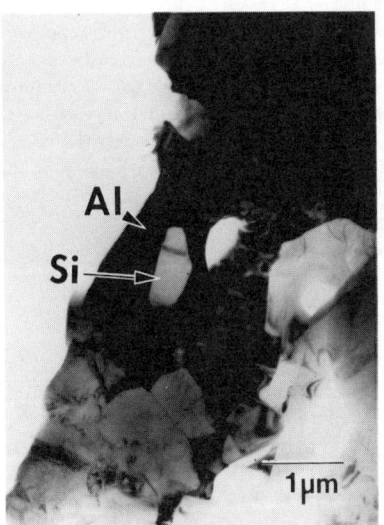

Figure 2(b)

Figure 2. (a) TEM micrograph of interface region in Al-SiC particulate specimen. The phase distribution is indicated. (b) Large Al grain surrounding Si grain and adjacent to Al_4C_3 grains.

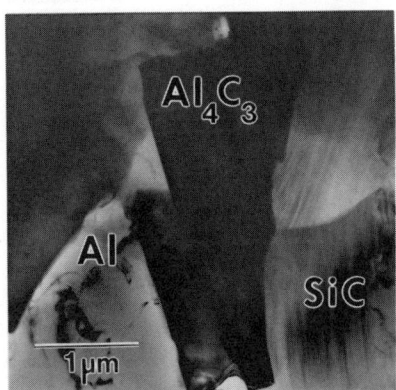

Figure 3. $Al/Al_4C_3/SiC$ grains showing faceted interface morphologies.

free energy of formation from the carbides Al_4C_3 and SiC is only -0.75 kJ/mole.[2] Within the limits of the calculation technique, therefore, it is not possible to determine whether Al_4SiC_4 is stable with respect to the Al_4C_3 and SiC. An analysis is underway to determine how diffusion kinetics affects the formation of Al_4SiC_4.

In the present study, a small amount of Al was detected in the SiC by EDX and a small amount of Si was detected in the Al_4C_3. Arsenault and Pande have suggested that significant diffusion of Al into SiC occurs at 500°C [12]; however, as Arsenault and Pande pointed out, bulk diffusion is exceedingly slow at low temperatures. Further analytical electron microscopy is required to sort out questions of solid solution formation and diffusion from those of specimen contamination during polishing, ion milling, or fracture.

High dislocation densities were observed in some Al and Al_4C_3 grains; in particular the Al grains had a high dislocation density near the Al/Al_4C_3 interface, as shown in Figure 3. High dislocation densities resulting from the thermal expansion difference between Al and SiC have been observed at the Al-SiC interface.[13,14] The thermal expansion coefficients of Al_4C_3 and SiC are the same order of magnitude, $8 \times 10^{-6}/K$ [15] and $3 \times 10^{-6}/K$ [13] respectively, whereas the thermal expansion coefficient of Al is approximately $30 \times 10^{-6}/K$.[13] Dislocation generation at the Al-Al_4C_3 interface due to thermal stresses is therefore expected.

Figure 3 shows an Al_4C_3 grain between Al and SiC with a faceted interface separating the Al and Al_4C_3. This interface is faceted on a scale of $1\mu m$; SAD patterns indicate that the large facets are parallel to the basal plane of the carbide. The Al_4C_3 and SiC interfaces in the particulate were also facetted, in some cases with easily identifiable facets corresponding to the basal plane of SiC, Figure 3, and in other cases, with many different interface facet planes, as shown in Figure 4a. In the particulate composite, facetted SiC surfaces were observed in contact with Al. The orientation relationships and facet planes of adjacent Al_4C_3-SiC grains and of Al_4C_3 and SiC in contact with Al remain to be determined.

The Al_4C_3 grains were sometimes twinned as indicated by the letter "T" in Figure 4a. A SAD pattern from the twinned region (Figure 4b) shows that the spots in some of the closely spaced rows are split by one third of the spot periodicity along the row and that every third row is not split. This diffraction pattern is characteristic of basal plane twinning in rhombohedral structures, in which the plane stacking sequence changes from *ABCA* to *ACBA* at the interface.

Major differences in the microscopic morphologies of the Al_4C_3-SiC interface were observed between the particulate and the fiber samples, primarily resulting from the different size scale of grain structures in the two types of SiC. Growth of Al_4C_3 on the fiber occurred by the apparent attack of the SiC between the individual fiber subgrains with intrusions of Al_4C_3 on the same scale as the columnar subgrains (80-100 nm in diameter), as shown in Figure 5a and 5b. The faulted SiC subgrains, delineated by the dashed lines in Figure 5b, are separated by regions of Al_4C_3. In the particulate sample, the SiC grains, while faulted, are large and equiaxed and the individual Al_4C_3 grains at the SiC-Al_4C_3 interface also appear to be at the same scale as the parent SiC grains. In both the fiber and the particulate composites, the Al_4C_3-SiC interface becomes rough but at different size scales. Although the bonding between SiC and Al_4C_3 appears to be good, the roughness can affect the mechanical properties of the SiC. The interpenetrated structure of Al_4C_3 and SiC, which is shown at higher magnification in Figure 5b, can then lead to regions of stress localization and can substantially weaken the fiber, as noted previously in mechanical testing of reacted fibers.[10]

The Al (which is molten at the heat treatment temperature) is found in direct

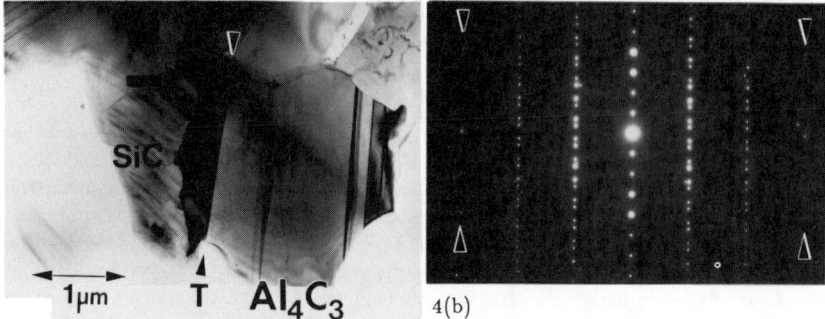

Figure 4. (a) Twinned grain of Al_4C_3, and faceted interface with SiC. (b) SAD pattern from twinned region in (a). Arrows indicate lines of unsplit spots.

Figure 5. (a) TEM micrograph of roughness of interface in fiber specimen showing interpenetration of Al_4C_3 and SiC. (b) Higher magnification micrograph showing details of Al_4C_3 penetration between faulted, columnar SiC grains.

contact with both SiC and the Al_4C_3 precipitates formed in the reaction zone. The coexistence of Al-SiC and Al-Al_4C_3 interfaces suggests that growth of Al_4C_3 is not limited by solid state diffusion and formation of Al_4C_3 can continue until much of the reinforcing SiC is undermined by Al_4C_3 formation.

SUMMARY

The Al-SiC system has been used in an examination of phase stability in the presence of a reactive liquid phase and microstructure development in a model metal-matrix composite system. The Al-Si-C phase diagram has been calculated for temperatures between 500°C and 1500°C. A complete phase analysis of interface regions in fiber and particulate composites has been carried out for composites heated to 920°C for times up to 8 days, and the morphologies of the Al-Al_4C_3, Al-SiC, and Al_4C_3-SiC interfaces have been examined. The only phases found were Al, Al_4C_3, SiC and Si. The SiC grain structure was found to influence strongly the morphologies of the Al_4C_3-SiC and Al-SiC interfaces.

ACKNOWLEDGEMENTS

Helpful discussions with Dr. S. R. Nutt regarding specimen preparation are gratefully acknowledged. The support of the Office of Naval Research (Contract N00014-87-F-0023) is also gratefully acknowledged.

REFERENCES

1. R. Warren and C.-H. Andersson, Composites, 15 101 (1984).
2. P. Dörner, "Konstitutionsuntersuchungen an Hochtemperaturkeramiken des Systems B-Al-C-Si-N-O mit Hilfe thermodynamischer Berechnungen," Ph.D Thesis, Universität Stuttgart, F.R. Germany, 1982.
3. B. Sundman, B. Jansson, and J.-O. Andersson, CALPHAD 9 153 (1985).
4. L.L. Oden and R.A. McCune, Met. Trans. A. 18A 2005 (1987).
5. S.R. Nutt, and F.E. Wawner, J. Mat. Sci. 20 1953 (1985).
6. V.M. Bermudez, Appl. Phys. Lett. 42 70 (1983).
7. T. Iseki, T. Kameda, and T. Maruyama, J. Mat. Sci., 19 1692 (1984).
8. S.R. Nutt and R.W. Carpenter, Mat. Sci. Eng., 75 169 (1985).
9. S.R. Nutt, in: Interfaces in Metal Matrix Composites, ed. A.K. Dhingra and S.R. Fishman, AIME, 1986.
10. S. Kohara and N. Muto, J. Jpn. Inst. Met. 45 411 (1981).
11. F.E. Wawner, A.Y. Teng, and S.R. Nutt, SAMPE Quarterly, 14 39 (1983).
12. R. J. Arsenault and C. S. Pande, Scr. Metall. 18 1131 (1984).
13. M. Vogelsang, R.J. Arsenault and R.M. Fisher, Met.Trans. 17A 379 (1986).
14. Y. Flom and R. J. Arsenault, Mat. Sci. Eng. 75 151 (1985).
15. T. Iseki, T. Kameda and T. Maruyama, J. Mat. Sci. Lett. 2 675 (1983).

PART VI

Composite Structures

NOVEL METHOD FOR CONSTRUCTING TETRAHEDRAL FRAMES

JOHN J. GILMAN
Center for Advanced Materials, Lawrence Berkeley Laboratory, University of California, Berkeley, CA 94720

INTRODUCTION

This paper is concerned with the design and construction of a frame that makes optimal use of modern materials together with optimal structural design. It is modular in format and can be constructed using puckered rings which combine to make a frame with tetrahedral nodes and cubic symmetry; analogous with the crystal structure of diamond. This frame can be extended indefinitely in three dimensions by adding modules to its core. Such a frame has high capacity to bear loads relative to its weight. For some types of loading it has the highest possible specific-structural strength (for a frame of struts). It is a structural analog of a foam.

The struts of a frame are commonly straight and joined together at nodes by means of various types of male and female coupling devices. Thus the struts are discontinuous at the nodes unless the material of construction can be welded. But some of the materials with the highest specific strengths are composites consisting of strong filaments embedded in less strong matrices, and such materials cannot be welded. They are difficult to join using coupling devices without adding substantial extra weight. Therefore, in building lightweight structures, it is desirable to be able to use continuous strut-like members that do not need nodal coupling devices.

The frame described here uses continuous puckered rings as strut-like members. These rings provide maximum structural efficiency, and can be readily constructed from materials that have maximal specific stiffnesses and strengths, including filamentary composites.

In two dimensions the minimum number of struts needed to fully determine the position of a point is three. Therefore, frames are usually based on triangular configurations. But in three dimensions the minimum number of struts becomes four. For maximum symmetry, and therefore maximum specific-structural-strength, these four struts must be arrayed with equal angles between them at each node. Thus, they must have tetrahedral angles between them (109° 28'). This allows the nodes to be connected in one of two ways yielding frames of either cubic or hexagonal symmetry. Of these two configurations, the cubic one has the most symmetry; and it is the one that will be emphasized here.

A tetrahedral frame constructed from puckered rings is shown in Figure 1. It is periodic in three dimensions. As a result of this periodicity the frame can be built up in a repetitive way by adding fundamental modular units to it until a desired set of dimensions is obtained. Its geometric form causes it to have three-dimensional stability without depending on lateral stabilizing members or complex networking. Also, this design takes advantage of the inherent rigidities of skeletal tetrahedra. Since the design achieves these features with maximum geometric efficiency (it uses the minimum number of struts per node to have stability) it has a maximum amount of open space within it; thereby improving its usefulness. The open space is in the form of quasi-cylindrical channels that pass through it along three orthogonal directions. A sketch showing one set of these channels is presented in Figure 2.

Figure 1. Model of tetrahedral frame constructed of puckered rings. Since the frame is periodic it can be extended indefinitely in three-dimensions. Note the continuity of the material at the nodes.

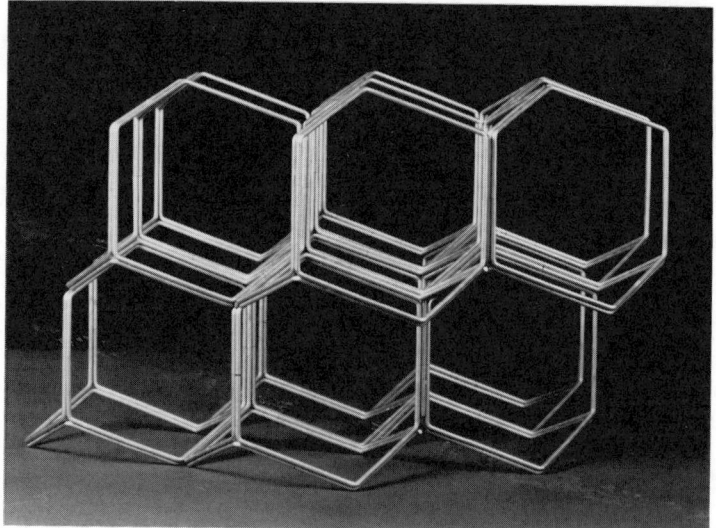

Figure 2. One set of channels through a tetrahedral frame. There are six equivalent sets through the structure. On each of four planes, sets of three channels from equilateral triangular arrays.

FRAME CONSTRUCTION

The standard approach to constructing a tetrahedral frame would be to connect a set of straight struts of equal length with nodal connectors, or with welds (1). Welding could be used for some metals or alloys, but cannot be applied effectively to ceramics, aligned polymers, or composites. In general, there are two types of nodal connectors to be considered: male and female as illustrated in Figure 3. The female connector (Figure 3A) has the disadvantage of requiring excess materials on the outside of the strut which adds to the weight of the system. The male one (Figure 3B) is less weighty, but tends to splay the end of the hollow strut if there is a bending moment present. This is especially serious for filamentary composites because they have relatively little transverse strength, and is also a problem with brittle, notch-sensitive materials.

The approach proposed here uses puckered rings (2). The rings have six sides and have trigonal symmetry as shown in Figure 4. They can be constructed by winding up fibers to form skeins which can then be appropriately puckered and set into shape with an impregnated adhesive (or wound with the adhesive already on the fiber). Thus there need be no discontinuities in the rings.

The puckered rings stack well, so they can be stored compactly until the time comes to construct the frame. Then they can be assembled using adhesives, brazements, mechanical connectors such as splines, or weldments. Even if they are not used for strength, splines may be used to aid in assembly because they minimize disruption of the continuity in the rings.

To maximize the packing density prior to assembly, standard elbows may be the structural unit (Figure 5). Six of them will form a ring, and such rings can be assembled into a frame.

The structural units can be made of metals, glasses, polymers, ceramics, or composites. Their cross-sections may be circular, but they will pack more densely if they are hexagonal or triangular. Also, they may be hollow, solid, or channels.

The first step in constructing a frame is to assemble four rings into a tetroid space-filling unit as shown in Figure 6. Then, either by joining tetroids together (Figure 7), or by adding rings to a tetroid, a frame can be built up (Figure 1). Such a frame uses a minimum amount of material to sustain a given set of applied forces. This was proven long ago by Michell (3).

Sometimes it is desirable to have a higher frame density on one or more sides of an underlying less dense frame. One example arises when a large beam is to have a frame as its substructure so more strength is needed at the outer surfaces than in the

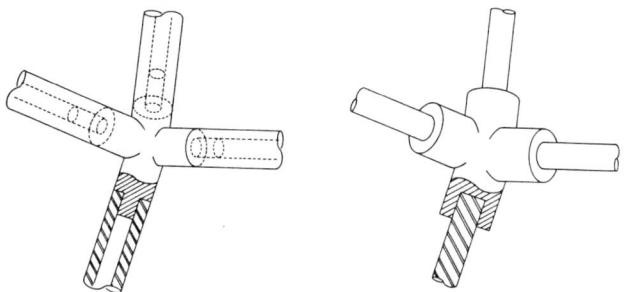

Figure 3. Schematic nodal couplers for the struts of a tetrahedral frame.

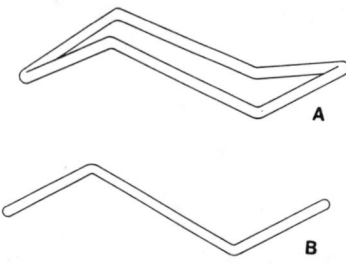
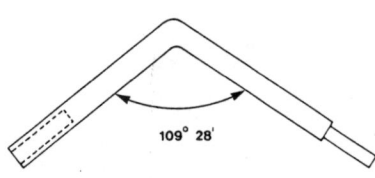

Figure 4. Puckered ring that can be used to build up a tetrahedral frame. The sides are all of equal length and the angles of the bends are 109° 28. Two views are shown: A - angled top view; B - side view.

Figure 5. A standard elbow from which the ring of Figure 4 can be built up.

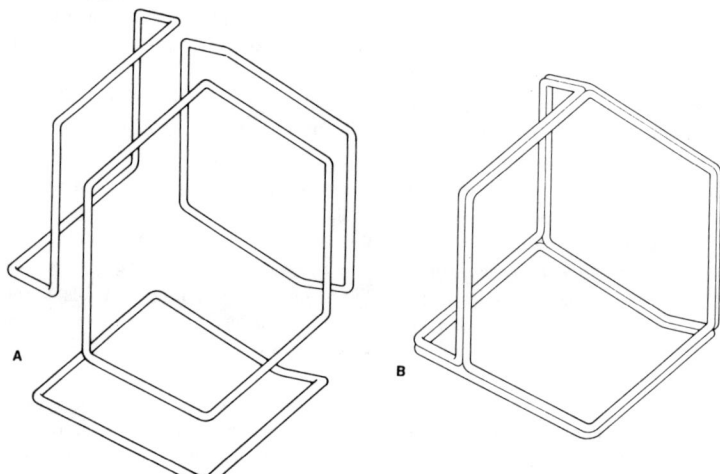

Figure 6. Assembly of four puckered rings into a tetroid unit. The view at A shows the rings prior to assembly;

interior. The same may be true for large plates or shells. Another example arises when a thin skin is to be supported by a frame, so an increased density of support points is desired at the surface of an otherwise low density frame. Figure 8 shows how increased frame density can be obtained with the present method of construction. In order for the smaller and larger rings to mesh coherently, the sizes of the smaller ones need to be integral sub-multiples of the larger ones.

Another method that can be used to "close-off" a surface on a tetrahedral frame is to use triangular rings in a two-dimensional array. This is the most efficient two-dimensional frame.

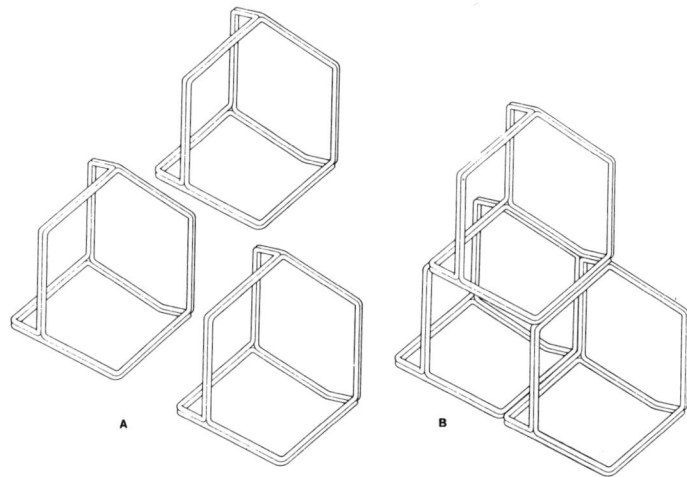

Figure 7. Assembly of three tetroid units into a section of frame. At A the three tetroids approach each other; at B the completed frame section is shown.

The tetroid units of Figure 6 are related to the "diamond saddle polyhedra" described by Pearce (4). However, they differ in being skeletal rather than blocky. Also, the method of constructing them from rings as described here is novel.

MECHANICAL RESPONSE OF TETRAHEDRAL FRAMES

Tetrahedral frames are not isotropic. They may have either cubic or hexagonal symmetry. In the case of the cubic ones, their structural stiffnesses in response to distributed applied loads, require three stiffness constants for their description; thus they are orthotropic. Three constants analogous with the elastic stiffnesses of cubic crystals can be used (5). Their conventional designations are: C_{11}, C_{12}, and C_{44}. The resistance of the frame to a uniform dilatational strain (its bulk modulus) can be described in terms of these constants. Its resistance to uniaxial stretching along its three characteristic directions: $<100>$, $<110>$, and $<111>$ can also be described in terms of them. The first of these characteristic directions lies along any edge of the reference cube sketched in Figure 9. The second lies along any face-diagonal; and the third along any body-diagonal.

Stiffness

In order to develop a description of the frame stiffness, the first step is to define strut constants for the elements that connect the nodes. These elements consist of bundles of six of the sides of adjacent puckered rings. The strut lengths, l approximately equal the distances between nodes. The ring sides may be assumed to be cylindrical, but they might have hexagonal or other more complex cross-sections. Their cross-sectional shapes will determine their section moduli and hence their resistance to bending. Then, if the axial stiffness of the material in the struts is

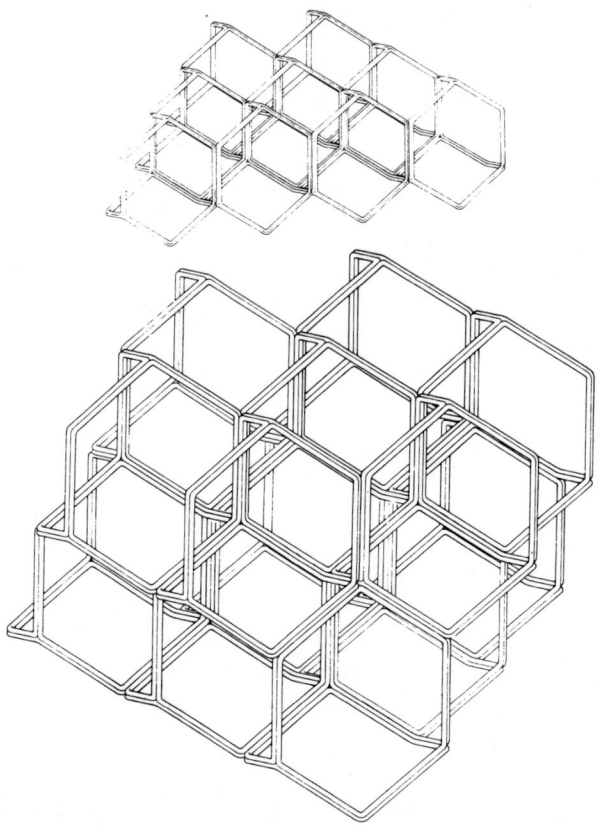

Figure 8. Schematic drawing of a higher density frame that meshes coherently with the surface of a lower density frame. The ratio of the ring sizes must be an integer.

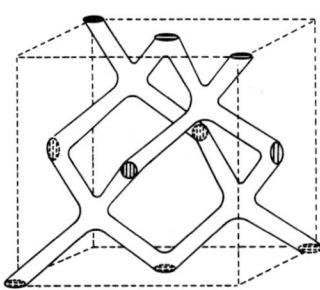

Figure 9. Coordinate frame of reference for a tetrahedral frame.

described by Young's Modulus, E, and the cross-sectional area is A, the stretching constant for a strut is:

$$C_s = AlE \tag{1}$$

and the total strain energy in a strut with a strain, ϵ is:

$$C_s \epsilon^2 / 2$$

In shear a strut is like two face-to-face cantilevers with each end "built-in". If the strut has an effective diameter, d and d/l is small, the strain-energy can be determined for a given deflection angle, θ(7). Then, since the total energy is:

$$C_b \theta^2 / 2$$

the bending constant becomes

$$C_b = \frac{6EI}{l} \tag{2}$$

where I is the section modulus for a circular bar and equals: $\pi d^4/64$.

To design a frame in terms of the stretching and bending constants, expressions that relate them to the stiffness constants, C_{ij} are needed. Then a frame with desired properties can be designed through proportioning of the strut dimensions, and selection of the materials of construction. The analysis used here parallels that for tetrahedrally-bonded crystals given by Harrison (6). It will not be repeated here except for the results.

The elastic stiffness constants in terms of the strut constants are given by the following expressions:

$$C_{11} = D(C_s/6 + 2C_b) \tag{3}$$

$$C_{12} = D(C_s/6 - C_b) \tag{4}$$

$$C_{44} = 3DC_sC_b/(C_s + 8C_b) \tag{5}$$

where $D = \sqrt{3}/2l^3$

In terms of these Cij, the bulk stiffness, B which measures the resistance of the frame to uniform volume changes is:

$$B = \frac{1}{3}(C_{11} + 2C_{12})$$

or, in terms of the stretching constant:

$$B = DC_s/6$$

and substituting for Cs

$$B = (d/l)^2 E/9 \qquad (6)$$

This expression indicates a reduction from the Young's modulus of the material in proportion to the slenderness ratios of the struts. This is just as expected from qualitative considerations.

If B is divided by the density of the structure, the specific bulk modulus of the frame is obtained. The frame density in terms of the material density, ρ is:

$$\rho^* = (d/l)^2 \rho$$

and this can be very low. For example, if the slenderness ratio is 0.1, the frame density is only one per cent of the material density.

The specific bulk modulus for the frame is $B^* = B/\rho^*$

or

$$B^* = (E/\rho)/9 \qquad (7)$$

Note that this is independent of the slenderness ratio (for moderately high ratios). Also note how relatively high the specific modulus is; or in other words how efficient this design is in using the specific stiffness of the material to full advantage. Another way of looking at this is that this frame design allows very large and quite open frames to be constructed with minimum sacrifice of specific stiffness. As Gordon (8) has pointed out, this is the optimum approach to the design of large structures. It may be noted in this connection that a tetrahedral fame is a kind of foam built of linear struts rather than thin films.

In shear this frame is also efficient because no diagonal stiffeners are needed. A simple cubic frame is much heavier because it requires shear stiffeners. For the case of dilatation, the bulk modulus is the same for both the simple cubic configuration and the tetrahedral one. However, there are six struts at each node so the weight is 50% more. Adding diagonal shear stiffeners increases the weight even more.

Compliance

The stiffness constants, C_{ij} when multiplied by a set of strains yield the corresponding stresses. But frames are usually subjected to tractions (stresses) rather than strains, and one wishes to know the corresponding strains (deflections). Such responses are conveniently described by a set of compliance constants, S_{ij} which yield the strains when they are multiplied by the stresses. They are inversely related to the C_{ij}'s (5).

Anistropy

Since a tetrahedral frame does not have isotropic properties in general, its resistance to shear depends on the direction and plane of shear. Two shear constants are required. For convenience, one of these is C_{44} which describes shearing parallel to the faces of the cubic unit cell along the directions of the cube edges. The other is $1/2\,(C_{11} - C_{12})$ which describes shearing parallel to diagonal planes of the unit cell containing a cube edge and a face-diagonal. The shear direction is parallel to the face-diagonal. This constant, in terms of the strut constants is: $3DC_b/2$.

The ratio of the two shear constants measures the elastic anistropy, of the frame. It is:

$$A = 2C_{44}/(C_{11}-C_{12}) = GC_s/(C_s+8C_b)$$

If the anistropy ratio equals unity, the frame is elastically isotropic. This condition is realized if:

$$C_s = 8C_b/5$$

Thus, if an isotropic frame is desired, it can be obtained by adjusting the strut design to satisfy this condition. Similarly, by adjusting the ratio of C_s to C_b the Poisson contractions of the frame can be designed to be positive, negative, or zero.

Strength

Of prime importance for a frame is its strength. This determines the load at which it begins to have excessive deflections, or at which collapse begins. One virtue of tetrahedral frames is that because of their high symmetry only one element needs to be designed for strength; namely, the strut that connects two nodes. However, these units must be designed to resist: buckling, inelastic bending, tension, and compression. If the frame is subject to fluctuating loads they must also resist fatigue; and if impacts are applied resistance to dynamic fracture also becomes important. Little can be said in general about these resistances until particular materials of construction have been specified, so this topic will not be pursued here.

APPLICATIONS

Tetrahedral frames can be used in the same situations as other space frames, but they have two advantages that may make them especially attractive for some kinds of aerospace systems. One is their exceptionally high specific structural stiffness and strength. The other is the simple module that can be used repetitively to construct them. Among the systems for which they might be useful are: the frames of lighter-than-air vehicles; space colony frames; the support for a solar sail; and the frames for large optical systems.

Pertinent to the last of these is the fact that graphite reinforcing fibers can be aligned with the axes of all of the struts if they are aligned in the puckered rings. Then the thermal expansion coefficient along the strut axes will be very low. In turn, by symmetry, the thermal expansion coefficient for the frame as a whole will be low. This will help provide the stability needed for optical systems.

Some other structures that might be built from these frames are: large boats; houses and other buildings; unsupported domes; large beams or plates; climbing frames for recreation; and large storage tanks.

It is not necessary to completely fill a section of space with a tetrahedral frame. They can be constructed containing internal cavities or channels. Thus they can be a variety of shapes, including cylinders, plates, and tubes. Just as exterior surfaces of them can be closed off with triangular rings, so can internal surfaces. Plates or membranes can be readily attached to their external surfaces.

A rather different application opportunity is that of embedded frames. For example, steel reinforcing bars might be cut and welded in the puckered ring configuration of Figure 4. These might then be assembled as a tetrahedral frame inside a mold which is to be filled with concrete. The re-bar frame would efficiently reinforce the subsequent concrete casting (in terms of the amount of steel needed to obtain a given stiffness of strength). Ceramic and polymeric objects might similarly be reinforced. Also, one metal might be used to reinforce another.

ACKNOWLEDGMENT

This work was supported in part by the Allied-Signal Corporation, the Advanced Defense Research Projects Agency, and the U.S. Department of Energy under contract No. DE-AC03-76SF00098.

REFERENCES

1. R. W. Kraft, "Construction Arrangement", U.S. Patent #3,139,959.
2. J. J. Gilman, "Tetrahedral Truss", U.S. Patent #4,446,666
3. A. G. M. Michell, "The Limits of Economy of Material in Frame-Structures", Phil. Mag. Ser. 6, 8, 589 (1904).
4. P. Pearce, Structure in Nature as a Strategy for Design, The MIT Press, Cambridge, MA (1978).
5. R. P. Feynman, R. B. Leighton, and M. Sands, The Feynman Lectures on Physics, Vol. II, p. 39-40, Addison-Wesley, Reading, MA (1964).
6. W. A. Harrison, Electronic Structure and the Properties of Solids, Chap. 8, pp. 180-202, W. H. Freeman and Co., San Francisco (1980).
7. S. Timoshenko, Theory of Elasticity, Article 41, p.145 McGraw-Hill, New York (1934).
8. J. E. Gordon, Structures, pp. 294-299, 385-387, Plenum Press, New York 1978).

Author Index

Abdullah, M.J., 205
Anton, D.L., 57
Armstrong, J.H., 111
Awerbuch, J., 121

Becher, P.F., 271
Biancaniello, F.S., 35
Bieler, T.R., 137
Bischoff, E., 333
Boisvert, R.P., 157
Bose, A., 51
Boulanger, Christophe H., 279
Bour, J.S., 351
Bourell, D.L., 23
Burkland, C.V., 163

Chiang, Yih-Cherng, 279
Chou, J.-C., 163
Chou, Tsu-Wei, 185, 279
Christodoulou, L., 29
Chumbley, L.S., 45
Clough, R.B., 35
Cranmer, D.C., 253
Crowe, C.R., 29

Das-Gupta, D.K., 205
Deshmukh, U.V., 253
Dick, C.M., 247
Diefendorf, R.J., 157
Dimiduk, D., 103
Downing, H.L., 45
Drescher-Krasicka, E., 341

Eliezer, Z., 23
Elliott, C.K., 95
Evans, A.G., 213, 293

Fareed, A., 121
Farmer, S.C., 169
Fishman, S.G., 285
Freiman, S.W., 253

Gac, F., 313
German, R.M., 51
Gilman, John J., 369
Gonsalves, Kenneth E., 199

Handwerker, Carol A., 357
Hasson, D.F., 285
Henager, C.H., 313
Heuer, A.H., 169

Johnson, H.H., 193
Jones, R.H., 313

Kanei, A., 253
Kattner, Ursula R., 357
Kembaiyan, K.T., 199
Kerr, W., 103
Koczak, M.J., 121
Krempl, E., 129
Krotz, P.D., 45

Lancin, M., 351
Langdon, Terence G., 265
Lee, B.-H., 23
Lee, Doh-Jae, 357
Lee, Hae-Weon, 175
Lee, K.D., 129
Lewandowski, J.J., 103
Lin, F., 323
Lipetzky, P., 271
Lucas, G.E., 95
Luh, E.Y., 333

Madhukar, M.S., 121
Majidi, Azar P., 279
Mannan, S.K., 89
Marcus, H.L., 23
Marieb, T., 323
Marshall, D.B., 213
McLean, Malcolm, 67
Mehrabian, Robert, 3
Mendiratta, M.G., 103
Misra, M.S., 111
Moore, B., 51
Morrone, A., 323
Mukherjee, A.K., 137

Nieh, T.G., 137
Nix, W.D., 247
Nutt, S.R., 271, 323

Odette, G.R., 95

Parrish, P.A., 29
Persad, C., 23
Pirouz, P., 169
Prewo, Karl M., 145

Raghunathan, S., 23
Rawal, S.P., 111
Ritchie, R.O., 81
Rojas, Oswaldo E., 175
Rühle, M., 293

Sacks, Michael D., 175
Sbaizero, O., 333

Schilling, C.H., 313
Schoenlein, L.H., 313
Shang, Jian Ku, 81
Sheckherd, J.W., 95
Simmons, J.A., 341
Singh, Raj N., 259
Spitzig, W.A., 45
Sprissler, B., 89
Steinwall, James E., 193
Stoloff, N.S., 51

Tai, Nyan-Hwa, 185
Thibault-Desseaux, J., 351
Thouless, M.D., 333

Vaudin, Mark D., 357
Verhoeven, J.D., 45
Viswanadham, R.K., 89

Wadley, H.N.G., 35, 341
Wadsworth, J., 137
Weihs, T.P., 247
Whittenberger, J. Daniel, 89

Xia, Kenong, 265

Yang, J.-M., 163

Subject Index

acoustic emission, 35
adhesion, 293
Al_3Ta, 57
Al/Al_2O_3, 2, 4, 6, 7, 9
Al_2O_3/SiC, 175
$Al_2O_3/ZrO_2/SiC$, 175
Al/B_4C, 12, 13
Al/SiC, 15
$Al/Nicalon^{TM}$, 18
Al/SiC_p, 81
Al/SiC, 121
Al/B, 121
Al-SiC, 348
Al-Zn-Mg-Cu/SiC, 81
amorphous matrix, 193
anisotropy, 129, 288, 377

bonding, 299
bridge zone toughening, 95

ceramics
  borosilicate glass, 255
  composites, 259, 271
  monolithic, 15
chemical vapor infiltration, 163, 185
composites
  constitutive equations, 70, 78
  continuous, 72
  damage mechanics, 70
  glass ceramic, 145
  metal matrix, 67, 111, 121, 340, 357
  micro, 193
  particulate, 3, 15, 68, 89
  pathways, 4
  polymer ceramic, 205
  routes, 3
C/borosilicate glass, 286
C/glass, 148
C/Mg, 111
C/MgO.6Zr(KIA), 111
C/Mg1.0Mn(MIA), 111
C/SiC, 148
crack
  blunting, 105
  driving force, 81
  nucleation, 323
  opening displacement, 84
  tip, 215
creep, 89, 271
  anisotropy, 69
  behavior, 67
  fracture, 72
  primary, 64
  steady state, 268
  stress exponent, 252, 269
  tertiary, 70
  testing, 248, 265
Cu-20% Nb, 45

damage development, 75
  grain boundary sliding, 137
damping capacity, 114
deformation processing, 45
  draw ratio, 46
  high-energy/rate, 23
deposition film growth, 185
dielectric properties, 205
diffusion reactions, 357
  Knudsen, 185
  multicomponent, 185
dislocation densities, 48
  Granato-Lucke model, 111, 114
ductile phase toughening, 103
dynamic recovery, 73

elastic moduli, 147
  nonlinear, 222
electrical resistivity, 46
elevated temperature
  mechanical properties, 89, 121
environmental stability, 146

fatigue crack growth, 81, 83
fiber
  aspect ratio, 73
  matrix compatibility, 60
  pull out, 218, 253
  reinforced composites, 35
  volume fraction, 78
  whisker pullout, 280
filaments; Nb, 45-46
  F.P. Al_2O_3/Al_2O_3, 148
  F.P. alumina, 57
fracture mechanics, 214, 287
  mechanisms, 214, 279
  toughness, 163
frame stiffness, 373
frictional resistance, 247
frictional stress, 247, 253

Hashin theory, 115

impact properties, 121
  toughness, 145, 285
inelastic deformation, 129

in-situ composites, 45, 67, 103
interfaces, 35, 296, 340, 351
  chemistry, 299, 313
  cracks, 214
  debonding, 215
  fiber/matrix, 158
  roughness, 358, 333
  shear stress, 259, 335
intermetallic compounds, 29, 103
internal friction, 105, 111, 114

lanxide, 4
lithium aluminosilicate, 152

microfabrication, 193
micro-indentation, 193
micromechanical testing, 35
  4-pt flex, 61
micromechanics, 3, 214
  viscoplastic analysis, 129
micromechanisms, 81
microstructural design, 3, 225
  degradation, 67

Nb_3Al, 60
Nb-Si, 103
Nb_5Si_3, 103
NiAl, 89
Ni_3Al/Al_2O_3, 51
Ni_3Al + B, 51
Ni_3Al, Cr + B, 51
Nicalon™, 4
Nicalon™/Al_2O_3, 148
Nicalon™/carbon coated, 158
Nicalon™/glass, 148
Nicalon™/polyvinylsilane, 161
Nicalon™/SiC, 148, 163
Nicalon™/Si_3N_4, 320

organotitanium, 199
oxidation, 323
  melt, 17
oxide dispersion strengthened, 137

PAN fibers, 201
precision space structures (LPSS), 111
phase stability, 357
ply constitutive equation, 130
polyvinylsilane/SiC, 158
processing, 5
  HIP, 17, 19, 49, 52, 159
  hot pressed, 314
  liquid metal, 5
  melt infiltration, 8
  melt oxidation, 5, 17

plasma deposition, 13
polymer pyrolysis, 157
powder, 23
vacuum investment casting, 112
puckered rings, 369
pyroelectric properties, 205

rapid solidification, 13
reaction rates, 188
Reuss averaging, 72
reverse sliding, 230, 247

SEM, 48, 104
shear properties, 121
SiC, 157, 163
SiC/Al, 357
SiC_f/Ti, 351
SiC/Al_2O_3, 265, 271, 279, 323
SiC_w/Al_2O_3, 148
SiC/lithium-aluminosilicate, 248
SiC/mullite, 259
SiC/Si_3N_4, 169, 313
SiC/zircon, 242, 259
sintering, 181, 301
solute drag, 142
structural stiffness, 377
  rate sensitivity, 137
  softening, 71
stress, 67, 220
  intensity range, 86
  internal, 71
  redistribution, 129
  residual, 223
  thermal, 133
substructure, 45
superplasticity, 137

Tateho SiC whiskers, 169
TEM, 48, 276, 306, 307, 308, 327, 329, 334, 354, 355, 356
tetrahedral frames, 369
tetroid space filing unit, 371
thermal expansion mismatch, 253
  coefficient, 147
thermal shock resistance, 163
Ti-45 a/o Al, 29
Ti_3Al, 31
Ti_3Al + Nb/SiC, 25
Ti6AlV, 95, 121
TiAl, 4, 16, 19, 60, 95
TiB_2, 89
TiB_2/TiAl, 30
Ti-C-B-N fibers, 201
Ti composites, 199
Ti/SiC, 127

Toughness, 145, 213
transverse strength, 121

ultrasonic propagation, 340

Voigt average, 73

Weibull statistics, 333
wetting behavior, 176
W-N-Fe/B C, 25

XD^{TM} composites, 5
XD^{TM} titanium aluminides, 29